Processing Technology of Brittle Crystal Materials

Processing Technology of Brittle Crystal Materials

Editors

Chen Li
Chongjun Wu
Binbin Meng
Shanshan Chen

MDPI

Basel • Beijing • Wuhan • Barcelona • Belgrade • Novi Sad • Cluj • Manchester

Editors

Chen Li
Harbin Institute of Technology
Harbin
China

Chongjun Wu
Donghua University
Shanghai
China

Binbin Meng
Soochow University
Suzhou
China

Shanshan Chen
Xi'an Jiaotong University
Xi'an
China

Editorial Office
MDPI
St. Alban-Anlage 66
4052 Basel, Switzerland

This is a reprint of articles from the Special Issue published online in the open access journal *Crystals* (ISSN 2073-4352) (available at: https://www.mdpi.com/journal/crystals/special_issues/brittle_crystal).

For citation purposes, cite each article independently as indicated on the article page online and as indicated below:

Lastname, A.A.; Lastname, B.B. Article Title. *Journal Name* **Year**, *Volume Number*, Page Range.

ISBN 978-3-0365-8866-7 (Hbk)
ISBN 978-3-0365-8867-4 (PDF)
doi.org/10.3390/books978-3-0365-8867-4

Contents

About the Editors

Chen Li

Dr. Chen Li, Associate Professor and doctoral supervisor of School of Mechatronics Engineering, Harbin Institute of Technology. He received Bachelor degree from Northeastern University in 2015. Later he received PhD degree from Harbin Institute of Technology in 2019. From 2018 to 2019 he worked as a visiting scholar in The University of Queensland, Australia. His work was supported by 15 research projects, such as National key R & D program, 973 program, National Natural Science Foundation, etc. The total cost of the all Projects is approximately $2, 000, 000. He published more than 70 peer-reviewed scientific research papers, and the total citation number of these publications is more than 1500 in Google Scholar. His citation index (h-index) in Google Scholar is 20. These papers were published in International Journal of Machine Tools and Manufacture, Journal of the European Ceramic Society, International Journal of Extreme Manufacturing, Journal of Materials Processing Technology, International Journal of Mechanical Sciences, and so on. He was awarded the Hiwin Doctoral Dissertation Award in 2020, Young Elite Scientists Sponsorship Program by CAST in 2022, Outstanding Contribution Award (at CCAT2017), Excellent Young Researcher Award (at CJUMP2016), Excellent Paper Award (at CJUMP2015), Excellent Young Paper Award (at CCAT2017). He is a member of Extreme Manufacturing Committee, Production Engineering Committee of Chinese Mechanical Engineering Society, and Heilongjiang Standardization Technical Committee for Instrument. He is an editorial board member of several academic journals and peer reviewer of more than 30 academic journals indexed by SCI. His interests are ultra-precision machining of hard-brittle materials, nano-manufacturing, mechanical characterization of single crystal and ceramics, application of finite element simulation and numerical simulation in machining technology.

Chongjun Wu

Dr. Chongjun Wu is serving as an Associate Professor at School of Mechanical Engineering, Donghua University. He has published more than 50 high-level SCI/EI papers as the first/corresponding author with about 600 citations in Scopus and 3 ESI papers. He is currently a member of the Production Engineering Branch of the Chinese Mechanical Engineering Society (Abrasive technology field), a member of the Smart Factory Special Committee of the Chinese Graphic Society, and a member of the young editorial board of periodicals such as Journal of Advanced Manufacturing Science and Technology, Diamond and Abrasive Tools Engineering and Surface Technology. He has presided over more than 10 projects such as the National Nature Science Foundation, Shanghai Science and Technology Commission projects, and first-class grants for postdoctoral students. Dr. Wu has also won a number of provincial and ministerial teaching and technology awards.

Binbin Meng

Binbin Meng, Ph.D. in Engineering, has a primary research interest in precision machining technology for hard and brittle materials, deformation and separation mechanisms at the micro- and nano-scales, and surface micro- and nano-structure machining under composite energy fields. As a post-doctor, he joined the Southern University of Science and Technology and was mainly involved in developing new methods for the ultra-precision machining of third-generation semiconductor materials under composite energy field effects. Following his doctoral research, he assumed the role of research assistant professor, with a primary focus on analysis and experimental research in the directions of AIMD and QM, specifically the microscopic behaviors of atomic/near-atomic-scale material migration and separation, with the ultimate goal of achieving sub-nano-meter-level quality machining for hard and brittle single-crystal and amorphous materials.

Shanshan Chen

Shanshan Chen, assistant professor, is mainly engaged in the research of ultra-precision machining techniques and measuring techniques for complex surfaces, received PhD degrees from Hong Kong Polytechnic University and Harbin Institute of Technology in April 2018 and June 2019, respectively, has been engaged in post-doctoral research at Xi'an Jiaotong University since June 2019, and was funded by the China Scholarship Council to conduct research at the University of Nottingham from October 2020 to January 2022. Being committed to the research of machined surface characterization technology and the grinding surface formation mechanism of optical difficult-to-machine materials, Shanshan Chen has participated in three national key basic research projects, and presided over the National Natural Science Foundation of China (NSFC)'s Youth Project, Chinese Postdoctoral Research Foundation Project, Science and Technology Planning Bureau Project of Xi'an Beilin District, and Basic Public Welfare Research Project of Zhejiang Province. As the first author, Shanshan Chen has published 9 academic papers in international journals such as Optics Express and Precision Engineering, won the Best Paper Award in two successive nanoMan International Conferences (2018 and 2021), and co-authored 1 foreign language book. Shanshan Chen is also a member of the International Association of Engineering and Technology (AET), a senior member of Chinese Society of Optical Engineering, a member of the Tribology Branch of Shaanxi Mechanical Engineering Society, a guest editor of International Journal Crystals, and a reviewer of International Journal of Machine Tools and Manufacture, Journal of Manufacturing Systems, Journal of Manufacturing Processes, International Journal of Nanomanufacturing, Nanomanufacturing and Metrology.

Preface

Brittle crystals and ceramics have been widely used in the fields of aerospace, 5G networks, and new energy vehicles owing to their excellent mechanical properties and steady chemical properties. For these advanced applications, brittle crystals and ceramics must be shaped into smooth substrates with high surface integrity using precision and ultra-precision processing technologies such as grinding, lapping, and polishing. However, these materials have high brittleness and hardness with low fracture toughness, which pose great challenges for efficient machining. Damanges due to brittleness, including fractures and cracks, are easily generated during the machining process, which inevitably shortens the service life of the crystal components and compromises further applications. Understanding the mechanical properties, revealing the the evolution of damange and the material removal mechanism at the micro- and nano-scales, exploring innovative machining technology, and optimizing machining process parameters are of great significance to realizing the high-efficiency and precision machining of brittle materials. This collection aims at summarizing the frontier research on the processing and surface integrity characterization of brittle crystals, ceramics, and composite materials.

Chen Li, Chongjun Wu, Binbin Meng, and Shanshan Chen
Editors

crystals

MDPI

Article

Surface Quality Experimental Study on Rotary Ultrasonic Machining of Honeycomb Composites with a Circular Knife Cutting Tool

Gang Liu, Jie Yang, Liqiang Zhang *, Qiuge Gao, Long Qian and Rongyao Zhang

School of Mechanical and Automotive Engineering, Shanghai University of Engineering Science, Shanghai 201620, China; liugang@sues.edu.cn (G.L.); xiaogegejie@outlook.com (J.Y.); 15836421741@163.com (Q.G.); Mofei0925@outlook.com (L.Q.); zry2276801367@163.com (R.Z.)

* Correspondence: zhanglq@sues.edu.cn

Abstract: Honeycomb composites (HCs) are diversely employed in aerospace, national defense and other fields owing to their remarkable spatial geometry and excellent mechanical properties. Their complex hexagonal cell structure and heterogeneous material properties cause major problems when implementing high-quality processing. Surface defects generated by processing will reduce the capability and service lifespan of the honeycomb sandwich structure. Therefore, the high quality of HCs is a topic of close attention for researchers. In this paper, the consequences of different cutting parameters of rotary ultrasonic machining (RUM) on surface quality with an ultrasonic circular knife (UCK) were studied through multiple groups of single-factor and orthogonal experiments with two-factors/four-levels and one-factor/three-levels. The single factor experiment was used to explain the effect that the degree of cutting parameters has on surface quality, and the orthogonal experiments were applied to explain the interaction between the processing parameters and the influence law of each factor on surface quality. Therefore, the reasonable cutting parameters of HCs were determined through experimental results to provide guidance for the realization of the precise and efficient machining of HCs. This study can provide a basis for the subsequent comprehensive consideration of various factors to achieve high-quality machining of HCs.

Keywords: Nomex honeycomb composites (NHCs); aluminum honeycomb composites (ALHCs); rotary ultrasonic machining (RUM) technology; surface quality; cutting parameters

Citation: Liu, G.; Yang, J.; Zhang, L.; Gao, Q.; Qian, L.; Zhang, R. Surface Quality Experimental Study on Rotary Ultrasonic Machining of Honeycomb Composites with a Circular Knife Cutting Tool. *Crystals* **2022**, *12*, 725. https://doi.org/10.3390/cryst12050725

Academic Editors: Binbin Meng, Anatoliy V. Glushchenko, Chen Li, Chongjun Wu and Shanshan Chen

Received: 21 April 2022
Accepted: 11 May 2022
Published: 19 May 2022

1. Introduction

Compared with other structures of the same material, the honeycomb sandwich structure has excellent spatial geometry and mechanical properties [1]. The honeycomb core sandwich structure is a sandwich composite material formed by connecting the honeycomb core and the upper and lower panels with adhesive and curing. It has high stiffness along the axial direction of the cell, therefore, it can bear an impact force on the sandwich structure [2]. However, honeycomb is a typical thin-walled porous structure. The particularity of its structure leads to low in-plane rigidity of the honeycomb core, which is prone to various defects during processing and reduces the performance and service life [3–9]. The mechanical properties largely rely on honeycomb composite core density, quality and the bonding strength between the core and the panel [1,10]. Therefore, the influence of machining defects on the honeycomb core and its mechanical properties is a key factor to be considered. Consequently, it is significant to research the morphology features of the honeycomb composite (HCs) core and the effect of cutting parameters on HCs core surface quality.

Due to the merits of high processing efficiency and convenient manufacturing of processing equipment, conventional high-speed CNC milling was widely used in honeycomb core material processing. However, traditional processing methods cause the following problems, such as poor processing quality, harsh processing environment, tool wear and so on [11,12]. Thence, rotary ultrasonic machining technology (RUM) was applied to solve the shortcomings of traditional technologies for HC core machining. As a rapidly developing processing technology, RUM technology has been more and more widely employed in the processing of various materials such as composites [13–16], crystal [17], ceramics [18,19], titanium alloys [20,21], glass and aluminum alloys [22–25]. In the past, most scholars have carried out research on RUM technology, mainly including the design of the ultrasonic system and ultrasonic cutting tools, the degree of effect of the independent variable cutting parameters on the surface quality of the dependent variable, etc. Xiang et al. [26] proposed that the ultrasonic longitudinal–torsional composite vibration machining method has excellent performance and better surface morphology compared with the transonic longitudinal vibration machining method by analyzing the kinematics of the two processing methods. Xia et al. [27] optimized the structure of ultrasonic cutting tools and explored the influence of distinct tool structure parameters on the machining efficiency of the Nomex honeycomb composite core. Sun et al. [28] concluded that the introduction of RUM technology can significantly reduce the cutting generation of tool cutting forces during machining, thereby reducing the deformation of aluminum honeycomb core walls, resulting in high-quality workpieces. Ahmad et al. [29] conducted a comprehensive review on machining technologies for Nomex HC and Ahmad et al. [30] discussed the influence of different structural parameters of the UCK cutting tool used in machining on ultrasonic amplitude as well as the resonant frequency. Ahmad et al. [31] also researched RUM characteristics of Nomex HCs by a series of UCK cutting tools through a series of controlled variable experiments and orthogonal experimental methods. Asmael et al. [32] reported a review on carbon-fiber-reinforced plastic composites by ultrasonic machining. Hu et al. [33] proposed a cutting force model of a triangular blade on honeycomb composites by ultrasonic machining. Cao et al. [34] constructed a three-dimensional finite element model through ABAQUS to analyze the ultrasonic vibration-assisted tool of the Nomex honeycomb core.

In this paper, a group of single-factor experiments and orthogonal experiments of RUM technology with a circular ultrasonic knife for HC core in terms of the aluminum honeycomb core and the Nomex honeycomb core workpiece were conducted to discuss the effect of different processing parameters, i.e., tool feed rate, tool cutting width, ultrasonic spindle speed on surface quality. Then, we further analyzed the significance of different cutting parameters on surface quality through orthogonal experiments developed in this study. Therefore, the framework of this paper is arranged as follows. The experimental synopsis is introduced in Section 2. Section 3 shows the experimental program. The phenomenon and discussion are presented in Section 4. The next section, namely, Section 5, summarizes this paper.

2. Experimental Synopsis

Most RUM systems are mainly composed of the following parts: ultrasonic power supply, ultrasonic converters, amplitude modifier (booster) and ultrasonic cutting tool. Ultrasonic power supply (ultrasonic generator) outputs high-frequency AC oscillating electrical signals. The ultrasonic transducer converts the high-frequency AC oscillating electrical signal output by the ultrasonic generator into a mechanical vibration signal with the same frequency, the ultrasonic transducer used in this paper is a piezoelectric ceramic sensor. The ultrasonic horn amplifies the amplitude of the weak mechanical vibration signal from the ultrasonic transducer to achieve the amplitude required for the actual cutting process, and at the same time, it can ensure that the entire acoustic system is always in a resonance state. The ultrasonic circular knife was used for processing the HC core. A schematic diagram of the RUM system for HC core workpiece machining was displayed in Figure 1.

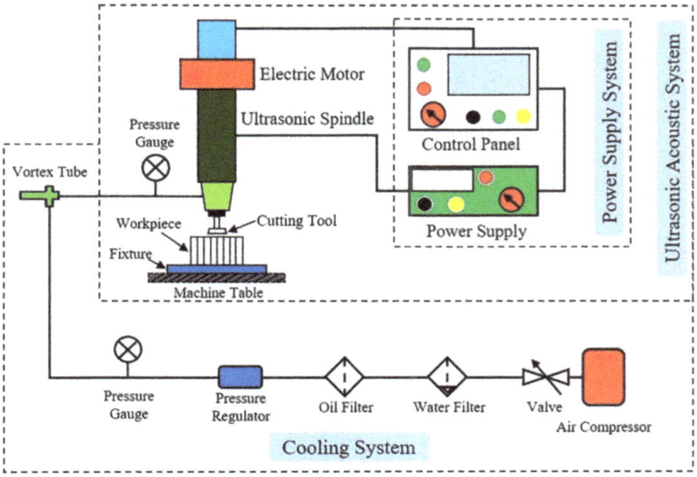

Figure 1. Abridged general view of RUM system for HCs core.

In this paper, the experimental devices for RUM of Aluminum honeycomb composites (ALHCs) core workpiece and Nomex honeycomb composites (NHCs) core workpiece were illustrated in Figures 2 and 3, respectively. Surface quality experiments for RUM of the HC core workpiece by UCK were executed on the ultrasonic machine tool (NO: IT5X111) produced by INNOTECH company. Ultrasonic equipment was installed on the machine tool, then ultrasonic cutting experiment was carried out by switching the power of the ultrasonic generator and air cooling was adopted for the cooling system in the machine tool, as shown in Figures 1–3. Ultrasonic generator (Model: GT2010-GT2015), ultrasonic converters, amplitude modifier and ultrasonic cutting tool were developed by SONIMAT (EUROPE TECHNOLOGIES) in France. The major parameters of the RUM system are ultrasonic frequency (20 KHz), ultrasonic power (Max 1500 w), power supply:230 V mono 50/60 Hz, ultrasonic spline speed (Max 1000 rpm) and vibration amplitude (Max 0.08 mm). ∅63 mm disc cutting tool was used in the experiment, as shown in Figure 4.

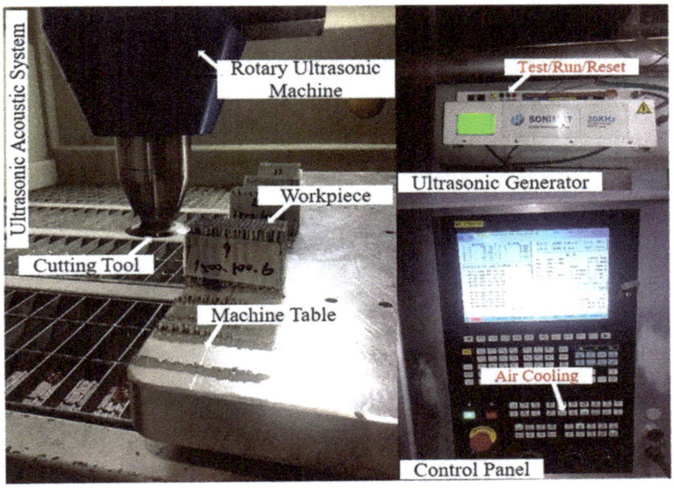

Figure 2. Experimental equipment of ultrasonic machining system for ALHCs core specimen.

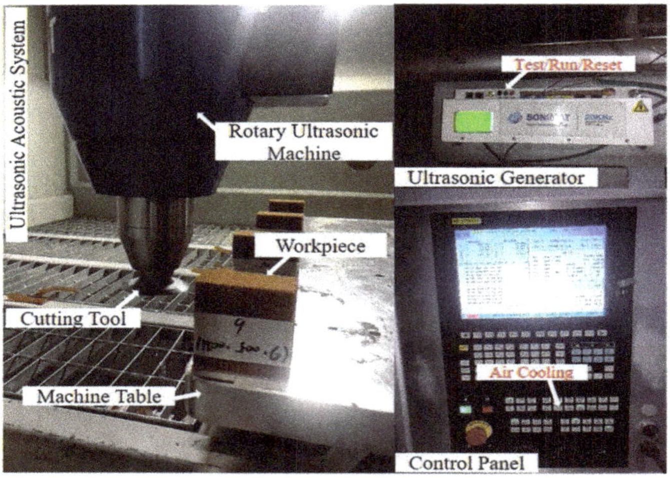

Figure 3. Experimental equipment of ultrasonic machining system for NHCs core specimen.

Figure 4. UCK (∅63 mm Disc) cutting tool used in RUM experimental for HCs core workpiece.

3. Experiment Program

3.1. Materials Used in Experiment

In this experiment, the Nomex honeycomb composite core workpiece (Type Designation: ACCH-2I-1/8-3.0) was produced by AVIC Composite Materials Co., Ltd. The aluminum honeycomb composite core workpiece (Type Designation: HexWeb® CR-PAA-5052-3/16-3.1-Non-p) was produced by HEXCEL company. NHCs core workpieces with dimensions of ribbon (L), a direction perpendicular to the ribbon (W) and cell depth (T) as 100 mm × 50 mm × 100 mm and ALHCs core workpieces with dimensions of ribbon (L), a direction perpendicular to the ribbon (W) and cell depth (T) as 100 mm × 50 mm × 80 mm were used in this experiment, as shown in Figure 5. Key parameters for the hexagonal cellular structure of the workpiece are cell size, single wall thickness, double wall thickness and cell edges, as shown in Figure 5 and specimen attributes of ALHCs core workpiece and NHCs core workpiece were demonstrated in Tables 1 and 2, respectively.

Double-sided adhesive tape and synthetic resin adhesive were used to fix ALHCs core and NHCs core workpiece (a group of four workpieces) on the machining table of the machine tool instead of fixing them on a special fixture due to the small dimensions of the workpiece. The fixing system of the specimen is shown in Figure 6.

Table 1. Material attributes and cell parameters of AL honeycomb core workpiece.

Material Properties	Value	Cell Parameters	Value
Compressive modulus	513 MPa	Cell size (c)	4.76 mm
Compressive strength	2.31 MPa	Cell signal wall thickness (t)	0.05 mm
Plate shear strength L direction	1.45 MPa	Cell edge (a)	1.37 mm
Plate shear strength W direction	0.86 MPa	Cell edge (b)	1.37 mm
Plate shear modulus L direction	307.8 MPa	Cell wall angle (φ.)	120°
Plate shear modulus W direction	150.48 MPa		
Nominal density	49.6 kg/m^3		

Table 2. Material attributes and cell parameters of Nomex honeycomb core workpiece.

Material Properties	Value	Cell Parameters	Value
Compressive modulus	131 MPa	Cell size (c)	3.17 mm
Compressive strength	2.15 MPa	Cell signal wall thickness (t)	0.13 mm
Plate shear strength L direction	1.28 MPa	Cell edge (a)	1.83 mm
Plate shear strength W direction	0.60 MPa	Cell edge (b)	1.83 mm
Plate shear modulus L direction	36 MPa	Cell wall angle (φ.)	120°
Plate shear modulus W direction	20 MPa		
Tensile strength	2.45 MPa		
Tensile modulus	148 MPa		
Nominal density	48 Kg/m^3		

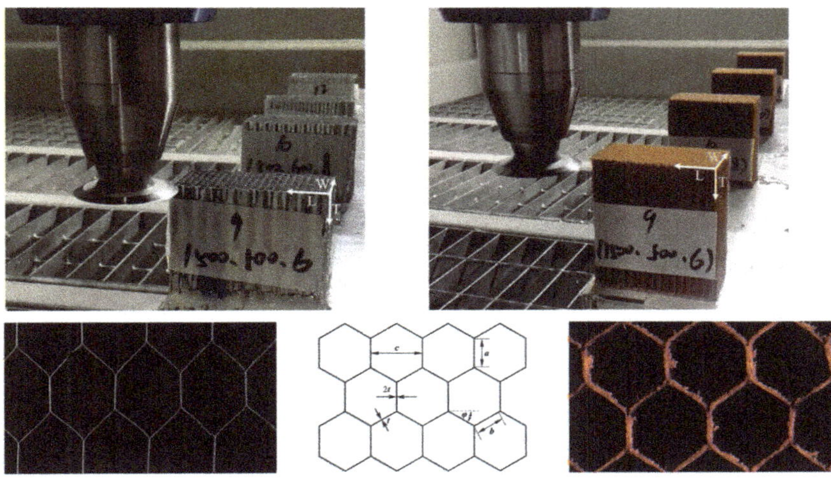

Figure 5. Honeycomb core workpieces and their structural parameters used in this experiment.

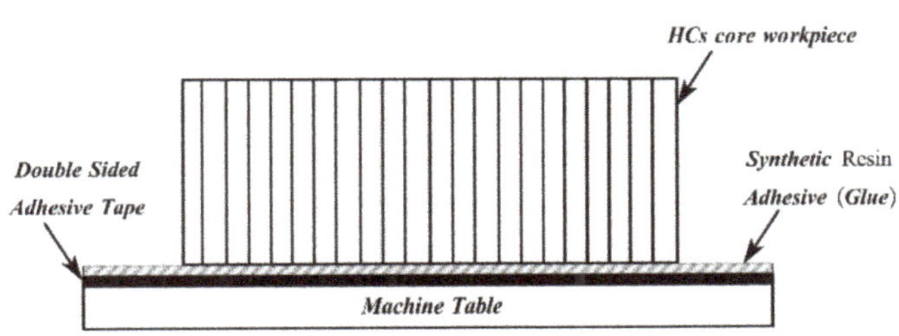

Figure 6. Schematic of workpiece clamping system.

3.2. *Measurement Methods for Surface Characteristics*

To observe the machining results, a high-definition video detector (CHECKER145Linear-M, STILL, Guangdong China) was used to measure the surface morphology of ALHCs core and NHCs core workpieces after RUM, then, to observe processing defects of the workpiece under different cutting parameters and the effect on surface quality. The concrete analysis is reflected in Section 4. Experimental setup for surface morphology of ALHCs core and NHCs core workpieces are displayed in Figure 7.

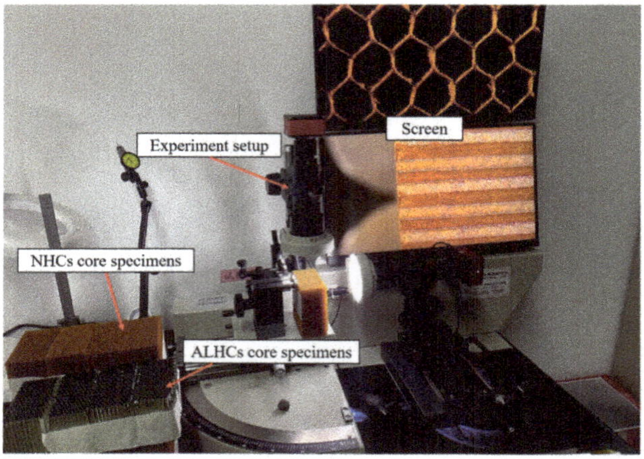

Figure 7. Experimental setup for surface morphology.

3.3. Processing Conditions in the Experiment

High quality has always been the core goal of processing. Compared with traditional technology, rotary ultrasonic machining reduces surface defects and improves efficiency. Therefore, the reasonable selection of processing parameters has become the core issue to improve efficiency and workpiece surface quality. Thus, in this research, a group of single-factor experiments and orthogonal experiments with two-factors/four-levels and one-factor/three-levels were used to discuss the effect of cutting parameters such as cutting tool feed rate, tool cutting depth and ultrasonic spindle speed on surface quality. In experiments for ALHC core workpieces, the cutting depth remained at 0.8 mm for each experiment due to material property and great cutting effects. In the experiments for NHC core workpieces, cutting depth remained at 2 mm for each experiment. Vibration amplitude was kept fixed at 0.08 mm according to the manufacturer. According to the parameters of the ultrasonic equipment itself and the performance parameters of the machine tool with ultrasonic equipment, the single-factor experiments designed are illustrated in Table 3.

Table 3. Control variables of experimental parameters.

Variable	Parameter Levels	Constant Parameters
Spindle Speed (rpm)	500, 600, 700, 800	Feed rate 500 mm/min, cutting width 8 mm
Feed Rate (mm/min)	500, 1000, 1500, 2000	Spindle speed 800 rpm, cutting width 8 mm
Cutting Width (mm)	4, 6, 8, 10	Feed rate 2000 mm/min, spindle speed 800 rpm

Through single-factor controlled variable method experiments, the effect of each parameter on the surface quality can be obtained. However, the interaction between different machining parameters cannot be illustrated in single-factor controlled variable method experiments. Therefore, two-factors/four-levels and one-factor/three-levels orthogonal experimental methods were designed to discuss the influence of distinct processing variables in this paper. Meanwhile, orthogonal experimental technology was used to obtain the optimal association of cutting parameters for rotatory ultrasonic machining of ALHC core workpieces and Nomex honeycomb core workpieces with a UCK cutting tool according to the orthogonal experiments. The parameter factors and levels in the orthogonal experiments of the RUM are displayed in Table 4. The two-factors/four-levels and one-factor/three-levels $L_{12}(3 \times 4^2)$ orthogonal experimental conditions of ALHC core and NHC core specimens are shown in Table 5.

Table 4. The relevant parameters used in orthogonal experiments.

Level	Spindle Speed (rpm)	Feed Rate (mm/min)	Cutting Width (mm)
1	500	500	4
2	600	1000	6
3	700	1500	8
4	800		10

Table 5. Two-factors/four-levels and one-factor/three-levels $L_{12}(3 \times 4^2)$ orthogonal table of ALHC and NHC core specimens.

Exp.	Processing Factors		
	Spindle Speed (rpm)	Feed Rate (mm/min)	Cutting Width (mm)
1	500	500	4
2	500	600	6
3	500	700	8
4	500	800	10
5	1000	500	6
6	1000	600	8
7	1000	700	10
8	1000	800	4
9	1500	500	6
10	1500	600	8
11	1500	700	10
12	1500	800	4

4. Results and Discussion

The honeycomb sandwich structure consists of a high-strength adhesive bonding a relatively thin skin panel to a comparatively thick lightweight honeycomb core, and the overall structure is like a honeycomb structure. As the key component of a sandwich structure, a honeycomb core can greatly increase stiffness and strength on the premise of slightly increasing the overall quality of the sandwich structure, and is lightweight. Therefore, processing quality directly affects the performance and service life. RUM technology is used in composite material machining to greatly reduce defects and improve surface quality. Therefore, it is worthwhile to further study the processing characteristics of RUM with UCK and the influence of different cutting parameters on the workpiece surface quality. However, the ultrasonic cutting tool used in this experiment was newly fabricated, so tool tear was ignored and dust, delamination and collapse of cells did not exist in this study due to the RUM technology. In these experiments, the processing principle of the RUM technology and the disc cutting tool ultrasonic vibration trajectory are displayed in Figure 8. In this figure, A represents ultrasonic amplitude, f represents ultrasonic frequency, n represents disc tool rotary velocity and P represents any point on the disc cutting tool. When the disc cutting tool is without ultrasonic vibration, the tangential velocity V_t was much greater than feed velocity V_f, therefore, tool rotational motion was the main motion, and V_t can be expressed as:

$$V_t = \pi D n \tag{1}$$

where D represents disc cutting tool diameter.

When the disc cutting tool used ultrasonic vibration, the cutting tool produced displacement S with periodic changes in magnitude and direction along the tool axis. S can be expressed as:

$$S = A \cdot \sin(2\pi f t + \theta) \tag{2}$$

where t represents the cutting time, θ represents the vibration initial phase angle. Differently to Equation 2, vibration cutting velocity along the tool axis (Z direction) can be obtained as:

$$V_s = 2\pi f A \cdot \cos(2\pi f t) \tag{3}$$

In the RUM technology, the vibration cutting velocity V_s was much greater than the tangential velocity V_t and feed velocity V_f, therefore, tool vibration motion along the tool axis became the main motion in processing. The magnitude and the direction of V_s changed periodically over time.

Then, the authors observed the effect of different processing parameters on the workpiece surface with a disc ultrasonic cutting tool. The surface morphology of the ALHC core and the NHC core workpieces under different processing parameters was analyzed in detail below.

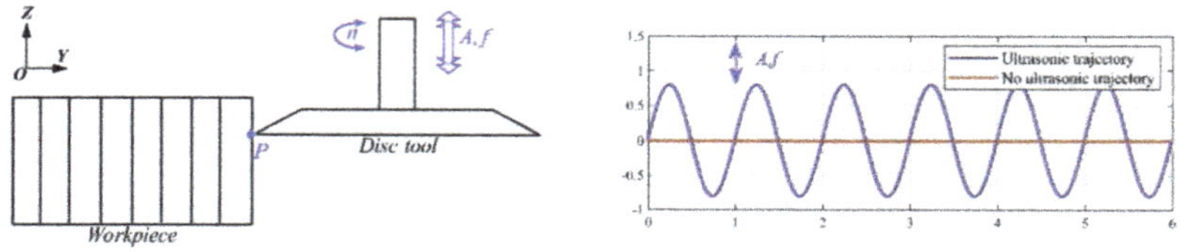

Figure 8. The RUM processing principle and disc tool trajectory.

4.1. ALHC Core Surface Quality Analysis

4.1.1. Results of Ultrasonic Spindle Speed on Surface Morphology

The effect degree of the ultrasonic spindle speed on the workpiece surface morphology was studied by rotatory ultrasonic machining tests on the ALHC core workpiece with a UCK cutting tool on the machine tool. The parameters of the cutting tool feed rate, tool cutting width, and tool cutting depth were fixed, and the spindle speed was gradually increased. The average length of tears L was used as the evaluation parameter to measure the surface quality due to tears that seriously affected the service life of the part. The processed surface was divided into three equal observation areas, two honeycomb cells can be observed in each observation area. The average tearing length was measured by the experiment setup in Figure 7, as shown in Figure 9. The surface quality experimental results of the RUM with the UCK cutting tool are shown in Figure 10.

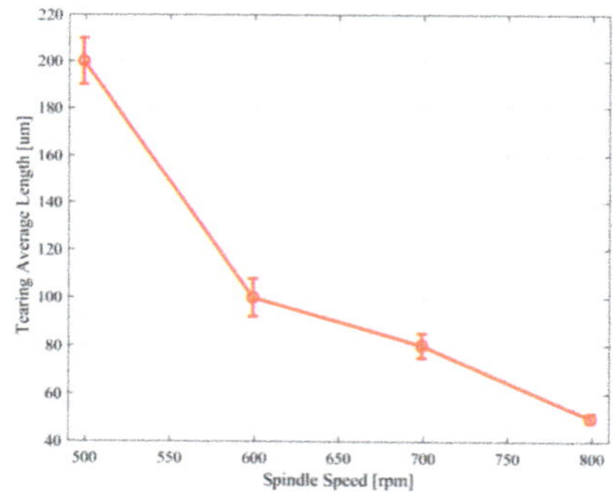

Figure 9. The effect of ultrasonic spindle speed on surface morphology.

As shown in Figures 9 and 10, the feed rate is fixed at 500 mm/min, the cutting width is fixed at 8 mm and the cutting depth is fixed at 0.8 mm. The range of spindle speeds is 500~800 rpm. The defect mainly included tearing, burr and crush fold on the surface morphology with the machined workpiece. The amount and the average length of tearing decreased with increased spindle speed during the ultrasonic machining experiment of the ALHC core workpieces with an ultrasonic circular knife cutting tool. The shearing effect on the material increased with an increased spindle speed, therefore, the shearing length decreased in this experiment at different spindle speeds from 500~800 rpm. Burr changed the same as tearing.

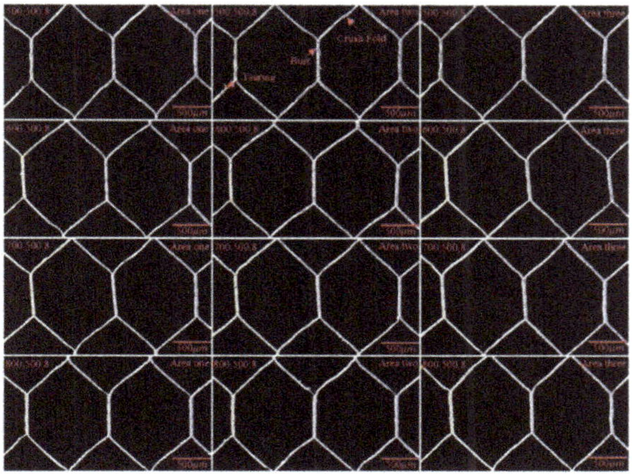

Figure 10. Surface morphology at different spindle speeds.

4.1.2. Results of Tool Feed Rate on Surface Morphology

The effect of the degree of tool feed rate on surface morphology was studied by rotatory ultrasonic machining tests on the ALHC core workpiece with a UCK cutting tool on the machine tool. The parameters of spindle speed, cutting width, and cutting depth were fixed, and the feed rate was gradually increased. The average tearing length was measured with the experimental setup in Figure 7, as shown in Figure 11. The surface quality experimental results of the RUM with a UCK cutting tool are shown in Figure 12.

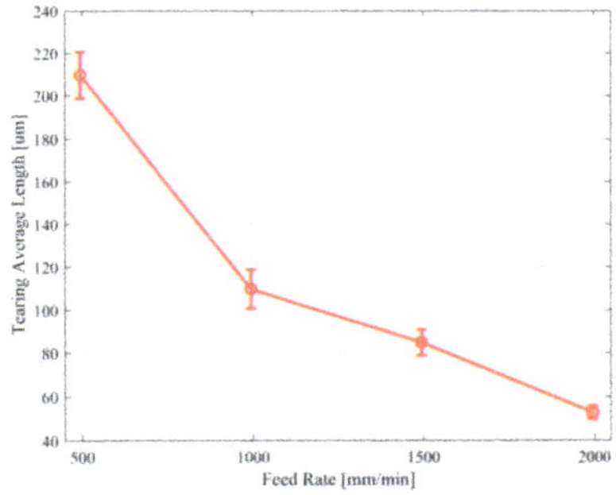

Figure 11. The effect of tool feed rate on surface morphology.

As shown in Figures 11 and 12, spindle speed is fixed at 800 rpm and cutting width is fixed at 8 mm. Both the number and length of tears decreased as the feed rate increased. The reason for this phenomenon was due to the increased feed rate speeding up the cutting process, shortening the touching time between the tool and the workpiece, thus improving the surface morphology to a certain extent, although the cutting force in the processing increased with an increased cutting tool feed rate.

4.1.3. Result of Tool Cutting Width on Surface Morphology

The effect of the degree of cutting width on surface morphology was researched by rotatory ultrasonic machining experiments on the ALHC core workpiece with a UCK cutting tool on the machine tool at different cutting widths from 4 μm to 10 μm. The parameters of the ultrasonic spindle speed, tool feed rate, tool cutting depth were fixed. The average tearing length was measured by the experimental setup in Figure 7, as shown in Figure 13. The surface quality experimental results of the RUM with a UCK cutting tool are shown in Figure 14.

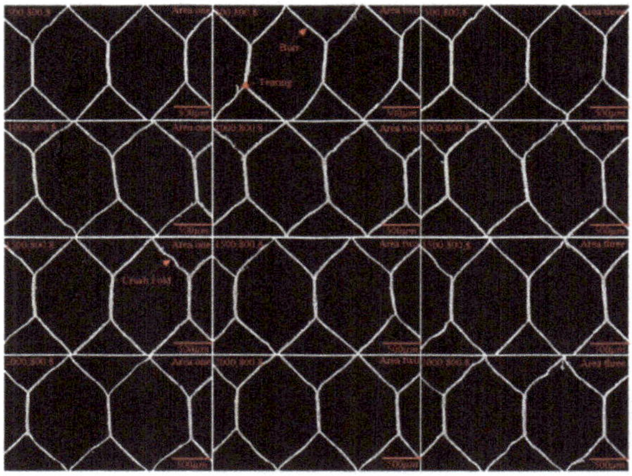

Figure 12. Surface morphology at different feed rate.

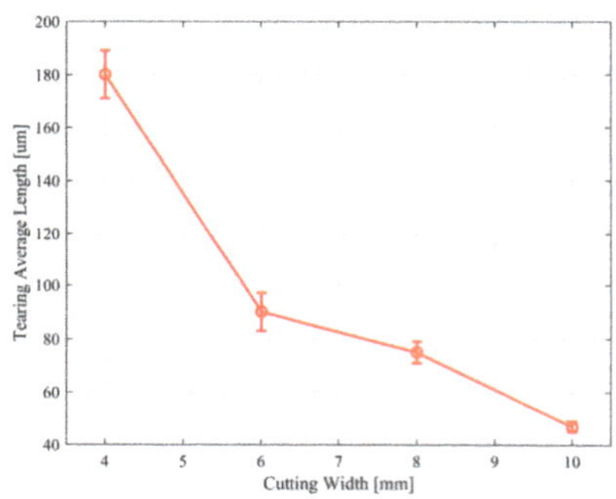

Figure 13. The effect of tool cutting width on surface morphology.

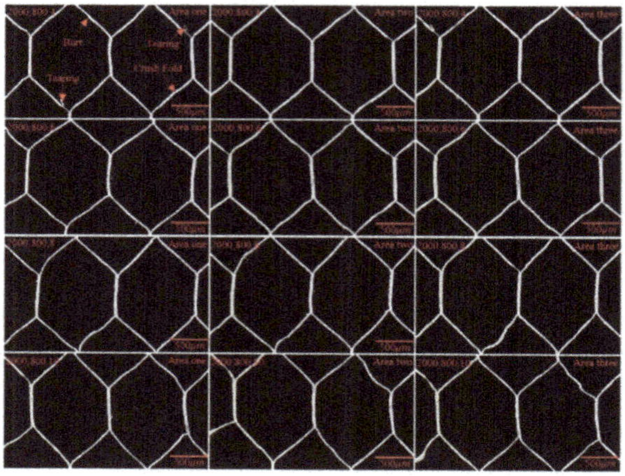

Figure 14. Surface morphology at different cutting widths.

As shown in Figures 13 and 14, the feed rate was fixed at 2000 mm/min and the spindle speed was fixed at 800 rpm. During the RUM experiments of the NCH core workpieces, tearing length decreased to a certain extent with an increased tool cutting width, but the effect was small. The authors found that the crush fold phenomenon was largely affected by the cutting width. The stiffness of the cell single wall was lower, therefore, the AL honeycomb cell single wall underwent plastic deformation due to the cutting force. The increase in tool cutting width increased the touching area between the cutting tool and the aluminum honeycomb, therefore, the material volume increased in the cutting process. The larger the cutting width, the more prone the aluminum honeycomb wall was to plastic deformation, which can lead to a crush fold defect.

4.1.4. Results of Cutting Parameters on Surface Morphology by Two-Factors/Four-Levels and One-Factor/Three-Factors $L_{12}(3 \times 4^2)$ Orthogonal Experiments

The influence of a single variable on the surface morphology of the ALHC core workpiece can be obtained through a group of single-factor experiments. However, the effect of a different combination of multiple variables on surface morphology cannot be determined. Therefore, according to the performance of the machine tool and ultrasonic device, two-factors/four-levels and one-factor/three-factors orthogonal experiments technology was proposed to study the effect of different cutting parameters on surface morphology. In this paper, the influence of various parameters, namely ultrasonic spindle speed, cutting tool feed rate and cutting width under the constant value of vibration amplitude 0.08 mm and cutting depth 2 mm were obtained by rotatory ultrasonic machining on an ALHC core specimen with an ultrasonic circular knife cutting tool. The authors only showed surface experimental results in area one to save space, as shown in Figure 15. In order to further analyze the significance of machining variables, the results of the orthogonal experiments were quantified by the evaluation parameter, namely L, of surface macro-topography proposed above. The orthogonal experimental significance analysis results in the average length L of tearing, as shown in Table 6.

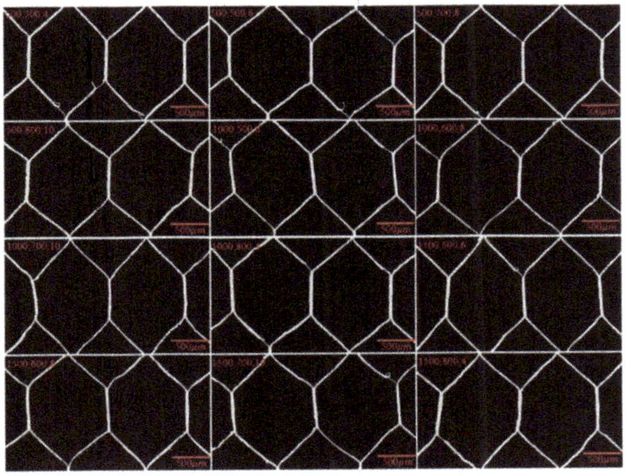

Figure 15. Surface morphology at different cutting parameters in orthogonal experiments.

Table 6. Two-factors/four-levels and one-factor/three-levels orthogonal experimental significance analysis of average length of tearing L (mm).

Level	Spindle Speed (rpm)	Feed Rate (mm/min)	Cutting Width (mm)
1	0.215	0.243	0.185
2	0.198	0.185	0.179
3	0.171	0.151	0.172
Range (R)	0.044	0.092	0.013

In conclusion from Table 6, the range variance of tearing average L among the three processing variables including ultrasonic spindle speed, tool feed rate and tool cutting width under the constant cutting depth and ultrasonic vibration amplitude during ultrasonic machining experiments of the Nomex honeycomb core workpiece by disc was:

$$R_{feed\ rate} > R_{spindle\ speed} > R_{cutting\ width} \quad (4)$$

It can be seen from Equation (1) that feed rate has the maximum effect on the tearing average length L and the cutting width has the minimum influence on the tearing average L.

4.2. NHC Core Surface Quality Analysis

4.2.1. Results of Ultrasonic Spindle Speed on Surface Morphology

The influence of the degree of ultrasonic spindle speed on surface quality was studied by rotatory ultrasonic machining experiments on a Nomex HC core workpiece with a disc cutting tool on the machine tool. The parameters of tool feed rate, tool cutting width, and cutting depth were fixed, and the spindle speed was gradually increased. The number of tears N (length more than 1 mm) and the average length of tears L were used as evaluation parameters to measure the surface quality as a surface with tears will seriously affect the service life of the part. The processed surface was divided into three equal observation areas, five honeycomb cells can be observed in each observation area, and the total number of tears and the average length of the three observation areas are counted. The surface quality experimental results of the RUM with a UCK cutting tool are shown in Figure 16. The average tearing length and amount were measured by the experimental setup in Figure 7, as shown in Figure 17.

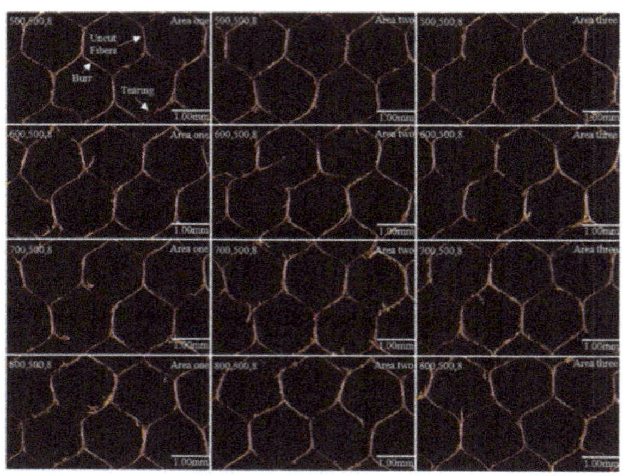

Figure 16. Surface morphology at different spindle speeds.

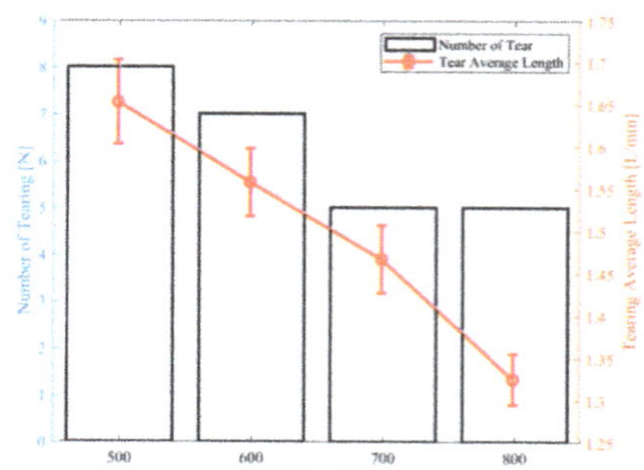

Figure 17. The effect of ultrasonic spindle speed on surface morphology.

As shown in Figures 16 and 17, area one, area two, and area three represent the observation areas, respectively. The feed rate was fixed at 500 mm/min, the cutting depth was fixed at 2 mm and the cutting width was fixed at 8 mm. The average length of tears L decreased when increasing the ultrasonic spindle speed and the total number of tears changes less during the experiments on the Nomex HC core specimen with a disc cutting tool.

The reason for this conclusion is that the high-frequency vibration of the cutting tool makes the vibration speed far greater than the tangential speed of the disc cutter during RUM, and the main cutting motion is the ultrasonic vibration of the tool. The main force for fracture separation of the honeycomb wall is not the shear force, but the high-frequency impact force of the disc cutter on the chip and the machined surface. Therefore, the change of spindle speed has little effect on the extrusion and the friction force of the cutting tool on the material along the feeding direction, and the amount of tearing does not change significantly. However, the increase in spindle speed can increase the shearing effect on the material to a certain extent, therefore, the shearing length decreases with an increase in cutting tool speed. The cutting force in processing decreases with ultrasonic spindle speed which increases in rotatory ultrasonic machining experiments of the NHC core specimens, therefore, the number of burrs decreases.

4.2.2. Results of Tool Feed Rate on Surface Morphology

The effect of the degree of tool feed rate on surface quality was studied with rotatory ultrasonic machining experiments on a Nomex HCs core workpiece with a disc cutting tool on the machine tool. The parameters of spindle speed, cutting width, and cutting depth were fixed, and the feed rate was gradually increased. The average tearing length and amount of tears were measured by the experimental setup in Figure 7, as shown in Figure 18. The surface quality experimental results of the RUM by disc cutting tool are displayed in Figure 19.

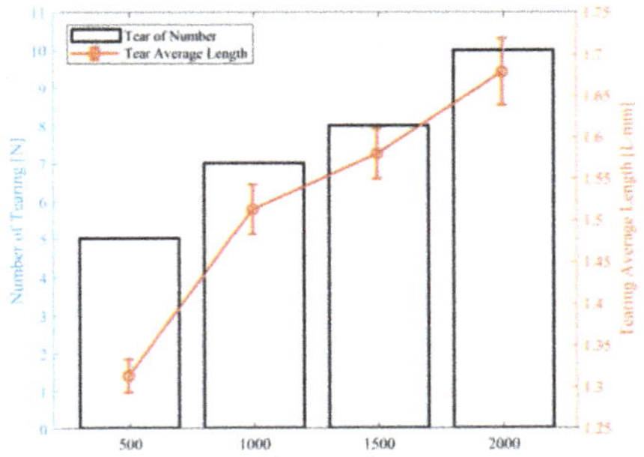

Figure 18. The effect of tool feed rate on surface morphology.

As shown in Figures 18 and 19, the spindle speed is fixed at 800 rpm and the cutting width is fixed at 8 mm. Both the number and length of tears increase as the feed rate increases. In ultrasonic cutting, the extrusion force of the cutting tool on the material decreases with a feed rate decrease. Therefore, the tearing defect generated by the cutting tool due to the compressive deformation of the honeycomb core wall increases when increasing the tool feed rate. However, RUM technology has a greater effect on reducing the extrusion force, therefore, the effect of feed rate on tearing tends to be gentle.

4.2.3. Results of Tool Cutting Width on Surface Morphology

The effect of the degree of tool cutting width on surface quality was studied by rotatory ultrasonic machining tests on a Nomex HC core workpiece with a disc cutting tool on the machine tool at different cutting widths from 4 μm to 10 μm. The parameters of the ultrasonic spindle speed, tool feed rate, and cutting depth were fixed. The average tearing length and amount of tears were measured by the experimental setup in Figure 7, as shown in Figure 20. The surface quality experimental results of RUM with the disc cutting tool are illustrated in Figure 21.

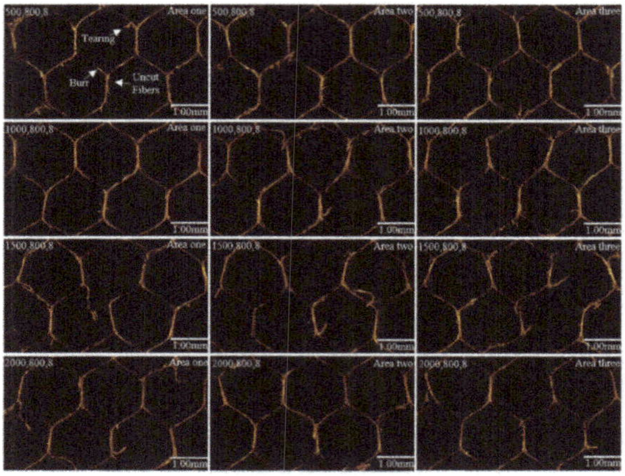

Figure 19. Surface morphology at different feed rates.

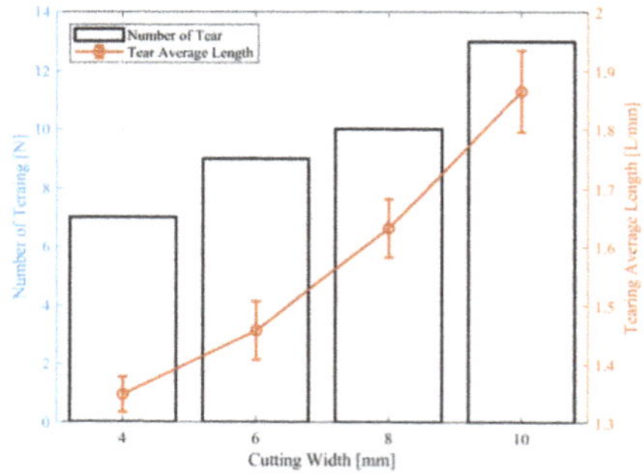

Figure 20. The effect of tool cutting width on surface morphology.

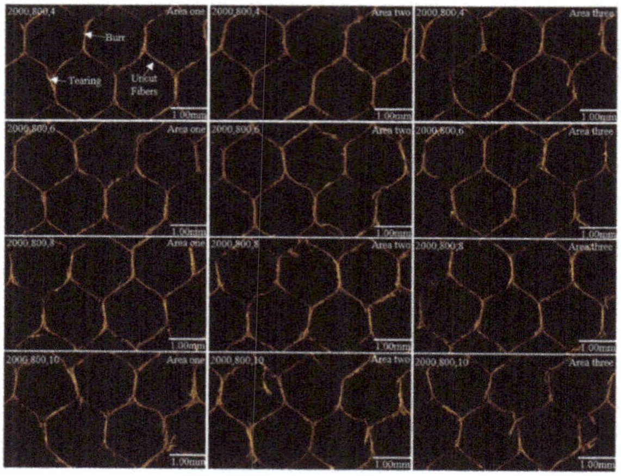

Figure 21. Surface morphology at different cutting widths.

As shown in Figures 20 and 21, the feed rate was fixed at 2000 mm/min and the spindle speed was fixed at 800 rpm.

Both the number and length of tears increase as the cutting width increases with the disc cutting tool during the RUM tests of the NCH core workpieces. The reason for this conclusion is that the touching area between the disc tool and the workpiece corresponding to the different cutting widths was also different, and the contact area was proportional to the cutting width, resulting in the larger the cutting width, the greater the friction between the tool and the workpiece in the cutting process. Burr changed the same as tearing.

4.2.4. Results of Cutting Parameters on Surface by Two-Factors/Four-Levels and One-Factor/Three-Factors $L_{12}(3 \times 4^2)$ Orthogonal Experiments

The influence degree of a single variable on the surface morphology of the Nomex HC core workpiece can be obtained through a group of single-factor experiments. However, the influence of a different combination of multiple variables on surface morphology cannot be determined. Therefore, according to the performance of the machine tool and the ultrasonic device, two-factors/four-levels and one-factor/three-factors orthogonal experiments technology were proposed to study the effect of different cutting parameters on surface morphology. In this study, the effect of cutting parameters on surface quality was investigated by establishing orthogonal experiments with different variable combinations. The authors only showed surface experiments in the results area one to save space, as shown in Figure 22. To further discuss the significance of cutting parameters, the results of orthogonal experiments were quantified by evaluation parameters, namely N, L, of the surface macro-topography, as proposed above. The orthogonal experimental significance analysis results of the total amount N and average length L of tearing were measured by the experimental device in Figure 7, as shown in Tables 7 and 8 separately.

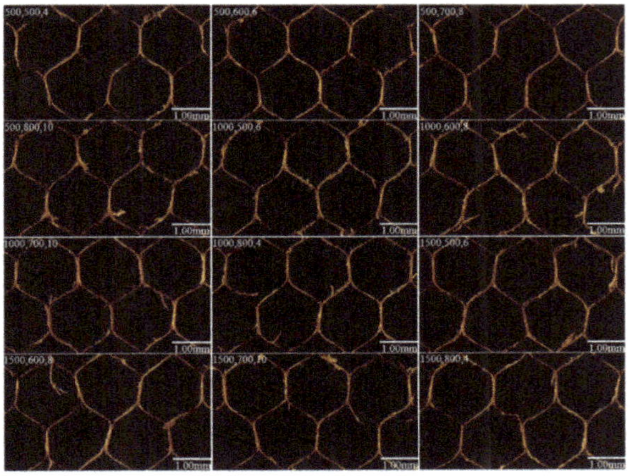

Figure 22. Surface morphology at different cutting parameters in orthogonal experiments.

Table 7. Two-factors/four-levels and one-factor/three-levels orthogonal experimental significance analysis of total amount of tearing N.

Level	Spindle Speed (rpm)	Feed Rate (mm/min)	Cutting Width (mm)
1	25	25	16
2	23	26	18
3	22	28	22
Range (R)	3	4	6

Table 8. Two-factors/four-levels and one-factor/three-levels orthogonal experimental significance analysis of average length of tearing L (mm).

Level	Spindle Speed (rpm)	Feed Rate (mm/min)	Cutting Width (mm)
1	1.724	1.417	1.347
2	1.538	1.515	1.632
3	1.482	1.628	1.769
Range (R)	0.242	0.211	0.434

In conclusion from Table 7, the range in the variance of tearing amount N among the three processing variables including ultrasonic spindle speed, tool feed rate and tool cutting width under the same depth and ultrasonic vibration amplitude of ultrasonic machining for the Nomex HC core workpiece with a disc cutting tool was:

$$R_{cutting\ width} > R_{feed\ rate} > R_{spindle\ speed} \tag{5}$$

From Equation (2) it can be observed that the cutting width has the maximum effect on the tearing amount N and the spindle speed has the least influence on the tearing amount N.

In conclusion from Table 8, the range in the variance of average tearing length L among the three processing variables including ultrasonic spindle speed, tool feed rate and tool cutting width under an identical depth and ultrasonic vibration amplitude of ultrasonic machining for the Nomex HC core workpiece with a disc cutting tool was:

$$R_{cutting\ width} > R_{spindle\ speed} > R_{feed\ rate} \tag{6}$$

From Equation (3) it can be seen that the cutting width has the maximum effect on the average tearing length L and the feed rate has the least influence on the tearing average L.

5. Conclusions

In this paper, the surface quality experiment for rotatory ultrasonic machining (RUM) of a honeycomb composite core (aluminum honeycomb and Nomex honeycomb) by ultrasonic circular knife (UCK) was developed to research the influence of different cutting variables on surface morphology for the aluminum honeycomb composite (ALHCs) core workpiece and the Nomex honeycomb composite (NHCs) core workpiece through a serious of single-factor experiments. Then, the optimal processing parameters of ultrasonic machining were received through orthogonal experimental technology. The results can be concluded as follows.

For the ALHC core workpiece, processing defects mainly include burr, tearing and crush fold of RUM with an ultrasonic disc cutting tool. Burr and tearing decrease with increasing ultrasonic spindle speeds and disc tool feed rates due to the increase in spindle speed increasing the shearing effect on the material to a certain extent and the increasing disc tool feed rate can speed up the cutting process, shortening the touching time between the disc cutting tool and the workpiece. Crush folds were largely affected by the cutting width through a group of single-factor experiments and these defects increase with the increase in cutting width due to the aluminum honeycomb being more prone to plastic deformation when there is a larger contact area between the cutting tool and the aluminum honeycomb. Feed rate has the greatest effect on the tearing average of surface morphology and cutting width has the smallest influence during the RUM experiment with a UCK cutting tool through significance analysis of two-factors/four-levels and one-factor/three-levels ($L_{12}(3 \times 4^2)$) orthogonal experimental results.

For the NHC core workpiece, burr, tearing and uncut fiber defects existed on the surface morphology of the RUM with a UCK cutting tool. Burr and tearing decrease when improving the ultrasonic spindle speed owing to the high-frequency impact force of the disc cutter on the chip and the machined surface. However, tearing increases with the improvement in disc tool feed rate because of an increase in extrusion force of the cutting tool on the material and the increase in material removal volume by the single-factor experiments. Therefore, burr and tearing increase when increasing the disc tool cutting width due to the increase in the touching area between the disc tool and the workpiece, resulting in an improvement in tool cutting force. Cutting width has the greatest effect on the tearing average and the number of tears to the surface morphology. Spindle speed has the smallest influence on the number of tears and the feed rate has the least effect on the tearing average during this experimental study by significance analysis of two-factors/four-levels and one-factor/three-levels ($L_{12}(3 \times 4^2)$) orthogonal experimental results.

The research in this paper can provide a basis for the subsequent comprehensive consideration of various factors to achieve high-quality processing of HC and it should be noted that the conditions of all single-factor experiments and orthogonal experiments in this paper were designed according to the actual performance of ultrasonic CNC machine tools in a factory, therefore, the results of this paper are generally applicable to most factories involved in the high surface quality processing of honeycomb composites.

Author Contributions: G.L. contributed to the conceptualization. J.Y. contributed to the experiment, writing and editing. L.Z. contributed to the methodology. Q.G. and L.Q. provided resources and performed experiment verification. R.Z. contributed to the English in the manuscript. All authors have read and agreed to the published version of the manuscript.

Funding: The research is sponsored by the National Natural Science Foundation of China (No. 51775328).

Institutional Review Board Statement: Not applicable.

Informed Consent Statement: Not applicable.

Data Availability Statement: The datasets generated and analyzed during the current study are available from the corresponding author on reasonable request.

Acknowledgments: The authors would like to thank Sichuan Innov Aviation Technology Co., Ltd., China and Suzou Interroll CNC Technology Co., Ltd., China for supporting the experiment equipment.

Conflicts of Interest: The authors declare no conflict of interest.

References

1. Harders, H.; Hupfer, K.; Rösler, J. Influence of cell wall shape and density on the mechanical behaviour of 2D foam structures. *Acta Mater.* **2005**, *53*, 1335–1345. [CrossRef]
2. Habib, F.N.; Iovenitti, P.; Masood, S.H.; Nikzad, M. Cell geometry effect on in-plane energy absorption of periodic honeycomb structures. *Int. J. Adv. Manuf. Technol.* **2018**, *94*, 2369–2380. [CrossRef]
3. Cvitkovich, M.K.; Jackson, W.C. Compressive Failure Mechanisms in Composite Sandwich Structures. *J. Am. Helicopter Soc.* **1999**, *44*, 260–268. [CrossRef]
4. Zhou, G.; Hill, M.; Loughlan, J.; Hookham, N. Damage Characteristics of Composite Honeycomb Sandwich Panels in Bending under Quasi-static Loading. *J. Sandw. Struct. Mater.* **2006**, *8*, 55–90. [CrossRef]
5. Kim, C.-G.; Jun, E.-J. Impact Resistance of Composite Laminated Sandwich Plates. *J. Compos. Mater.* **1992**, *26*, 2247–2261. [CrossRef]
6. Abrate, S. Localized Impact on Sandwich Structures with Laminated Facings. *Appl. Mech. Rev.* **1997**, *50*, 69–82. [CrossRef]
7. Lacy, T.E.; Hwang, Y. Numerical modeling of impact-damaged sandwich composites subjected to compres-sion-after-impact loading. *Compos. Struct.* **2003**, *61*, 115–128. [CrossRef]
8. Aktay, L.; Johnson, A.F.; Kröplin, B.-H. Numerical modelling of honeycomb core crush behaviour. *Eng. Fract. Mech.* **2008**, *75*, 2616–2630. [CrossRef]
9. Zhou, Z.; Wang, Z.; Zhao, L.; Shu, X. Experimental investigation on the yield behavior of Nomex honeycombs under combined shear-compression. *Lat. Am. J. Solids Struct.* **2012**, *9*, 515–530. [CrossRef]
10. An, Q.; Dang, J.; Ming, W.; Qiu, K.; Chen, M. Experimental and Numerical Studies on Defect Characteristics During Milling of Aluminum Honeycomb Core. *J. Manuf. Sci. Eng.* **2019**, *141*, 031006. [CrossRef]
11. Abbadi, A.; Tixier, C.; Gilgert, J.; Azari, Z. Experimental study on the fatigue behaviour of honeycomb sandwich panels with artificial defects. *Compos. Struct.* **2015**, *120*, 394–405. [CrossRef]
12. Meruane, V.; del Fierro, V. An inverse parallel genetic algorithm for the identification of skin/core debonding in honeycomb aluminium panels. *Struct. Control Health Monit.* **2015**, *22*, 1426–1439. [CrossRef]

13. Wang, J.; Feng, P.; Zhang, J.; Guo, P. Experimental study on vibration stability in rotary ultrasonic machining of ceramic matrix composites: Cutting force variation at hole entrance. *Ceram. Int.* **2018**, *44*, 14386–14392. [CrossRef]
14. Wang, H.; Cong, W.; Ning, F.; Hu, Y. A study on the effects of machining variables in surface grinding of CFRP composites using rotary ultrasonic machining. *Int. J. Adv. Manuf. Technol.* **2018**, *95*, 3651–3663. [CrossRef]
15. Yuan, S.; Li, Z.; Zhang, C.; Guskov, A. Research into the transition of material removal mechanism for C/SiC in rotary ultrasonic face machining. *Int. J. Adv. Manuf. Technol.* **2018**, *95*, 1751–1761. [CrossRef]
16. Chen, X.; Wang, H.; Hu, Y.; Zhang, D.; Cong, W.; Burks, A.R. Rotary ultrasonic machining of CFRP compo-sites: Effects of machining variables on workpiece delamination. In International Manufacturing Science and Engineering Conference. *Am. Soc. Mech. Eng.* **2019**, *58752*, V002T03A051.
17. Li, C.; Piao, Y.; Meng, B.; Hu, Y.; Li, L.; Zhang, F. Phase transition and plastic deformation mechanisms induced by self-rotating grinding of GaN single crystals. *Int. J. Mach. Tools Manuf.* **2022**, *172*, 103827. [CrossRef]
18. Sandá, A.; Sanz, C. Rotary ultrasonic machining of ZrO2-NbC and ZrO2-WC ceramics. *Int. J. Mach. Mach. Mater.* **2020**, *22*, 165–179.
19. Abdo, B.M.A.; El-Tamimi, A.; Alkhalefah, H. Parametric Analysis and Optimization of Rotary Ultrasonic Machining of Zirconia (ZrO_2) Ceramics. *IOP Conf. Series: Mater. Sci. Eng.* **2020**, *727*, 012009. [CrossRef]
20. Liu, J.; Jiang, X.; Han, X.; Zhang, D. Influence of parameter matching on performance of high-speed rotary ultra-sonic elliptical vibration-assisted machining for side milling of titanium alloys. *Int. J. Adv. Manuf. Technol.* **2019**, *101*, 1333–1348. [CrossRef]
21. Zhang, M.; Zhang, D.; Geng, D.; Shao, Z.; Liu, Y.; Jiang, X. Effects of tool vibration on surface integrity in rotary ultrasonic elliptical end milling of Ti–6Al–4V. *J. Alloy. Compd.* **2020**, *821*, 153266. [CrossRef]
22. Wang, J.; Zhang, J.; Feng, P.; Guo, P. Damage formation and suppression in rotary ultrasonic machining of hard and brittle materials: A critical review. *Ceram. Int.* **2018**, *44*, 1227–1239. [CrossRef]
23. Wang, J.; Feng, P.; Zhang, J. Reducing edge chipping defect in rotary ultrasonic machining of optical glass by compound step-taper tool. *J. Manuf. Process.* **2018**, *32*, 213–221. [CrossRef]
24. Zha, H.; Feng, P.; Zhang, J.; Yu, D.; Wu, Z. Material removal mechanism in rotary ultrasonic machining of high-volume fraction SiCp/Al composites. *Int. J. Adv. Manuf. Technol.* **2018**, *97*, 2099–2109. [CrossRef]
25. Dong, S.; Liao, W.; Zheng, K.; Liu, J.; Feng, J. Investigation on exit burr in robotic rotary ultrasonic drilling of CFRP/aluminum stacks. *Int. J. Mech. Sci.* **2019**, *151*, 868–876. [CrossRef]
26. Xiang, D.; Wu, B.; Yao, Y.; Liu, Z.; Feng, H. Ultrasonic longitudinal-torsional vibration-assisted cutting of Nomex®honeycomb-core composites. *Int. J. Adv. Manuf. Technol.* **2019**, *100*, 1521–1530. [CrossRef]
27. Xia, Y.; Zhang, J.; Wu, Z.; Feng, P.; Yu, D. Study on the Design of Cutting Disc in Ultrasonic-Assisted Machining of Honeycomb Composites. In *IOP Conference Series: Materials Science and Engineering*; IOP Publishing: Bristol, UK, 2019; Volume 611, p. 012032.
28. Sun, J.; Dong, Z.; Wang, X.; Wang, Y.; Qin, Y.; Kang, R. Simulation and experimental study of ultrasonic cutting for aluminum honeycomb by disc cutter. *Ultrasonics* **2020**, *103*, 106102. [CrossRef]
29. Ahmad, S.; Zhang, J.; Feng, P.; Yu, D.; Wu, Z.; Ke, M. Processing technologies for Nomex honeycomb composites (NHCs): A critical review. *Compos. Struct.* **2020**, *250*, 112545. [CrossRef]
30. Ahmad, S.; Zhang, J.; Feng, P.; Yu, D.; Wu, Z.; Ke, M. Research on Design and FE Simulations of Novel Ultrasonic Circular Saw Blade (UCSB) Cutting Tools for Rotary Ultrasonic Machining of Nomex Honeycomb Composites. In Proceedings of the 2019 16th International Bhurban Conference on Applied Sciences and Technology (IBCAST), Islamabad, Pakistan, 8–12 January 2019; pp. 113–119.
31. Ahmad, S.; Zhang, J.; Feng, P.; Yu, D.; Wu, Z. Experimental study on rotary ultrasonic machining (RUM) character-istics of Nomex honeycomb composites (NHCs) by circular knife cutting tools. *J. Manuf. Processes* **2020**, *58*, 524–535. [CrossRef]
32. Asmael, M.; Safaei, B.; Zeeshan, Q.; Zargar, O.; Nuhu, A.A. Ultrasonic machining of carbon fiber–reinforced plastic composites: A review. *Int. J. Adv. Manuf. Technol.* **2021**, *113*, 3079–3120. [CrossRef]
33. Hu, X.; Yu, B.; Li, X.; Chen, N. Research on Cutting Force Model of Triangular Blade for Ultrasonic Assisted Cutting Honeycomb Composites. *Proc. CIRP* **2017**, *66*, 159–163. [CrossRef]
34. Cao, W.; Zha, J.; Chen, Y. Cutting Force Prediction and Experiment Verification of Paper Honeycomb Materials by Ultrasonic Vibration-Assisted Machining. *Appl. Sci.* **2020**, *10*, 4676. [CrossRef]

crystals

Article

Surface Shape Evolution of Optical Elements during Continuous Polishing of Fused Quartz

Yiren Wang [1], Feihu Zhang [1,2,*] and Chen Li [1,2,*]

1 School of Mechatronics Engineering, Harbin Institute of Technology, Harbin 150001, China; 16b908085@stu.hit.edu.cn
2 State Key Laboratory of Robotics and System (HIT), Harbin Institute of Technology, Harbin 150001, China
* Correspondence: zhangfh@hit.edu.cn (F.Z.); lichen1992@hit.edu.cn (C.L.)

Abstract: Continuous polishing is the first choice for machining optical elements with a large aperture. The lubrication in the continuous polishing is an important factor affecting the surface quality of the optical elements. In this study, the lubrication system between the optic element and polishing lap was analyzed firstly and then was verified by the measurement experiment of the friction coefficient. In addition, the numerical simulation model of the mixture lubrication was established. The polishing pressure distribution and material removal distribution can be obtained by the model. The influences of the rotating speed, optical element load, and surface roughness of the polishing lap on polishing pressure were also analyzed. Finally, the influence rules of the lubrication on the surface shape of optical elements were revealed by the polishing experiments.

Keywords: continuous polishing; lubrication; optical elements; surface shape

Citation: Wang, Y.; Zhang, F.; Li, C. Surface Shape Evolution of Optical Elements during Continuous Polishing of Fused Quartz. *Crystals* **2022**, *12*, 736. https://doi.org/10.3390/cryst12050736

Academic Editor: Anton Meden

Received: 25 April 2022
Accepted: 17 May 2022
Published: 20 May 2022

1. Introduction

Abrasive machining technologies such as grinding and polishing are the necessary means to realize the precision machining of optical elements [1–3]. Continuous polishing technology is a full-aperture covering method for polishing optics. The research on continuous polishing is more qualitative than quantitative. The machining instability of the continuous polishing seriously restricts the efficiency of planar optical elements with a large aperture. To reduce the production cycle and processing cost, it is necessary to enhance the understanding of the continuous polishing technology and solve the key problems which restrict the development of the continuous polishing [4–8].

The lubrication in the continuous polishing area is an important factor affecting the surface quality of optical elements [9]. It affects the polishing pressure and material removal rate, and has a distinct influence on the surface shape of the substrates [10,11]. Therefore, the study on the lubrication of the machining area is of great significance to reveal the removal mechanism of the continuous polishing technology [12,13].

Chang et al. [14] studied the thickness of the liquid film between the wafer and polishing lap, and the results indicated that the thickness was directly proportional to the rotational speed and Hersey number of the wafer. The film thickness had distinct effect on the performance of the chemical-mechanical polishing (CMP) process, such as material removal and deformation behaviors. Controlling and optimizing the slurry film thickness were helpful to maintain the process stability. Runnels et al. [15,16] believed that the hydrodynamic effect in CMP polishing was an important factor affecting the uniform removal of the work materials. Revealing the hydrodynamic effect can improve the stability and surface integrity of the polishing process. Terrell et al. [17] summarized the research on the hydrodynamics in CMP, and the results demonstrated that the existing research results were imperfect, and did not point out the specific impact of the hydrodynamics during the polishing process. More in-depth research should be carried out on the hydrodynamics effect. Waechter et al. [18] found that with the increase of the relative speed, the contact between the optics and polishing lap were divided into four stages, namely low-speed solid contact friction, medium-speed solid contact, hydrodynamic mixture friction, and high-speed hydrodynamic friction.

Suadarajan et al. [19] used the lubrication theory model to solve the thickness, pressure, and velocity distributions of the liquid film between the polishing pad and optics. The model did not consider the elasticity of the polishing pad, and considered that the surfaces of the optics and polishing pad were smooth. In addition, the results showed that only the hydrodynamic pressure of the liquid film was not enough to support the optics [20], e.g., the protrusion of the polishing pad inevitably contacted with the optics surface. Tichy model described a one-dimensional contact model by using contact mechanics and lubrication theory [21]. Although this model considered the elasticity of the polishing pad and deformation of the surface bulge, it simplified the actual situation.

In conclusion, lubrication has a significant impact on the polishing of optical elements. However, there are few studies on the flow field in the continuous polishing process, and the evolution law of the optic surface shape has not been revealed. To optimize the surface accuracy of the optical substrates, this study is aimed to clarify the lubrication mechanism in continuous polishing and reveal the influence law of flow field on the surface accuracy of optical elements.

2. Lubrication Method in Continuous Polishing

2.1. Lubrication Status Classification in Continuous Polishing

The continuous polishing machine uses a large substrate platen made from granite [22], as shown in Figure 1. An annular pitch layer is prepared on the granite plate surface as the grooves are cut into different shapes on the polishing lap surface to improve its fluidity and the slurry transmission. There are two work rings on one side of the annulus, and the inner diameter is equal to the annulus width. The plane optical elements to be polished are placed in these rings. There is a large circular truing tool called the 'conditioner' on the remaining portion of the annulus, which is substantially wider than the radius of the polishing pad. During the continuous polishing processes, the polishing lap, work rings and conditioner are driven by a servo motor. Abrasive slurry is sprayed on the surface of the polishing lap through the nozzle, and it transported to the bottom of the element through the groove. The surface finishing can be realized by the mechanical action of the polishing particles.

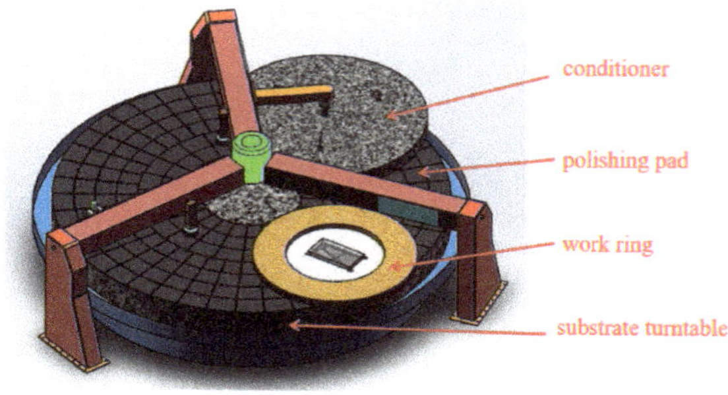

Figure 1. Continuous polishing machine.

Different contact modes between the optical element and polishing lap directly determine the different lubrication characteristics, as shown in Figure 2. As a result, the polishing fluid has different flow characteristics between the element and polishing lap. Different polishing fluid flow characteristics affect the removal mechanism, removal rate and accuracy of continuous polishing.

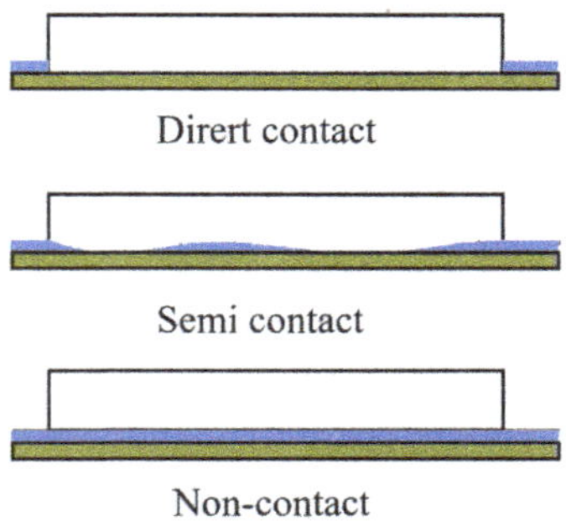

Figure 2. Contact modes between the optical element and polishing lap.

The transformation between lubrication state and several kinds of states is the basis of theoretical analysis of the fluid lubrication. According to the formation mechanism and characteristics of the fluid lubrication film, the lubrication state can be divided into boundary lubrication, fluid film lubrication and mixture lubrication.

(1) Boundary lubrication refers to the transition of the lubricant between the friction pairs from the internal friction between molecules to direct contact between two friction surfaces. In this lubrication state, the viscosity characteristics of the lubricant do not have effect on the lubrication. The lubrication function of the fluid film is weak and can be ignored, and there is more contact between the micro-convex bodies on the friction surface. The surface friction and wear characteristics are determined by the interaction among the lubricant, friction surface and properties of the boundary film. The load between the friction pairs is provided by the surface micro-convex body.
(2) Fluid lubrication refers to the lubrication state in which the friction surface is separated by a continuous fluid film. In this state, the surface of the friction pair does not directly contact with each other and the contact wear does not occur.
(3) Mixture lubrication refers to the coexistence of the boundary lubrication and fluid lubrication. In this state, the lubricating film is discontinuous, and is divided into fluid lubricating film and boundary lubricating film. On the one hand, the fluid lubrication film bears part of the load and produces viscous friction. On the other hand, the boundary lubrication film and the micro-convex body on the contact surface bear another part of the load, resulting in the dry friction on the contact surface. In addition, the mixture lubrication state is often unstable and fluctuates between the fluid film lubrication and boundary lubrication.

2.2. *Judgment Method of Lubrication State of Continuous Polishing*

In the state of the fluid film lubrication, when the load increases and the relative motion speed of the friction pair and lubricant viscosity decrease, the bearing capacity of the fluid film, thickness of the lubrication film and spacing between the friction surfaces decrease. If there is a certain amount of the micro-convex contact in the friction, this state may be transformed into the mixture lubrication. A judgment method is introduced below.

The calculation formula of the film thickness ratio is given in Equations (1) and (2).

$$\psi = h_m / R_q \tag{1}$$

$$R_a = 4\left(R_{a1}^2 + R_{a2}^2\right)^{\frac{1}{2}} / 3 \tag{2}$$

where, h_m is the minimum lubricating film thickness, R_q is the comprehensive surface roughness. R_{a1}, and R_{a2} are the average deviations of the arithmetic contour of the surface roughness.

According to the calculation formula, when $\psi \leq 1$, it is in the boundary lubrication state, and there are many micro-convex contacts in the friction pair. When $1 < \psi < 3$, the film thickness ratio increases within this range, the fluid transits from the boundary lubrication to fluid lubrication, and the fluid film thickness increases. When $\psi \approx 3$, for the high pair contact with a large load, it is in the state of full elastic hydrodynamic lubrication. When $\psi > 3$, it is in the state of complete fluid film lubrication. The thickness of the liquid film and roughness of the polishing lap are approximately 25 and 10 μm, respectively [23]. Therefore, the film thickness ratio of the continuous polishing ψ is equal to 2.5, which belongs to mixture lubrication. The numerical simulation model of the mixture lubrication will be established, and, then, the accuracy of the model will be verified by the experiments.

3. Numerical Model of Mixture Lubrication for Continuous Polishing

3.1. Load Sharing Theory

When the mixture lubrication is formed through continuous polishing, it is necessary to consider both the rough peak and liquid film in the contact area. According to the Johnson load thought [24], the total contact load F_N is shared by the liquid film pressure F_h and rough peak force F_a on the contact surface, as given in Equation (3). The friction F_f is composed of the lubricating liquid film friction F_{fh} and rough peak friction F_{fa}, as shown in Equation (4).

$$F_N = F_h + F_a \tag{3}$$

$$F_f = F_{fh} + F_{fa} \tag{4}$$

Introducing the scale factor r, the liquid film bearing ratio is $1/r_1$, and the micro-convex body bearing ratio is $1/r_2$. The bearing ratio of the liquid film refers to the proportion of the pressure carried by the whole liquid film in the total optic load. The relation between these parameters can be expressed by Equations (5) and (6).

$$F_h = F_N/r_1, F_a = F_N/r_2 \tag{5}$$

$$\frac{1}{r_1} + \frac{1}{r_2} = 1 \tag{6}$$

On the contact interface, different micro-convex bodies have the same friction coefficient, and the friction of micro-convex bodies is expressed by Equation (7).

$$F_{fa} = \mu_c F_a = \frac{1}{r_2} F_N \mu_c \tag{7}$$

The polishing fluid of the continuous polishing can be regarded as Newtonian fluid at a low speed [25]. The shear stress of the fluid is known from the Newton law of fluid viscosity, as shown in Equation (8).

$$\tau = \frac{\eta u}{h} \tag{8}$$

where τ is the shear stress, η is the viscosity of polishing solution, and u is the sliding speed.

Therefore, the friction coefficient of mixture lubrication can be calculated by Equation (9).

$$u_f = \frac{F_C}{F_N} = \frac{\sum\limits_{i=1}^{N} \iint\limits_{A_{Ci}} \tau_{Ci}(\dot{r}) dA_{Ci} + \frac{1}{r_2} F_N \mu_c}{F_N} = \frac{\eta u A_H}{h F_N} + \frac{1}{r_2}\mu_c \tag{9}$$

where μ_c is the boundary friction coefficient, lubrication area is $A_H = A_{mon} - A_r$, A_{mon} is the nominal contact area, A_r is the actual contact area, and h is the average liquid film thickness.

3.2. Calculation Method of Liquid Film Thickness

The basic concept of the lubrication is to solve Reynolds equation to reveal the distribution of the pressure in fluid lubrication film. The temperature difference in the annular polishing lubrication system is very small, so it is unnecessary to consider the change of the lubricant viscosity and density with the temperature. In addition to the isothermal conditions, the following assumptions should be used in the derivation:

(1) The effect of the volume force, such as gravity or magnetic force is ignored. In addition to the theory of the electronic fluid or magnetic fluid lubrication, this assumption is usually reasonable.
(2) The fluid has no sliding on the solid interface, that is, the velocity of the fluid particle attached to the interface is the same as that of the point on the interface.
(3) In the direction along the thickness of the lubricating film, the change of the pressure is ignored. Because the film thickness is only less than one micron to tens of microns, in such a thin range, the pressure cannot change significantly. From this hypothesis, it can be inferred that the viscosity and density of the fluid do not change in the direction of film thickness.
(4) Compared with the film thickness, the radius of the curvature of the support surface is large, and, accordingly, the change of velocity direction caused by the surface curvature can be ignored.
(5) The lubricant is Newtonian fluid.
(6) The flow is laminar, without vortex and turbulence.
(7) Compared with the viscous force, the influence of inertial force can be ignored, including the inertial force of the fluid acceleration and centrifugal force of the fluid film bending.

Based on the above assumptions, the Reynolds equation under the isothermal conditions is given in Equation (10).

$$\frac{\partial}{\partial x}\left(\frac{\rho h^3}{\eta}\cdot\frac{\partial p}{\partial x}\right)+\frac{\partial}{\partial y}\left(\frac{\rho h^3}{\eta}\cdot\frac{\partial p}{\partial y}\right)=12\frac{\partial}{\partial x}(\rho U h)+12\frac{\partial}{\partial y}(\rho V h)+12\frac{\partial(\rho h)}{\partial t} \tag{10}$$

where, ρ is the fluid density. U and V are the speeds of the optic and polishing lap, respectively.

$$U=(U_1+U_2)/2, V=(V_1+V_2)/2 \tag{11}$$

The pressure P is integrated in the whole lubricating film range, and the result is the bearing capacity of the lubricating film. For optical elements in the ring polishing, the bearing capacity is given in Equation (12).

$$W=\iint p dx dy \tag{12}$$

Two dimensionless parameters are defined as Equation (13).

$$\overline{W}=\frac{W}{E'R}, \overline{U}=\frac{\eta_0 U}{E'R} \tag{13}$$

where $\overline{W}$ is the load parameter, and $\overline{U}$ is the speed parameter. In the definition of $\overline{W}$, E' is the equivalent elastic modulus, and it can be calculated by the elastic modulus E_1, E_2 and Poisson's ratio v_1 and v_2 of the two surfaces, as given in Equation (14).

$$\frac{1}{E'}=\frac{1}{2}\left(\frac{1-v_1^2}{E_1}+\frac{1-v_2^2}{E_2}\right) \tag{14}$$

In the definition of $\overline{U}$, R is the equivalent radius of curvature, and it can be calculated by the radii R_1 and R_2 of the two contact surfaces, as given in Equation (15).

$$\frac{1}{R}=\frac{1}{R_1}+\frac{1}{R_2} \tag{15}$$

The schematic diagram of the liquid film is show in Figure 3, where the fluid density ρ is assumed as a constant. The Reynolds equation can be derived from Equation (10) by assuming that the viscosity and density of the fluid do not change in the direction of film thickness, as shown in Equation (16).

$$\frac{dp}{dx} = 12\eta_0 U \frac{h - h_0}{h^3} \tag{16}$$

where h_0 is the actual film thickness at $x = 0$.

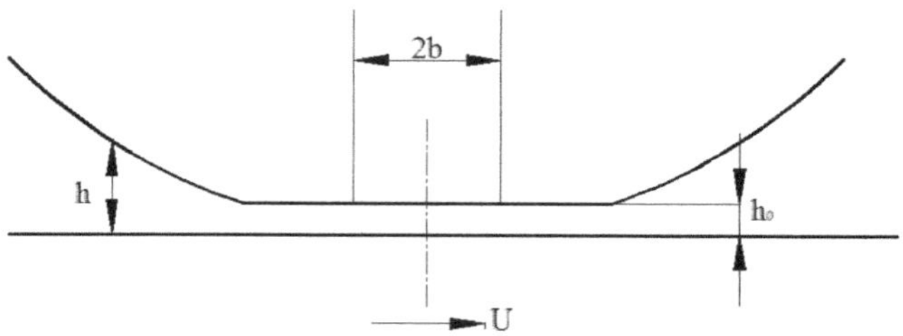

Figure 3. Schematic diagram of the liquid film.

When $\overline{h} = h/R, \overline{p} = p/E, \overline{x} = x/b$, Equation (16) can be rewritten into Equation (17).

$$\frac{d\overline{p}}{d\overline{x}} = 48\left(\frac{\overline{W}}{2\pi}\right)^{1/2} \overline{U}\frac{\overline{h} - \overline{h}_0}{\overline{h}^3} \tag{17}$$

The film thickness can be expressed by Equation (18).

$$\overline{h}_0 = 1.95\left(E'\overline{U}\right)^{8/11} / \overline{W}^{1/11} \tag{18}$$

The dimensional form is given in Equation (19).

$$h_0 = 1.95(a\eta_0 U)^{8/11} R^{4/11} \left(E'/W\right)^{1/11} \tag{19}$$

Equation (19) is the empirical formula of the average thickness of the liquid film in the contact area.

In the mixture lubrication problem, the actual thickness of the liquid film should also consider the elastic deformation and surface roughness of the polishing lap. Note that the sum of the elastic deformation of the two surfaces is δ and the roughness function is S(x). The calculation of the liquid film thickness is given in Equation (20).

$$h = h_0 + \frac{x^2}{2R} + \mathrm{S(x)} + \delta \tag{20}$$

It is known from the elasticity that, for a semi-infinite body, when a concentrated force P uniformly distributed in the Y direction acts on the z-axis, the normal elastic deformation $\delta(x)$ at any point M with abscissa x is given in Equation (21).

$$\delta(x) = -\frac{1 - v^2}{\pi E} P \ln x^2 - \frac{1 + v}{\pi E} p + C_0 \tag{21}$$

where E is the elastic modulus, v is the Poisson's ratio, and C_0 is the constant. C_0 and H_0 are combined into one term, and the liquid film thickness h is obtained in Equation (22).

$$h_c = h_0 + \frac{x^2}{2R} - \frac{1 - v^2}{\pi E} P \ln x^2 - \frac{1 + v}{\pi E} P \tag{22}$$

When two rough surfaces come into contact with each other, Greenwood and Trip [26] deduced the equation of the total load by the micro-convex body, as given in Equation (23).

$$F_h = \frac{8\sqrt{2}}{15}\pi n^2 \beta^{1.5} {\sigma_s}^{2.5} E' F(H) \tag{23}$$

where n is the number of micro-convex bodies per unit area, β is the radius of micro-convex body, σ is the root mean square value of the surface roughness. $F(H)$ can be calculated by Equation (24).

$$F(H) = 4.4086 \times 10^{-5}(4 - H)^{6.804} \tag{24}$$

$$H = \frac{h_c - d_d}{\sigma_s} \tag{25}$$

where d_d is the distance between the average surface of the micro-convex body and plane of the polishing lap, which is about 1.15 σ_s.

3.3. Numerical Simulation Method of Mixture Lubrication

In order to calculate the friction coefficient of the mixture lubrication, it is necessary to obtain the liquid film thickness h and bearing ratio of the micro-convex body $1/r_2$. The liquid film pressure is calculated according to the liquid film bearing ratio, and the liquid film thickness is calculated according to the liquid film pressure. Calculation steps are as follows.

(1) The initial parameters are set, including the viscosity η of the polishing solution and the characteristic values of the surface roughness of the polishing lap (n, β, σ).
(2) The initial values of the liquid film bearing ratio r_1 and micro-convex body bearing ratio r_2 are set, and h_0 is calculated according to the empirical formula.
(3) The liquid film load W and dimensionless parameter in Equation (1) are calculated.
(4) Calculate the liquid film thickness h_c and load on the micro-convex body F_h according to Equation (3).
(5) Calculate the load on the micro-convex body F_{h1} according to Equation (23).
(6) Judge whether the difference between the two ends of the load on the micro-convex body is less than 10^{-4}.
(7) If the friction coefficients r_1 and r_2 are substituted into the balance condition, the friction coefficient μ is obtained by substituting it into Equation (9). Otherwise, repeat steps (2)–(6) until the error converges.

3.4. Numerical Simulation Results of Mixture Lubrication

3.4.1. Calculation Results of Liquid Film Thickness

The relationship between the liquid film thickness and relative speed of the polishing lap is shown in Figure 4a. When the surface roughness of the polishing lap is constant, the thickness of the liquid film increases with the increase of the rotating speed of the polishing lap. This occurred because the increase of the polishing lap speed increases the flow rate of the polishing liquid. The relationship between the liquid film thickness and relative speed of the polishing lap is shown in Figure 4b. When the rotating speed of the polishing lap is constant, the thickness of the liquid film between the element and polishing lap increases with the increase of the surface roughness of the polishing lap. This occurred because the increase of the surface roughness of the polishing lap will make more polishing liquid flow between the element and polishing lap, resulting in the increase of the thickness of the liquid film.

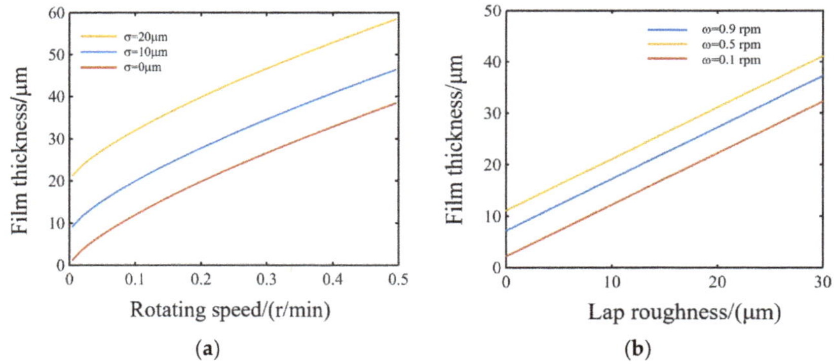

Figure 4. Influence of rotating speed and lap roughness on liquid film thickness. (**a**) Liquid film thickness vs. rotating speed (**b**) Liquid film thickness vs. lap roughness.

The Influence of the total load on liquid film thickness is shown in Figure 5. When the roughness of the polishing pad is a constant, the thickness of the liquid film decreases with the increase of the optic load. This is because the contact area between the element and polishing lap becomes larger, which squeezes the flow space of the polishing liquid. When the roughness increases and the total load is a constant, the liquid film thickness still maintains an increase trend.

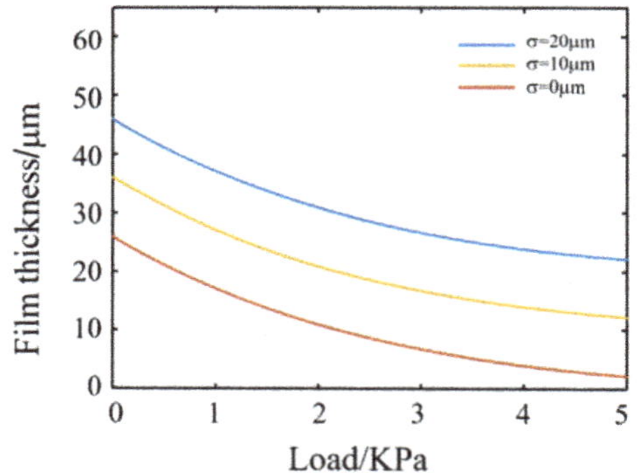

Figure 5. Influence of total load on liquid film thickness.

3.4.2. Calculation Results of Liquid Film Bearing Ratio

The relationship between the liquid film bearing ratio and polishing lap speed is shown in Figure 6, which indicates that, when the rotating speed is constant, the proportion of the liquid film load increases with the increase of the surface roughness of the polishing lap. This is because the rough surface increases the thickness of the liquid film.

When the surface roughness of the polishing lap is a constant, the proportion of the liquid film load will increase with the increase of the lap rotating speed. This is because the probability of the direct contact between the element and micro-convex body on the surface of the polishing lap decreases with the increase of the thickness of the liquid film. When the rotating velocity increases to a certain extent, the proportion of the liquid film load increases slowly and is close to 1. It indicates that the lubrication state is in hydrodynamic lubrication mode at this time. The element and polishing lap are completely separated by a continuous liquid film, and the load is completely borne by the liquid film.

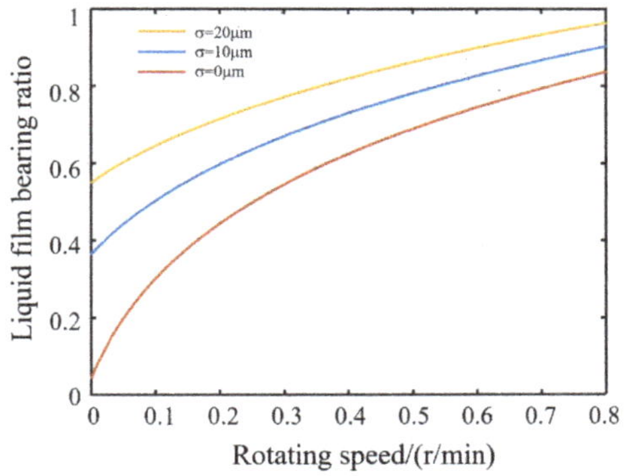

Figure 6. Relationship between liquid film bearing ratio and velocity under different roughness.

The relationship between the liquid film bearing ratio and polishing load under different roughness is shown in Figure 7. When the rotating speed of the polishing lap is a constant, the proportion of the liquid film decreases with the increase of the total load, and gradually approaches zero. This is because when the contact area between the element and polishing lap increases, it squeezes the flow space of the polishing liquid and reduces the thickness of the liquid film. When the total load is a constant and the roughness increases, the liquid film thickness still increases.

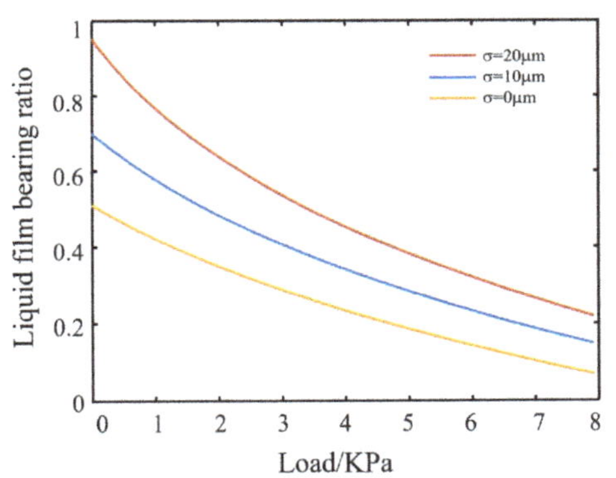

Figure 7. Relationship between liquid film bearing ratio and element pressure under different lap roughness.

3.4.3. Calculation Results of Pressure Distribution

In continuous polishing, the optical element is removed by the friction with the polishing particles on the surface of the polishing lap. It is assumed that the load of the element is borne by the micro-convex body on the polishing lap and polishing fluid. The actual material removal is only related to the contact pressure on the surface of the polishing lap. Therefore, the actual polishing pressure of the element in the mixture lubrication mode is obtained by subtracting the liquid film pressure from the load of the optic. According to the simplified Reynolds Equation (16), the liquid film pressure distribution under different total loads is shown in Figure 8a. The micro-convex body bearing pressure distribution under different total loads is shown in Figure 8b. It can be seen from the figure that the distribution of the liquid film pressure is symmetrical around the center of the element, and the pressure in the center of the element is the highest. The maximum liquid film pressure increases with the increase of the external load. This is because, on the one hand, the increase of the overall pressure of the element leads to the increase of the liquid film pressure. On the other hand, due to the increase of the external load of the element, the thickness of the liquid film decreases and the bearing capacity of the liquid film is weakened. As a result, the maximum liquid film pressure increases slowly when the external load becomes large.

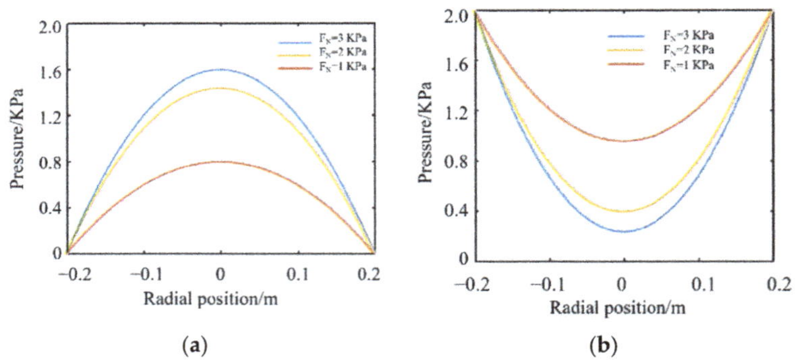

Figure 8. Distribution of liquid film pressure along horizontal direction under different external loads. (**a**) Liquid film pressure distribution (**b**) Polishing pressure distribution.

The distribution curve of the liquid film pressure along the horizontal direction is shown in Figure 9. It can be seen that the liquid film pressure increases with the roughness increase. This is because the roughness of the polishing lap surface enhances the flow of polishing liquid and increases the thickness of the liquid film. Thus, the loading force of the liquid film is increased, and the bearing capacity of the polishing liquid film is increased. At the same time, it can be seen that when the roughness increases, the maximum liquid film pressure increases faster, but at the same time, the liquid film pressure distribution becomes nonuniform, so the appropriate surface roughness of the polishing lap should be selected for polishing.

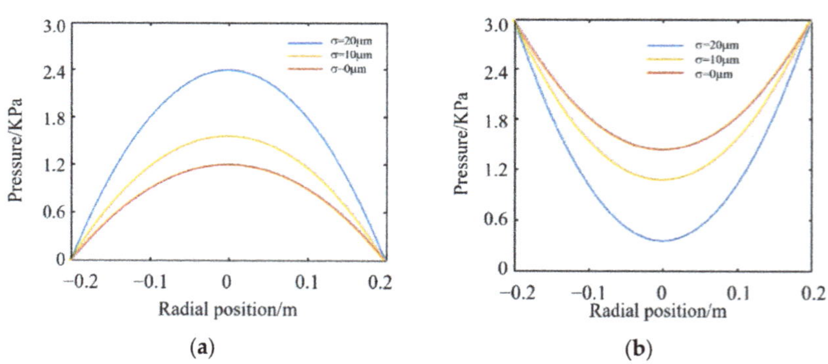

Figure 9. Liquid film pressure distribution along the radial direction of the element under different roughness. (**a**) Liquid film pressure distribution; (**b**) Polishing pressure distribution.

The liquid film pressure distribution under different lap rotating speeds is shown in Figure 10. It can be seen that the liquid film pressure is symmetrically distributed around the center of the element. The maximum liquid film pressure increases with the increase of the rotating speed of the polishing lap. It shows that higher polishing speed is not conducive to the uniform distribution of the liquid film pressure. When the polishing speed is large, the material removal rate of the optic is also larger, which is easy to damage the surface of the optical element. In order to ensure the surface quality, it is necessary to improve the roughness of the polishing pad and increase the average liquid film thickness to protect the surface of the optic.

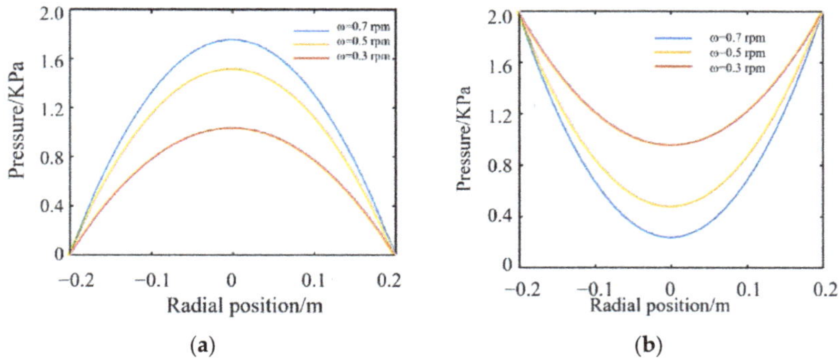

Figure 10. Relationship between liquid film pressure distribution and rotating speed. (**a**) Liquid film pressure distribution; (**b**) Polishing pressure distribution.

3.4.4. Friction Coefficient Calculation Results and Experimental Verification

To verify the accuracy of the numerical simulation results, the friction coefficient measurement experiments were carried out. Fpt-f1 friction stripping tester was used for the friction coefficient measurement test. The friction pair is fused quartz glass and asphalt block, and the polishing fluid is used for the lubrication. The friction coefficients of the friction pair at different rotating speeds are obtained. The numerical calculated and experimental results of the friction coefficient are as shown in Figure 11. The abscissa of the friction coefficient curve is the relative speed between the element and polishing lap, and the ordinate is the friction coefficient between the element and polishing lap. It can be seen that the friction coefficient curves obtained by the numerical calculation and experimental measurement are similar to the Stribeck curve. The friction coefficient decreases first and then increases with the increase of the relative speed. The numerical values of the two curves are also approximately equal in different lubrication zones, which verifies the correctness of the numerical results. At the same time, for the ring polishing process, the relative speed of the element and polishing lap is 0.3–0.5 rpm. In the figure, it belongs to the mixture lubrication area, which further proves that the lubrication mode of the annular polishing belongs to mixture lubrication.

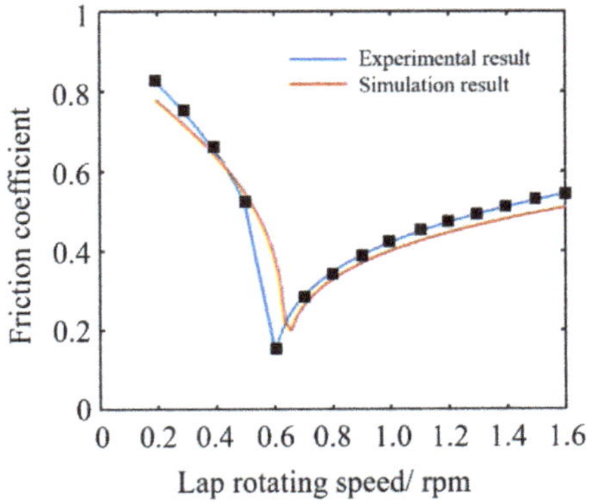

Figure 11. Experimental and calculated results of friction coefficient.

4. Numerical Model Verification Experiment of Mixture Lubrication

4.1. Experimental Method

The surface shape measurement experiments before and after the surface shape machining of the annular polishing element are carried out, the material removal distribution in the process of the element polishing is obtained, and the accuracy of the pressure distribution calculated by the numerical model is verified. First, the liquid film pressure and bearing pressure distribution of the polishing lap are obtained through the calculation of the mixture lubrication numerical model. The material removal distribution is obtained according to Preston material removal equation of Equation (26) [27,28]. A fused quartz optical element with a size of 430 mm × 430 mm × 80 mm is selected for the experiment, and the distance between the element center and center of the polishing lap is 1.6 m. The average polishing pressure of the optical element is about 2.0 KPa. Material removal coefficient in Preston equation is equal to $k = 4.73 \times 10^{-13}$/Pa, and polishing time is $t = 3$ h. Different process parameters are selected for rotating speed and polishing lap roughness.

$$MRR = K \cdot P \cdot V \tag{26}$$

4.2. Experimental Result

4.2.1. Influence of Different Rotating Speeds on Material Removal Uniformity

When the rotating speed of the polishing lap is changed, the result distribution of the material removal rate calculated according to the numerical simulation model is shown in Figure 12. When the rotating speed is 0.3 rpm, the difference of the material removal rate between the optic center and optic edge is 2.26×10^{-11} m/s. When the polishing time is 3 h, the material removal is 0.23 μm, which is about 0.36 λ (1 λ = 0.6328 μm). When the rotating speed is 0.7 rpm, the difference of the material removal rate between the optic center and optic edge is 9.6×10^{-11} m/s. When the polishing time is 3 h, the material removal is 1.03 μm, which is about 1.64 λ. When the polishing time is 3 h, the material removal is 0.23 μm, which is about 0.36 λ. When the rotating speed is 0.7 rpm, the difference of the material removal rate between the optic center and optic edge is 9.6×10^{-11} m/s. When the polishing time is 3 h, the material removal is 1.03 μm, which is about 1.64 λ.

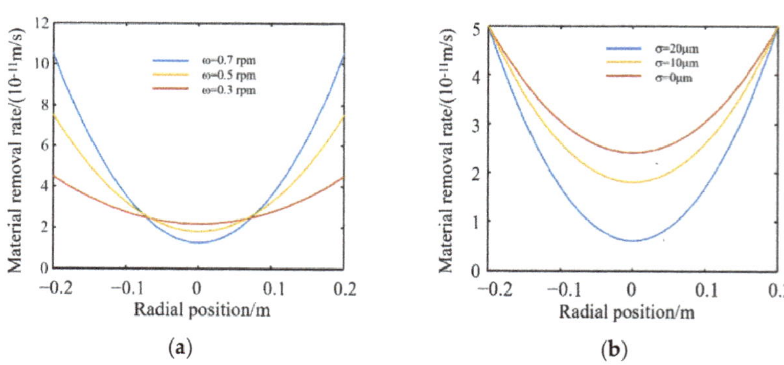

Figure 12. Factors affecting material removal rate. (**a**) Effect of rotating speed on material removal rate (**b**) Effect of roughness on material removal rate.

In the experiment, two optics with concave initial surface shape were polished at 0.3 and 0.7 rpm, respectively. Then, the surface shapes of the elements were measured after polishing for 3 h. The lowest point of the element is taken as the reference, and the difference of the material removal amount between the center and edge of the element is the subtraction of the surface shape before and after polishing. The initial surface shape and the processed surface shape of the elements are shown in Figure 13. It can be seen that when the rotating speed is 0.3 rpm, the optic edge material is removed by 0.33 λ more than the center, which is close to the simulated result. When the rotating speed is 0.7 rpm, the edge material of the optic is removed by 1.56 λ more than the center, which is also approximately equal to the calculated result. Thus, for the optical elements with concave shape, increasing the rotating speed can improve the surface flatness.

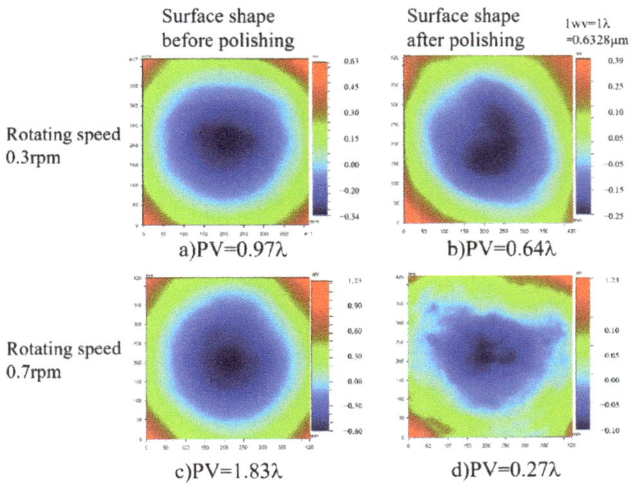

Figure 13. Experiment of influence law of rotating speed on element surface shape. (**a**) Rotating speed is 0.3 rpm before polishing, (**b**) rotating speed is 0.3 rpm after polishing, (**c**) rotating speed is 0.7 rpm before polishing and (**d**) rotating speed is 0.7 rpm after polishing.

In conclusion, the use of low-speed process parameters is conducive to the uniformity of the material removal distribution, but is not conducive to the processing efficiency. High rotating speed will increase the material removal rate and make the surface shape of the element become more convex, which is suitable for the element whose initial surface shape is concave. Therefore, in the actual processing, it is necessary consider the surface shape of the initial element and select the appropriate speed parameters [29–31].

4.2.2. Influence of Different Polishing Lap Surface Roughness on Material Removal Uniformity

According to the calculation by the simulation model, when the surface roughness of the polishing lap is 10 μm, the difference of the material removal on the surface of the component after machining for 3 h is about 0.35 μm, which is about 0.53 λ. When the surface roughness of the polishing lap is 20 μm, the difference of element surface removal is about 0.49 μm, which is about 0.8 λ.

The surface roughness *Ra* of the polishing lap selected in the experiment is about 9.8 and 19.4 μm respectively. When the surface roughness of the polishing lap is 9.8 μm, the material removal difference between the optic edge and optic center is 0.33 μm, which is about 0.52 λ. When the surface roughness of the polishing lap is 19.4 μm, the material removal difference between the optic edge and optic center is 0.48 μm, which is about 0.76 λ. The surface shapes of the optic before and after polished are shown in Figure 14. Thus, when it is necessary to improve the material removal rate at the center, the polishing lap roughness should be increased [32].

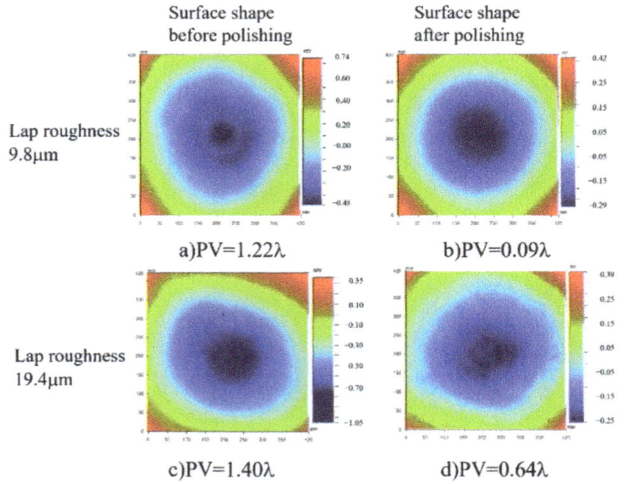

Figure 14. Influence law of roughness on element surface shape. (**a**) Lap roughness is 9.8 μm before polishing, (**b**) lap roughness is 9.8 μm after polishing, (**c**) lap roughness is 19.4 μm before polishing and (**d**) lap roughness is 19.4 μm after polishing.

4.2.3. Optic Surface Shape Optimization Experiment

The final surface shape of the element can be actively controlled by adjusting the process parameter according to above experiments. A fused quartz optical element with a size of 430 mm × 430 mm × 80 mm is selected in the experiment, and its initial surface shape is 0.97 λ, as shown in Figure 15a. According to the calculation, when the material removal required for the element edge is 0.97 λ more than the center, the rotating speed of the polishing lap is selected as 0.5 rpm, and the surface roughness of the polishing lap is controlled to be 9.8 μm. The surface shape of the element obtained after polishing for 3 h is shown in Figure 15b. The material removal amount of the optic center is 0.88 λ, which is approximately equal to the calculated result, and the final surface shape PV is better than 0.1 λ.

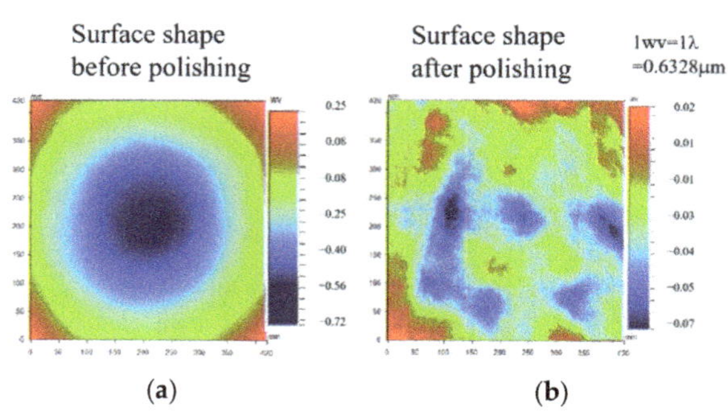

Figure 15. Experimental results of element surface shape measurement. (**a**) PV = 0.9 7λ; (**b**) PV = 0.09 λ.

5. Conclusions

Three possible contact modes between the optical elements and polishing lap in the annular polishing were analyzed. The lubrication mode in the continuous polishing belongs to mixture lubrication judged by calculating the film thickness ratio.

Based on the hydrodynamics theory, a numerical calculation model of the mixture lubrication in continuous polishing was developed. The influences of the polishing lap roughness, polishing speed and external load on the liquid film thickness, bearing ratio of the liquid film and polishing pressure distribution were revealed. The numerical simulation results indicated increasing the surface roughness of the polishing lap, polishing speed and optic load was not conducive to the uniformity of the pressure distribution. The numerical calculated and experimental results of the friction coefficient indicated that the friction coefficient decreased and then increased as rotating speed was increased.

The continuous polishing experiments of the fused quartz were carried out, and the influences of the surface roughness of the polishing lap and rotating speed on material removal rate of the optics were analyzed. The shape evolution rule of the optical element during the continuous polishing process was revealed. For fused silica elements, the material removal amount was calculated according to its initial surface shape. The surface shape accuracy was improved by selecting appropriate process parameters, such as the surface roughness and rotating speed of the polishing lap. By optimizing the process parameters, the surface PV of optical elements can be better than 0.1 λ.

Author Contributions: Conceptualization, F.Z. and C.L.; methodology, Y.W.; software, Y.W.; validation, Y.W., F.Z. and C.L.; formal analysis, Y.W.; investigation, Y.W. and C.L.; resources, Y.W.; data curation, Y.W.; writing—original draft preparation, Y.W.; writing—review and editing, C.L. and F.Z.; visualization, Y.W.; funding acquisition, C.L. and F.Z. All authors have read and agreed to the published version of the manuscript.

Funding: This research was funded by the National Natural Science Foundation of China (52005134, 51975154), Self-Planned Task (No. SKLRS202214B) of State Key Laboratory of Robotics and System (HIT), China Postdoctoral Science Foundation (2020M670901), Heilongjiang Postdoctoral Fund (LBH-Z20016), Open Fund of ZJUT Xinchang Research Institute, and Fundamental Research Funds for the Central Universities and The APC was funded by the National Natural Science Foundation of China (51975154).

Institutional Review Board Statement: Not applicable.

Informed Consent Statement: Not applicable.

Data Availability Statement: The study did not report any data.

Conflicts of Interest: The authors declare no conflict of interest.

References

1. Zhang, Y.; Wang, Q.; Li, C.; Piao, Y.; Hou, N.; Hu, K. Characterization of surface and subsurface defects induced by abrasive machining of optical crystals using grazing incidence X-ray diffraction and molecular dynamics. *J. Adv. Res.* **2022**, *36*, 51–61. [CrossRef] [PubMed]
2. Qu, S.; Yao, P.; Gong, Y.; Yang, Y.; Chu, D.; Zhu, Q. Modelling and grinding characteristics of unidirectional C–SiCs. *Ceram. Int.* **2022**, *48*, 8314–8324. [CrossRef]
3. Li, C.; Piao, Y.; Meng, B.; Zhang, Y.; Li, L.; Zhang, F. Anisotropy dependence of material removal and deformation mechanisms during nanoscratch of gallium nitride single crystals on (0001) plane. *Appl. Surf. Sci.* **2022**, *578*, 152028. [CrossRef]
4. Liao, D.; Xie, R.; Hou, J.; Chen, X.; Zhong, B. A polishing process for nonlinear optical crystal flats based on an annular polyurethane pad. *Appl. Surf. Sci.* **2012**, *258*, 8552–8557. [CrossRef]
5. Atuchin, V.V.; Maklakova, N.Y.; Pokrovsky, L.D.; Semenenko, V.N. Restoration of KTiOPO4 surface by annealing. *Opt. Mater.* **2003**, *23*, 363–367. [CrossRef]
6. Atuchin, V.V.; Kesler, V.G.; Kokh, A.E.; Pokrovsky, L.D. X-ray photoelectron spectroscopy study of β-BaB2O4 optical surface. *Appl. Surf. Sci.* **2004**, *223*, 352–360. [CrossRef]
7. Ramana, C.V.; Atuchin, V.V.; Becker, U.; Ewing, R.C.; Isaenko, L.I.; Khyzhun, O.Y.; Zhurkov, S.A. Low-energy Ar+ ion-beam-induced amorphization and chemical modification of potassium titanyl arsenate (001) crystal surfaces. *J. Phys. Chem. C* **2007**, *111*, 2702–2708. [CrossRef]
8. Korobeishchikov, N.G.; Nikolaev, I.V.; Roenko, M.A.; Atuchin, V.V. Precise sputtering of silicon dioxide by argon cluster ion beams. *Appl. Phys. A* **2018**, *124*, 833. [CrossRef]
9. Xie, R.; Zhao, S.; Liao, D.; Chen, X.; Wang, J.; Xu, Q. Numerical simulation and experimental study of surface waviness during full aperture rapid planar polishing. *Int. J. Adv. Manuf. Technol.* **2018**, *97*, 3273–3282. [CrossRef]
10. Luo, H.; Ajmal, K.M.; Liu, W.; Yamamura, K.; Deng, H. Polishing and planarization of single crystal diamonds: State-of-the-art and perspectives. *Int. J. Extrem. Manuf.* **2021**, *3*, 022003. [CrossRef]

11. Zhang, Z.; Jin, Z.; Mu, Q.; Yang, H.; Han, X. Optimization of CMP solution for yttrium aluminum garnet crystal. *Diam. Abras. Eng.* **2021**, *41*, 82–88.
12. Li, C.; Li, X.; Huang, S.; Li, L.; Zhang, F. Ultra-precision grinding of $Gd_3Ga_5O_{12}$ crystals with graphene oxide coolant: Material deformation mechanism and performance evaluation. *J. Manuf. Processes* **2021**, *61*, 417–427. [CrossRef]
13. Li, C.; Piao, Y.; Hu, Y.; Wei, Z.; Li, L.; Zhang, F. Modelling and experimental investigation of temperature field during fly-cutting of KDP crystals. *Int. J. Mech. Sci.* **2021**, *210*, 106751. [CrossRef]
14. Chang, A. The effect of slurry film thickness variation in chemical mechanical polishing (CMP) of patterned oxide wafers. In Proceedings of the VMIC Conference 2000, Santa Clara, CA, USA, 2000; pp. 112–114. Available online: https://jglobal.jst.go.jp/en/detail?JGLOBAL_ID=200909054958270070 (accessed on 16 May 2022).
15. Runnels, S.R.; Renteln, P. Modeling the effect of polish pad deformation on wafer surface stress distributions during chemical-mechanical polishing. In Proceedings of the Third International Symposium on Interconnects, Contact Metallization and Multilevel Metallization, 183rd Meeting of the Elecrochemical Society, Honolulu, HI, USA, 19–21 May 1993.
16. Li, C.; Piao, Y.; Meng, B.; Hu, Y.; Li, L.; Zhang, F. Phase transition and plastic deformation mechanisms induced by self-rotating grinding of GaN single crystals. *Int. J. Mach. Tools Manuf.* **2022**, *172*, 103827. [CrossRef]
17. Terrell, E.J.; Higgs, C.F. Hydrodynamics of slurry flow in chemical mechanical polishing: A review. *J. Electrochem. Soc.* **2006**, *153*, K15. [CrossRef]
18. Waechter, D.; Dambon, O.; Klocke, F. Analysis and modeling of tribology effects in conventional glass polishing. In *Optical Manufacturing and Testing X*; International Society for Optics and Photonics: Bellingha, EA, USA, 2013; Volume 8838, p. 88380V.
19. Sundararajan, S.; Thakurta, D.G.; Schwendeman, D.W.; Murarka, S.P.; Gill, W.N. Two-dimensional wafer-scale chemical mechanical planarization models based on lubrication theory and mass transport. *J. Electrochem. Soc.* **1999**, *146*, 761. [CrossRef]
20. Kim, A.T.; Seok, J.; Tichy, J.A.; Cale, T.S. A multiscale elastohydrodynamic contact model for CMP. *J. Electrochem. Soc.* **2003**, *150*, G570. [CrossRef]
21. Tichy, J. *Contact Mechanics and Lubrication Hydrodynamics of Chemical-Mechanical Planarization*; Tribology Series; Elsevier: Amsterdam, The Netherland, 2001; Volume 39, pp. 63–68.
22. Liao, D.; Ren, L.; Zhang, F.; Wang, J.; Xu, Q. Kinematic model for material removal distribution and surface figure in full-aperture polishing. *Appl. Opt.* **2018**, *57*, 588–593. [CrossRef]
23. Qin, K.; Moudgil, B.; Park, C.W. A chemical mechanical polishing model incorporating both the chemical and mechanical effects. *Thin Solid Film.* **2004**, *446*, 277–286. [CrossRef]
24. Johnson, K.L.; Greenwood, J.A.; Poon, S.Y. A simple theory of asperity contact in elastohydro-dynamic lubrication. *Wear* **1972**, *19*, 91–108. [CrossRef]
25. Habchi, W.; Eyheramendy, D.; Bair, S.; Vergne, P.; Morales-Espejel, G. Thermal elastohydrodynamic lubrication of point contacts using a Newtonian/generalized Newtonian lubricant. *Tribol. Lett.* **2008**, *30*, 41–52. [CrossRef]
26. Greenwood, J.A.; Tripp, J.H. The contact of two nominally flat rough surfaces. *Proc. Inst. Mech. Eng.* **1970**, *185*, 625–633. [CrossRef]
27. Guo, J.; Shi, X.; Song, C.; Niu, L.; Cui, H.; Guo, X.; Kang, R. Theoretical and experimental investigation of chemical mechanical polishing of W–Ni–Fe alloy. *Int. J. Extrem. Manuf.* **2021**, *3*, 025103. [CrossRef]
28. Xu, J.; Kang, R.; Dong, Z.; Wang, Z. Review on chemical mechanical polishing of silicon wafers. *Diam. Abras. Eng.* **2020**, *40*, 24–33.
29. Li, C.; Wu, Y.; Li, X.; Ma, L.; Zhang, F.; Huang, H. Deformation characteristics and surface generation modelling of crack-free grinding of GGG single crystals. *J. Mater. Process. Technol.* **2020**, *279*, 116577. [CrossRef]
30. Sun, Y.; Jin, L.; Gong, Y.; Wen, X.; Yin, G.; Wen, Q.; Tang, B. Experimental evaluation of surface generation and force time-varying characteristics of curvilinear grooved micro end mills fabricated by EDM. *J. Manuf. Processes* **2022**, *73*, 799–814. [CrossRef]
31. Li, C.; Hu, Y.; Huang, S.; Meng, B.; Piao, Y.; Zhang, F. Theoretical model of warping deformation during self-rotating grinding of YAG wafers. *Ceram. Int.* **2022**, *48*, 4637–4648. [CrossRef]
32. Cheng, Z.; Gao, H.; Liu, Z.; Guo, D. Investigation of the trajectory uniformity in water dissolution ultraprecision continuous polishing of large-sized KDP crystal. *Int. J. Extrem. Manuf.* **2020**, *2*, 045101. [CrossRef]

crystals

MDPI

Article

Splitting Opaque, Brittle Materials with Dual-Sided Thermal Stress Using Thermal-Controlled Fracture Method by Microwave

Xiaoliang Cheng [1,*], Zongyang He [2], Hailong Wang [2] and Yang Wang [2]

[1] School of Mechanical Engineering, Anhui Polytechnic University, Wuhu 241000, China
[2] School of Mechatronics Engineering, Harbin Institute of Technology, Harbin 150001, China; hit_hezongyang@126.com (Z.H.); whl19850505@163.com (H.W.); yangwanghit@163.com (Y.W.)
* Correspondence: chengxiaoliang@ahpu.edu.cn

Abstract: The thermal-controlled fracture method has been increasingly focused upon in the high-quality splitting of advanced brittle materials due to its excellent characteristics related to the fact that it does not remove material. For opaque, brittle materials, their poor fracture quality and low machining capacity resulting from their single-sided heat mode is a bottleneck problem at present. This work proposed the use of dual-sided thermal stress induced by microwave to split opaque, brittle materials. The experimental results indicate that the machining capacity of this method is more than twice that of the single-sided heat mode, and the fracture quality in splitting opaque, brittle materials was significantly improved by dual-sided thermal stress. A microwave cutting experiment was carried out to investigate the distribution characteristic of fracture quality by using different workpiece thicknesses and processing parameters. A dual-sided thermal stress cutting model was established to calculate the temperature field and thermal stress field and was used to simulate the crack propagation behaviors. The accuracy of the simulation model was verified using temperature measurement experiments. The improvement mechanism of the machining capacity and fracture quality of this method was revealed using the fracture mechanics theory based on calculation results from a simulation. This study provides an innovative and feasible method for cutting opaque, brittle materials with promising fracture quality and machining capacity for industrial application.

Keywords: opaque; brittle materials; thermal-controlled fracture method; dual-sided thermal stress; microwave cutting; machining capacity; fracture quality; simulation model; crack propagation behavior

Citation: Cheng, X.; He, Z.; Wang, H.; Wang, Y. Splitting Opaque, Brittle Materials with Dual-Sided Thermal Stress Using Thermal-Controlled Fracture Method by Microwave. *Crystals* **2022**, *12*, 801. https://doi.org/10.3390/cryst12060801

Academic Editor: Umberto Prisco

Received: 11 May 2022
Accepted: 30 May 2022
Published: 6 June 2022

1. Introduction

Opaque, brittle materials, mainly including advanced ceramics, have been widely used in industrial fields due to their excellent properties of high-temperature performance, low expansion, high hardness and corrosion resistance. However, these intrinsic characteristics induce processing difficulties, especially in the splitting process of separating a whole blank into small pieces. The conventional cutting method depends on removing surplus materials between the parts by contact force, melting, ablation, etc. Due to its brittleness, these methods tend to damage fresh surfaces and cause serious fracture quality problems such as micro-cracks, stripes and heat-affected zones [1].

In 1968, Lumley proposed the method of guiding crack propagation by thermal stress to split brittle material. According to its processing principle, it is called the thermal-controlled fracture method (TCFM) [2]. The TCFM has the advantages of no material removal, no high temperature effect (under 500 °C) and no force impact. It can achieve high splitting efficiency (its splitting speed can reach more than 10 mm/s) and good fracture quality (its arithmetic mean deviation of contour (Ra) can reach less than 10 nm) [3]. Therefore, the TCFM has obvious advantages compared with conventional cutting techniques such as abrasive wheel cutting, diamond saw cutting, abrasive water jet machining, ultrashort pulse laser beam ablation methods and other processing method [1,4–7], and has become a promising machining method, attracting the interest of many researchers in the field of brittle material processing [8–12]. The thermal stress used to split material in the TCFM is produced by a heat source on the surface or in the body of the workpiece (as shown in Figure 1).

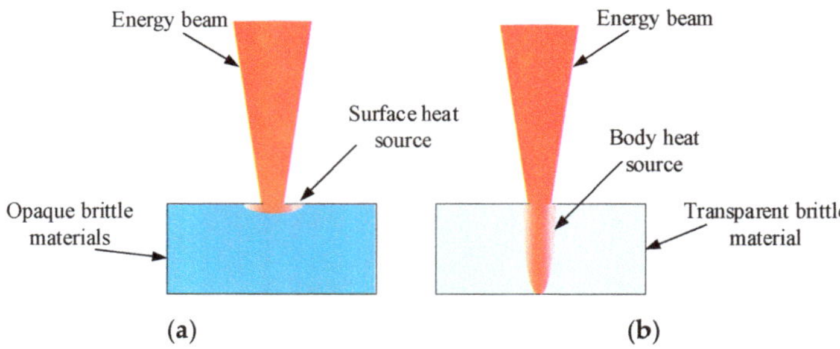

Figure 1. Schematic diagram of heat source type in thermal-controlled fracture method. (**a**) Surface heat source, (**b**) body heat source.

The site of a heat source in a material is mainly determined by its absorption type for energy sources such as laser beam. The research regarding splitting glass using the TCFM have proven that the body absorption type which produces body heat sources can produce better processing quality than that of surface heat sources [3,9]. Using body heat sources, the process of splitting LED glass using the TCFM has been realized in industrialization [13]. However, a body heat source is mainly formed in transparent, brittle materials such as glass. For opaque, brittle materials, a thermogenic beam such as a laser beam can only be absorbed by the surface of the material to form a surface heat source [14–17]. Since the heat source is formed on a single surface of the material in these cases, this cutting mode is called the single-sided thermal stress method (SSTM).

The SSTM is mainly used for cutting thin brittle materials such as silicon wafer, sapphire and ceramic substrate (generally no more than 1 mm thickness). Ueda used the SSTM to cut silicon wafers and ceramic sheets, and the results indicated that the surface roughness can reach 100 μm for Si_3N_4, 1.3 μm for Al_2O_3 ceramic and 0.7 μm for crystalline silicon; however, the thickness of these specimens was only 0.5 mm [8]. Although it has ideal processing quality for thin materials, it has been reported that the cutting quality by the single-sided cutting mode is remarkably worse than that of the full-body cutting form for thick brittle materials [13,17]. The work of Saman indicated that the worse fracture quality by the SSTM is because the effect of thermal stress on thickness is not adequate for the material to be cut, and its maximum tensile stress is in the reverse side of the pre-crack [18]. Cai used a laser to cut thick Al_2O_3 ceramic materials, and the results indicated that it was difficult to split thick ceramic materials with the single-sided heat mode, and its processing quality was poor [19]. Thus, the machining capacity and processing quality of the SSTM hinder the application of the TCFM in the field of cutting opaque, brittle materials.

Due to the high dielectric properties, some opaque, brittle materials have good absorption capacity for microwaves. This means these ceramics can form body heat sources via microwave beams. Wang successfully used a microwave of 2.45 GHz to treat SiC ceramics in the full-body cutting form using the TCFM [20]. However, for those other opaque, brittle materials such as Al_2O_3 and Si_3N_4 ceramics, which have low dielectric coefficients, heating in the body form via microwave is as difficult as it is with laser beams. Cheng and Wang used microwaves to cut Al_2O_3 ceramics coated with graphite, which has good absorptivity for microwaves; this method is an SSTM [21,22].

To overcome the defects of the SSTM, Cai used dual laser beams to cut glass–silicon–glass plates based on the TCFM [23]. However, the dual laser beams method requires relatively complex alignment technology, and the two laser beams need to have exactly the same operation states, which results in a great increase in processing costs. Furthermore, the minimum temperature needed to split brittle material using laser beams in the TCFM is higher than that needed for microwaves, which would induce unnecessary thermal damage [20].

In this work, microwave-induced dual-sided thermal stress (MIDT) was proposed for splitting opaque, brittle materials to improve the machining capacity and processing quality of the SSTM. Compared with the dual laser beams method, MIDT needs just one energy source. The high dielectric loss material of graphite was used to coat the upper and lower surfaces of the processing area of the opaque, brittle plates to form a dual heat source. The microwave absorbed by the upper surface was used to form an upper heat source, and the remainder of the energy penetrating the ceramic plate was absorbed by the lower surface and used to form lower heat source. Cutting experiments were conducted to study the machining capacity and the effects of processing parameters on fracture quality. An MIDT cutting simulation model was developed to analyze the improvement mechanism of the machining capacity and fracture quality using this method. This method demonstrated significant potential for the precise cutting of thick opaque, brittle materials using the thermal-controlled fracture method.

2. Experiment

2.1. Experimental Principle

Al_2O_3 ceramic is an opaque, brittle material that has a low dielectric coefficient. The cutting of this ceramic with the TCFM is usually performed in the single-sided heat mode. Figure 2 shows the schematic diagram of heating low-dielectric-coefficient materials on their dual-sided surface via microwaves. In this work, high-dielectric-coefficient graphite material was used to coat the upper and lower surfaces of the low-dielectric-coefficient Al_2O_3 ceramics to absorb microwaves and generate upper and lower surface heat sources. To control the distribution of the heat sources, the coating material was restricted within a certain range along the cutting line.

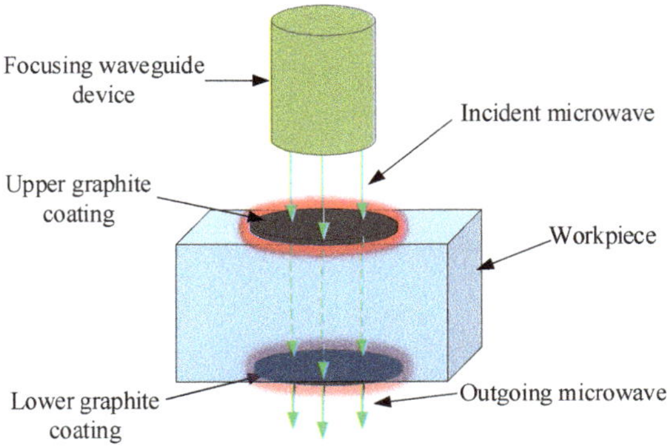

Figure 2. Schematic diagram of single microwave beam heating upper and lower surface of the material.

The cutting of ceramic via MIDT was realized by the heat generated by high electromagnetic loss on the dual side of the coating material to induce thermal stress and guide crack propagation. It is feasible to use one waveguide output to realize dual-sided thermal stress because only part of the energy was lost when the microwave irradiated the upper coating materials. The remaining microwave energy could pass through the dielectric ceramic without much electromagnetic loss onto the lower surface coating material to generate heat. In this way, the upper and lower coating materials directly transferred the heat generated by electromagnetic loss to the workpiece and produced dual-sided thermal stress. Since the upper and lower heat sources were generated by the same microwave beam, the neutrality of the heat source did not need to be adjusted. Compared with the dual laser beam heating scheme adopted by Zhao and Cai, this project greatly reduces the equipment and technical costs [9,10,23].

A schematic diagram of cutting ceramic via MIDT is shown in Figure 3. As is shown, the heat sites in both sides were located directly below the waveguide output (Figure 3a). The uneven heat distribution produced a thermal stress field in the material. The tip of prefabricated crack could produce a stress amplification effect and was the weakest position at the same time. When the tensile stress at the tip exceeded the fracture limit of the material, the crack system reached the cracking condition. Based on the appropriate moving speed between the material and the waveguide generated by the motion device, the crack propagated forward at the appropriate speed, so as to realize the cutting of the material (Figure 3b,c). Since the thermal stress was induced at the dual side of the material at the same time, this is called the microwave-induced dual-sided thermal stress cutting method.

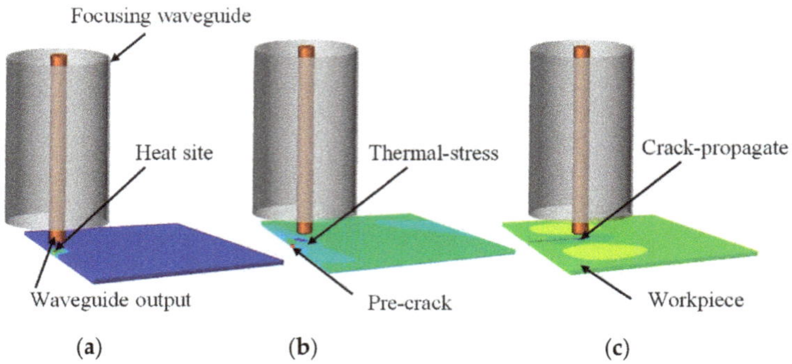

Figure 3. Schematic diagram of cutting ceramic by using microwave-induced dual-sided thermal stress. (**a**) Heating stage; (**b**) thermal stress stage; (**c**) crack propagation stage.

2.2. Experimental Material and Apparatus

Figure 4 shows the schematic diagram of the microwave-induced dual-sided thermal stress cutting system and the cutting machine. The microwave generated by the microwave source was modulated by the Bj26 waveguide device and was guided into the focusing equipment. Then, the focusing equipment output focused the microwave for processing from the inner conductor. The workpiece with a highly dielectric coating was directly below the inner core of the waveguide output. In order to obtain the accurate cutting position and movement, the workpiece was placed on an NC motion device which could realize x-y-z three-direction movement. The function of the z-axis was to adjust the distance between the inner core of the waveguide and the workpiece to achieve specific power density. The microwave cutting machine tool was equipped with safety facilities such as a safety filter screen, microwave absorption vessel and microwave safety door to ensure safety and reliability. The microwave frequency used in this study was 2.45GHz, and the maximum output power of the equipment was 1500 W.

Figure 5 shows the workpiece of Al_2O_3 ceramic, in which the black part in the middle is the coated graphite which had a thickness of 0.1 mm and a width of 1 mm. The dimensions of the ceramic material were 100 mm $\times$ 100 mm in size, and the thickness had different specifications of 1 mm, 2 mm, 4 mm, 6 mm and 8 mm. The coating position was symmetrically distributed on the upper and lower surfaces along the cutting line. A pre-crack was fabricated on the end of the workpiece with a diamond wire saw. The graphite was a micron-sized powder material (mesh of 8000 and particle size of 1.6 μm), evenly prepared with alcohol and then coated onto the surface of workpiece with precision powder spreading equipment.

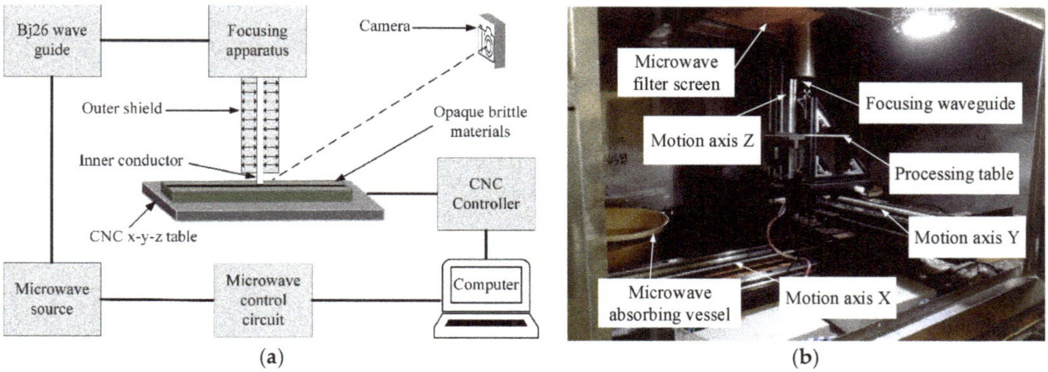

Figure 4. The microwave-induced dual-sided thermal stress cutting machine. (**a**) Schematic of the experimental apparatus; (**b**) microwave-induced dual-sided thermal stress cutting machine.

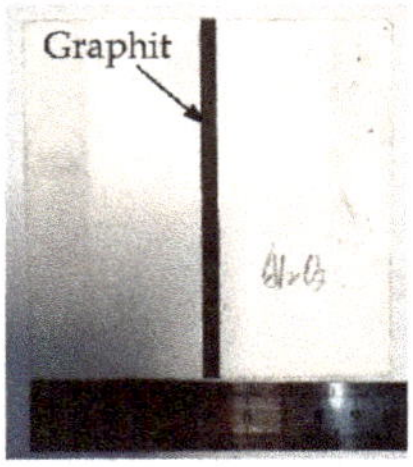

Figure 5. Al_2O_3 ceramic plate coated with graphite.

Table 1 shows the range of processing parameters used in the experiment. The main processing parameters which could be controlled were microwave power and scanning speed.

Table 1. Processing parameters of in microwave cutting of Al_2O_3 ceramic using TCFM.

Test Group No	Workpiece Thickness (mm)	Single-Sided Heat Source		Dual-Sided Heat Source	
		Microwave Power (W)	Scanning Speed (mm/s)	Microwave Power (W)	Microwave Power (W)
NO.1	1	600–1200	2.0–3.5	400–700	2.0–3.5
NO.2	2	900–1500	0.5–2.0	800–1100	2.0–3.5
NO.3	4	1200–1500	0.3–0.6	1200–1500	1.0–2.5
NO.4	6	1200–1500	0.1–0.4	1200–1500	0.5–2.0
NO.5	8	1200–1500	0.1–0.4	1200–1500	0.25–1.00

In order to study the effect of dual-sided heating method on the machining ability, experiments with both single-sided and dual-sided heat sources for cutting ceramic using TCFM were carried out. Each test was repeated four times. The mean values were selected as the final results. These processing parameter ranges were determined in advance by a thermal stress cutting simulation. The guidance for determining the processing parameter range from the simulation was that the cutting process could be realized and had acceptable crack propagation quality. To investigate the distribution characteristics of the fracture quality via MIDT, the surface roughness of the fracture surface was measured at 10 mm, 50 mm and 90 mm along the cutting direction, with three measurement positions along the thickness direction at each site. The distribution of these measurement sites is shown in Figure 6.

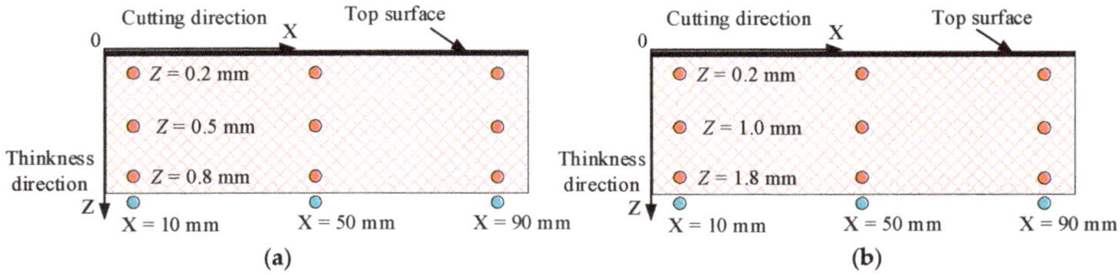

Figure 6. Location of surface roughness measurement points in fracture surface of the workpiece with different thickness of material. (**a**) 1 mm; (**b**) 2 mm.

3. Results

3.1. Machining Capacity

Table 2 shows the experimental results of machining capacity in the cutting of Al_2O_3 ceramic by using the TCFM induced by single-sided and dual-sided heat sources. The processing parameters shown in Table 1 could realize the cutting of ceramic materials and achieve acceptable cutting quality at the same time. It can be seen from Table 2 that when the single-sided heating method was adopted, the maximum processing thickness was 2 mm, while the dual-sided heat source could realize 8 mm. It is notable that the cutting speed could be more than 100 times higher than that of the diamond wire saw cutting method, which was about 0.01 mm/s [24].

Table 2. Experimental results of machining capacity in cutting of Al_2O_3 ceramic by using TCFM.

Workpiece Thickness (mm)	Single-Sided Heat Source		Dual-Sided Heat Source	
	Microwave Power (W)	Cutting Speed (mm/s)	Microwave Power (W)	Cutting Speed (mm/s)
1	1000	3	600	3
2	1500	0.5/1.0	1000	3
4	——	——	1500	2
6	——	——	1500	1
8	——	——	1500	0.5

Figure 7 shows the experimental results of cutting Al_2O_3 ceramic with thickness of 8 mm with MIDT. They indicate that the MIDT cutting method can achieve an excellent fracture surface and an approximately straight crack propagation path.

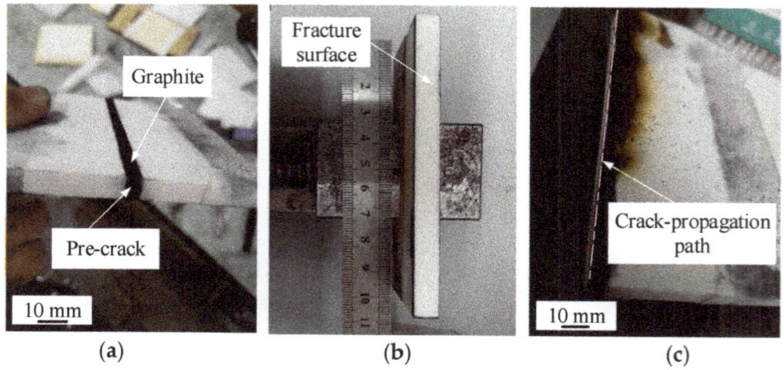

Figure 7. Experimental results of cutting Al_2O_3 ceramic with thickness of 8 mm with MIDT. (**a**) The workpiece before cutting, (**b**) fracture surface after cutting, (**c**) top surface and the crack propagation path after cutting.

3.2. Fracture Quality

Figure 8 shows the fracture surface micrograph and 3D outline figure of the workpiece after cutting Al_2O_3 ceramic materials with thicknesses of 2 mm and 8 mm by using MIDT. The sampling location of the 8 mm thickness was at the middle depth of the material with the same scale of the 2 mm thickness. The processing parameters for these two thicknesses of ceramic materials are shown in Table 1. From Figure 8b, it can be seen that most areas in the middle depth were relatively flat and were much better than those near the surface. At the middle depth of 8 mm thickness, the contour fluctuation was greater than that of the material with 2 mm thickness with larger irregularity. The arithmetic average surface roughness Ra at the middle depth after cutting could achieve 4 μm when cutting the workpiece with 8 mm thickness. This was much better than that of the wire saw cutting method, which is the most popular technique used to cut ceramics [1].

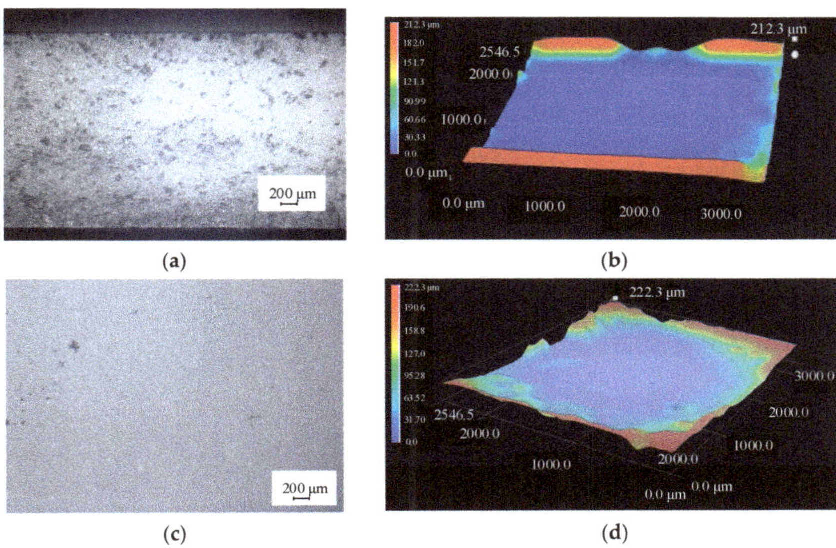

Figure 8. Fracture surface in cutting Al_2O_3 ceramic using MIDT. (**a**) The micrograph of fracture surface with thickness of 2 mm, (**b**) 3D figure of fracture surface of material with thickness of 2 mm, (**c**) the micrograph of fracture surface with thickness of 8 mm, (**d**) 3D figure of fracture surface of material with thickness of 8 mm.

Figure 9 shows a 3D figure of the section profile of Al_2O_3 ceramic with a thickness of 2 mm cut with a microwave-induced single-sided heat source using the TCFM. The processing parameters used are shown in Table 1, where the cutting speed used was 1.0 mm/s. Comparing Figures 8 and 9, it is notable that dual-sided induced thermal stress could obtain better fracture quality than that of single-sided induced thermal stress.

Figure 10 shows the effect of microwave power on surface roughness Ra in cutting Al_2O_3 ceramics using MIDT. It can be seen that the fracture quality of the material 1 mm thickness was better than that with 2 mm thickness. The surface roughness Ra increased dramatically with the increase in microwave power, which was independent of material thickness. It is noteworthy that the fracture quality in the middle section was much better than that in the inlet and outlet of the material. The fracture quality in the middle section, shown in Figure 10c,d, indicated that the fracture quality at the middle depth is better than that near the surface. The fracture quality shown in Figure 10c was better than that achieved with the laser-induced single-sided thermal stress method reported by [8], in which the Ra was 1.3 μm and the thickness was 0.5 mm, which was thinner than that in this study.

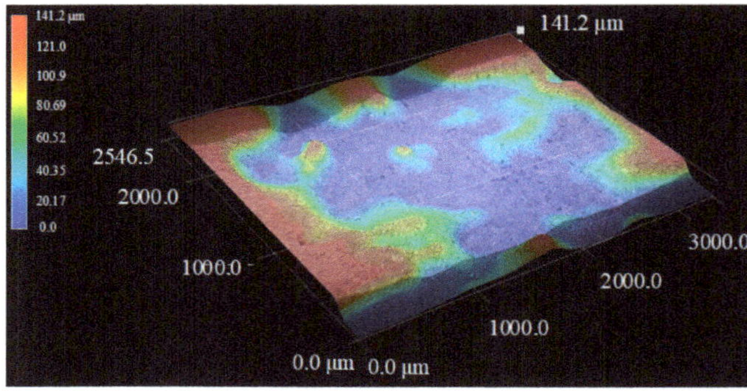

Figure 9. Fracture surface of Al_2O_3 ceramic with thickness of 2 mm cut with microwave-induced single side surface heat source.

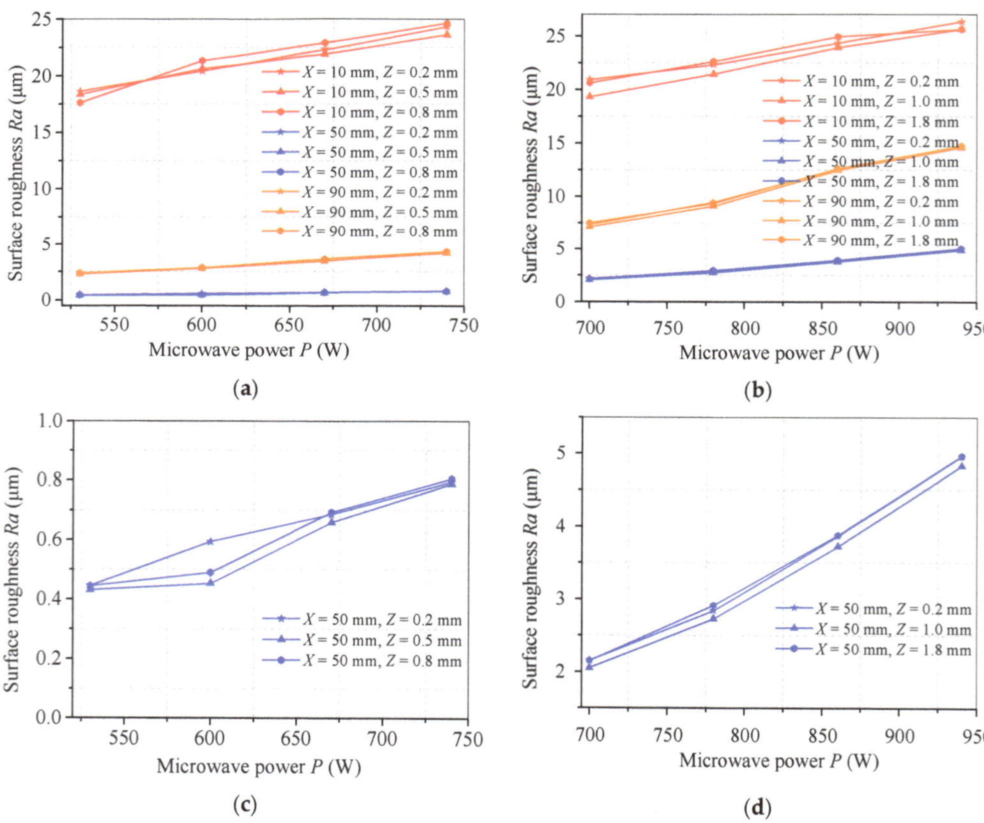

Figure 10. The influence of microwave power on the surface roughness Ra of fracture surface in cutting Al_2O_3 ceramic using MIDT. (**a**) Fracture quality along the cutting direction with thickness of 1 mm, (**b**) fracture quality along the cutting direction with thickness of 2 mm, (**c**) fracture quality distribution along thickness at the middle section with thickness of 1 mm, (**d**) fracture quality distribution along thickness at the middle section with thickness of 2 mm.

4. Discussion

4.1. Finite Element Modeling of the Cutting Process

The main physical processes of cutting opaque, brittle materials with MIDT are as follows:

(1) Absorption of microwave and heat generation.

The microwave is absorbed at the upper coating material, and then it penetrates the ceramic material body almost without loss and is absorbed by the lower coating material. In this way, the microwave is absorbed by the upper and lower coating materials and dual-sided surface heat is generated. The thermal power density on the upper and lower surface can be calculated by

$$P_Z = 2\pi f \varepsilon_0 \varepsilon_r \tan\delta |E_t(x, y, z)|^2 \quad (1)$$

where P_Z (W/m^3) is the thermal power density, f (GHz) is the microwave frequency, ε_0 (8.85×10^{-12} F/m) is the vacuum permittivity, ε_r is the relative permittivity of graphite, δ is dielectric loss angle and $E_t(x, y, z)$ (V/m) is the electric field intensity.

(2) Heat transfer process.

The heat generated at the upper and lower surfaces is transmitted from the surface to the ceramic material body. This process is accompanied by changes in heat convection and heat radiation between the material and the environment.

According to Fourier heat transfer and energy conservation law, taking the plane where the surface heat source is located as the X-Y plane and the direction which is perpendicular to the plane as the Z axis in the Cartesian coordinate system, the heat transfer differential equation inside the workpiece in a non-equilibrium state can be expressed as

$$\rho c \frac{\partial T}{\partial t} = \frac{\partial}{\partial x}\left(\lambda \frac{\partial T}{\partial x}\right) + \frac{\partial}{\partial y}\left(\lambda \frac{\partial T}{\partial y}\right) + \frac{\partial}{\partial z}\left(\lambda \frac{\partial T}{\partial z}\right) + \dot{Q}(x, y, z, t) \quad (2)$$

where ρ is material density ($Kg{\cdot}m^{-3}$), c is specific heat (J/Kg·°C), λ is thermal conductivity (W/m·°C) and $\dot{Q}$ is heat production per unit volume ($J{\cdot}m^{-3}$), which is related to P_Z in Equation (1). The heat convection and the heat radiation between the workpiece and the environment also needs to be considered.

(3) Thermal stress generation.

According to the thermal stress theory, the uneven heat distribution induces thermal stress in the material. The calculation of a thermal stress field is mainly based on the following constitutive equation:

$$\sigma_{ij} = D_{ijmn}\varepsilon_{mn} - \beta_{ij} T \quad (3)$$

where σ_{ij} is the stress component (MPa), D_{ijmn} is the elastic coefficient (MPa), ε_{ij} is the strain component, T is the temperature distribution in Equation (3) and β_{ij} is the thermal stress coefficient (MPa). The solution of thermal stress needs to combine geometric equations and equilibrium differential equations.

(4) Crack propagation process.

When the tensile stress acts on the prefabricated crack and exceeds the fracture strength of the material, the pre-crack can propagate. This process includes the mutual transformation of elastic energy and surface energy in the crack system.

When the vertical tensile stress component of the crack tip σ_A is greater than the critical stress σ_F, the crack begins to propagate, and the material begins to be cut. σ_F can be expressed as

$$\sigma_F = \left[2E'\gamma/(\pi c_0)\right]^{1/2} \quad (4)$$

where c_0 is the size of the pre-crack (m), E' is the equivalent elastic modulus (Pa) and γ is free surface energy per unit area (J/m^2). σ_A can be obtained by Equation (3) using the finite element method (FEM).

The cutting of opaque, brittle materials via thermal stress is an uncoupled thermoelastic physical process, in which the calculation of the temperature field, thermal stress field and crack propagation requires finite element modeling (FEM) technology. Based on the above theoretical modeling, the calculation of the temperature field, stress field and crack propagation in cutting opaque, brittle materials via MIDT was realized by using the finite element simulation software ABAQUS. The mechanical and thermophysical properties of Al_2O_3 ceramic materials are shown in Table 3.

Table 3. Mechanical and thermophysical properties of Al_2O_3 ceramic.

Physical Parameters	Value
Thermal conductivity (W/m·°C)	25
Density (g/cm^3)	3.9
Specific heat (J/Kg·°C)	880
Expansion coefficient (10^{-6}/°C)	7.5
Young's modulus (G Pa)	370
Poisson ratio	0.22

Figure 11 shows the workpiece mesh model used in the simulation. The geometric dimension of the model was 100 mm × 100 mm × 1 mm. Due to the large size of the workpiece, the balance between calculation accuracy and calculation efficiency needed to be considered. Therefore, mesh refinement was carried out around the crack propagation zone, and coarse mesh was used in the other sites. The refinement of the middle part, which had a 10 mm width, was the scanning mesh type, and the coarse mesh at other positions was of the structured mesh type. The minimum mesh size was 0.071 mm and was along the thickness direction of the workpiece.

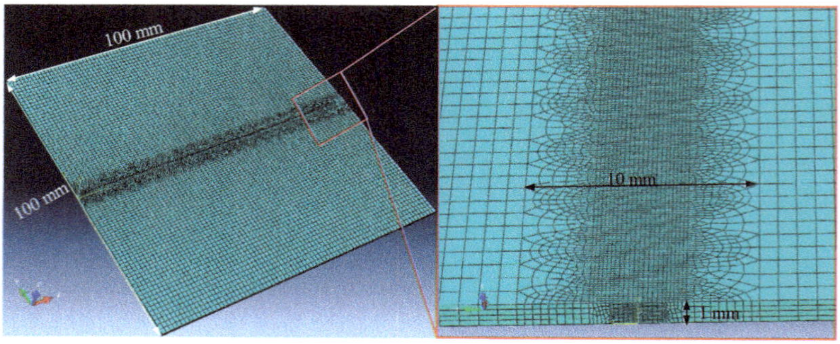

Figure 11. Workpiece mesh model.

Figure 12 shows the finite element simulation flow. The reasonable construction and setting were mainly carried out from three aspects: geometric modeling, physical modeling and the calculation method. The temperature was the loading condition for subsequent thermal stress and crack propagation simulation. The accuracy of the calculation results was related to the reliability of the whole model.

Thus, the simulated temperature needed to be verified by experiments to check the accuracy of the model. The optical fiber temperature measurement experiment was carried out to verify the temperature simulation results. Figure 13 shows the comparation between the temperament measurements results and the prediction results from the simulation. From Figure 13, it can be seen that the simulation results had good agreement with the experimental results. This indicates that the model was established correctly and could be used to analyze the cutting mechanism in this process.

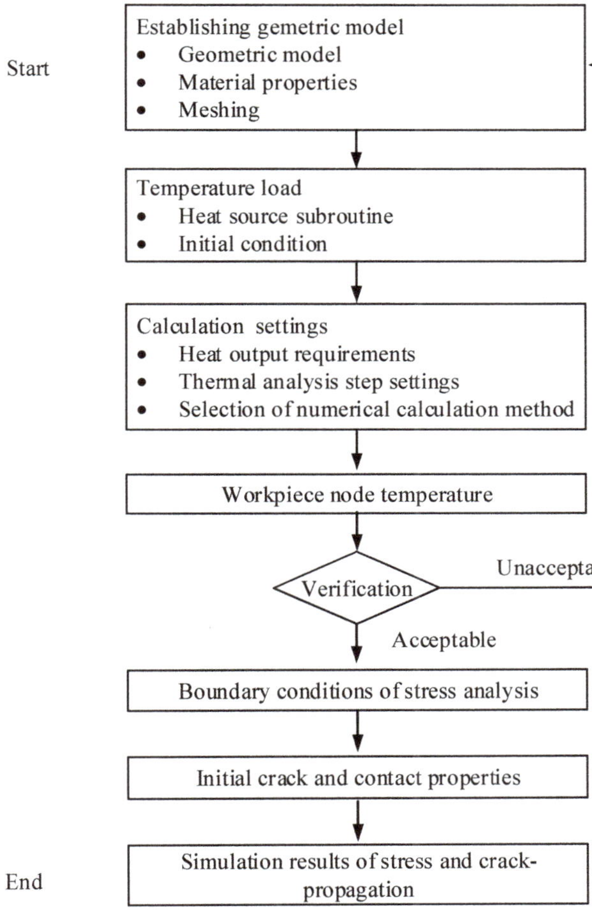

Figure 12. Flow chart of simulation process.

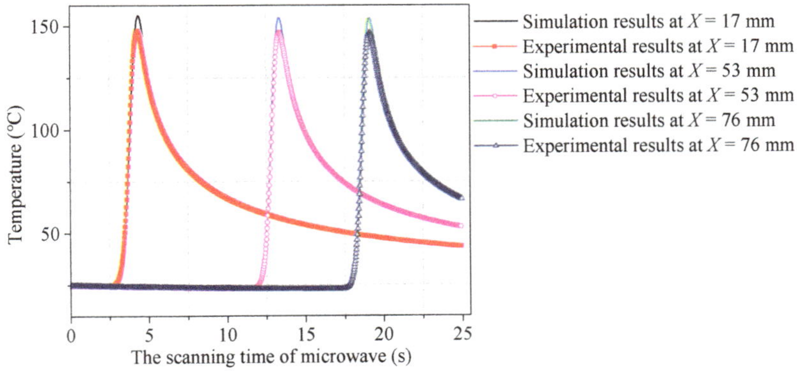

Figure 13. Comparation between the temperature simulation results and the experimental results.

4.2. Mechanism of the High Machining Capacity

Figure 14 shows the evolution of the temperature field induced by the dual-sided heat method. The scanning speed of the heat source was 3 mm/s, the microwave power was 600 W, the thickness of the workpiece was 1 mm and the dimensions of the other two directions were 100 mm. From Figure 14, it can be seen that the upper and lower heat-affected zone induced by the two heat sources kept approaching and finally connected as a thin waist shape. It can be derived that the deeper heating zone (at least twice as much) produced by the dual-sided heat mode was a direct reason for the improvement in the machining capacity. The effective heating depth enhanced by the dual-sided heat mode was more than twice that of the single-sided heat mode, because there was no low-temperature convection in the case of the dual-sided heating mode. This is the reason why the machining capacity of the dual-sided heating method was more than twice that of the single-sided heating method.

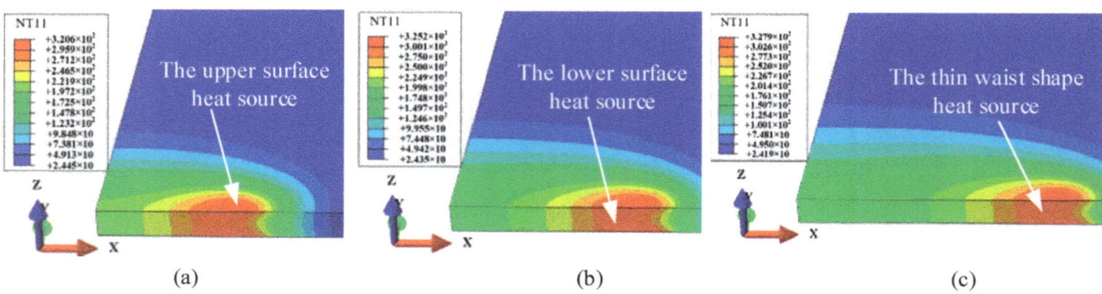

Figure 14. Evolution of temperature field in cutting Al_2O_3 ceramics with dual-sided heat mode at (**a**) 2.433 s, (**b**) 3.033 s and (**c**) 4.033 s.

Figure 15 shows the evolution of the transverse tensile stress induced by the dual-sided heat method. From Figure 15, it can be seen that the maximum compressive stress was divided into upper and lower areas at first and then kept approaching and connected as a thin waist shape. It could be found that the tensile stress zone in front of the compressive stress zone, which was used to cut the workpiece, had an integrated shape. Figure 16 shows the contrast of the distribution of the thermal stress of the single and dual-sided heat modes and their effect on the on the maximum cutting thickness. It is notable that the depth of the tensile stress had a positive correlation with the machining capacity.

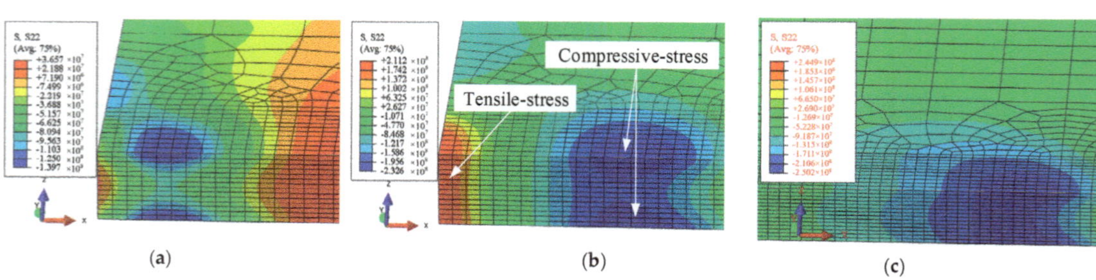

Figure 15. The distribution of transverse tensile stress in the workpiece heated by the dual-sided heat mode at (**a**) t = 0.9000 s, (**b**) t = 1.767 s and (**c**) t = 4.030 s.

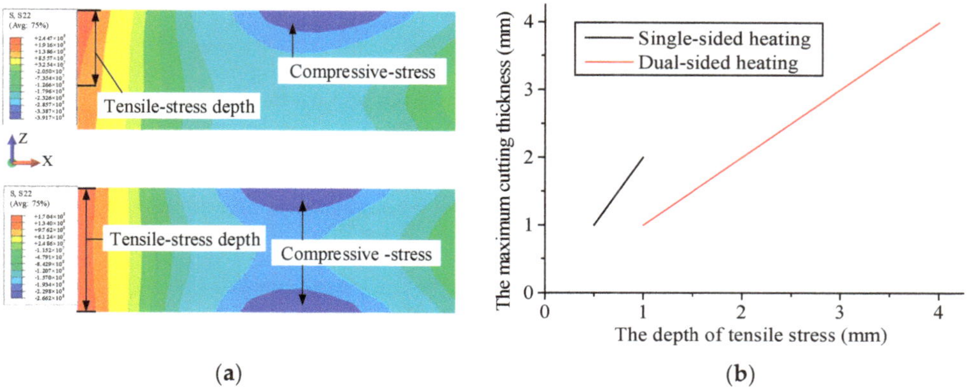

Figure 16. (**a**) Comparison of the distribution of thermal-stress between the single and dual-sided heat modes. (**b**) The effect of heat mode on the maximum cutting thickness.

4.3. Analysis of the Fracture Quality Distribution

Figure 17 shows the tensile stress evolution and crack propagation mode in the crack initiation stage in cutting Al_2O_3 ceramics using the thermal cracking method via a microwave-induced dual-sided heat source. As is shown in Figure 17a, the pre-crack first experienced the action of tensile stress as the upper and lower heat source were loaded on the edge of the workpiece.

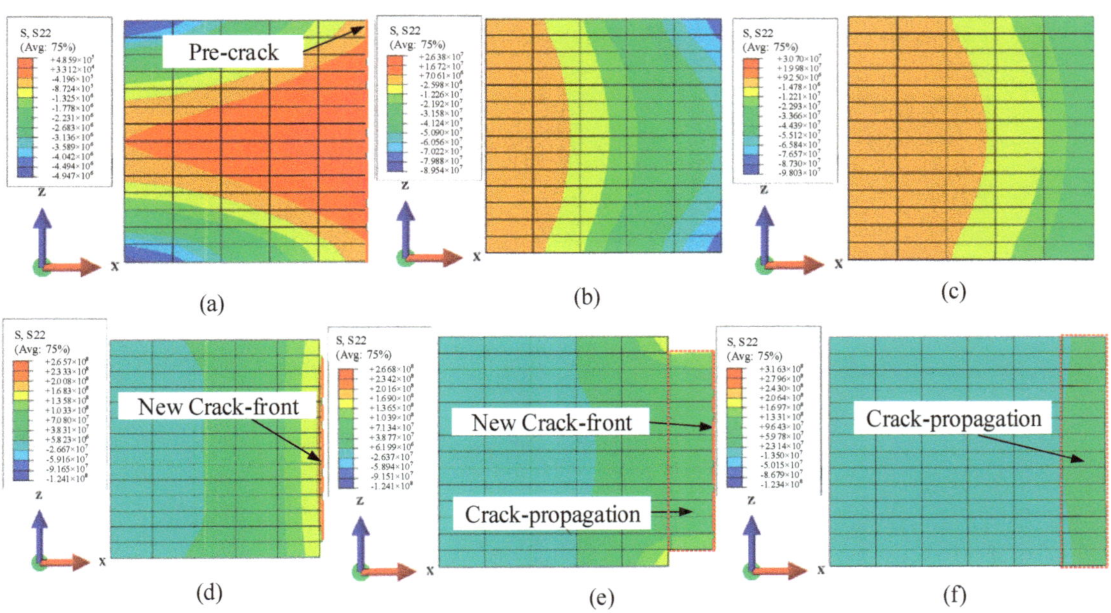

Figure 17. Transverse tensile stress evolution and initial crack behavior in cutting ceramics using thermal cracking method with microwave dual-surface heat source at (**a**) 0.06667 s, (**b**) 1.17083 s, (**c**) 1.33750 s, (**d**) 2.40857 s, (**e**) 2.42267 s and (**f**) 2.42860 s. (The microwave power is 600 W, the scanning speed is 3 mm/s, the dual-sided coated graphite thickness is 100 μm and the effective power radius 2 mm).

Then, with the dual-sided heat source moving to the site shown in Figure 17b, which was just above the pre-crack, and moving further, the tensile stress continually gathered toward the pre-crack again. At the stage shown in Figure 17d, when the tensile stress at the center thickness of the material exceeded the fracture strength, the new crack front formed. Finally, the new crack front at the center thickness propagated and drove the whole thickness to extend forward.

The initial cracking process under the action of bulk heat source was simulated to compare it with the crack initiation characteristic of dual-sided heat sources. Figure 18 shows the transverse tensile stress distribution and initial crack propagation evolution when the glass was cut by a bulk heat source. The geometric model used was the same as that used in Al_2O_3 ceramics, and the crack propagation simulation calculation was also carried out by using extended finite-element technology. The remarkable difference from the dual-sided loading mode was that the compressive stress zone, which represented the maximum temperature zone, was on the center of the material. This difference led to crack initiation near the upper and lower surface of the material, shown in Figure 18, which was also reported in Zhao's research [9].

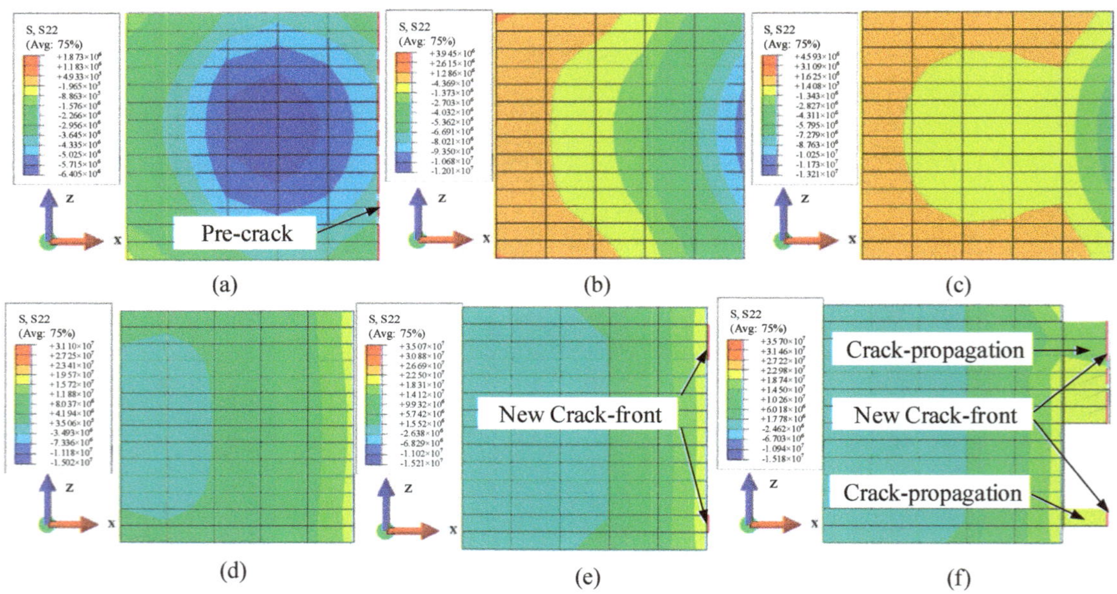

Figure 18. Transverse tensile stress evolution and crack behavior in the process of crack initiation of cutting glass using thermal cracking method via bulk heat source at (**a**) 0.8193 s, (**b**) 1.3189 s, (**c**) 1.5194 s, (**d**) 2.0206 s, (**e**) 2.1527 s and (**f**) 2.1860 s.

Figure 19 shows the transverse tensile stress distribution along the pre-crack at the time when the crack initiation occurred in the thermal cracking of ceramic materials via a dual-surface heat source (Figure 19a) and a bulk heat source (Figure 19b). From Figure 19, it can be seen that the transverse tensile stress distribution characteristic along the pre-crack in the dual-sided heat mode and the bulk heat mode was different, but together, they were more significantly different from the single-sided heat mode, which is shown in Figure 20. In the thermal cracking of glass, it was reported that the body heat mode has better cutting quality than that of the single-sided heat mode [3]. It is the similarity in the transverse tensile stress distribution style between the dual-sided heat mode and the bulk heat mode that results in their better fracture quality than the single-sided heat mode.

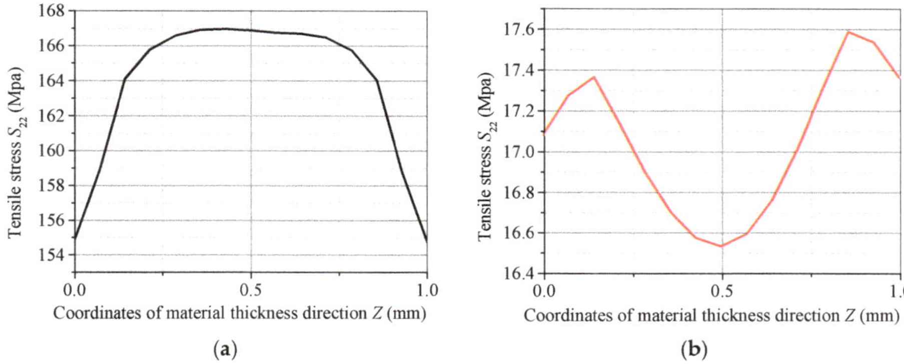

Figure 19. Transverse tensile stress distribution along crack length direction before crack initiation under the action of dual-surface heat source and comparison with that under the action of body heat source. (**a**) Dual surface heat source; (**b**) body heat source.

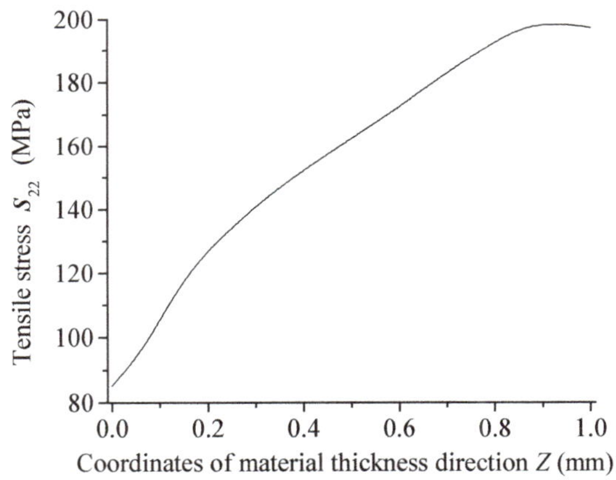

Figure 20. Transverse tensile stress distribution at the bottom of pre-crack along the direction of material thickness when the crack is expanding under single-sided heat source.

In fracture mechanics of brittle materials, a crack would propagate when the external tensile stress exceeds the fracture strength of the material. From Figure 19, it can be seen that the center thickness of the material in the dual-sided surface heat mode had approximately equal tensile stress. The workpiece would fracture first at these high-external-stress and low-gradient sections, which was accurately predicted in the crack propagation simulation in Figure 17. Because this section in the thickness had a proportion beyond 70% and this state could maintain at more than 80% of the cutting periods, the pre-crack almost synchronously propagated in a large proportion of the middle section, shown in Figure 17. Compared with the crack propagation style in the bulk heat mode, shown in Figure 18, in which the fracture first occurred near the upper and lower surface, the dual-sided surface heat mode would produce better cutting quality.

It is notable that the monotonous distribution of transverse tensile stress caused the crack to initiate from the surface with high tensile stress and then expand to another surface, shown in Figure 21, resulting in poor fracture quality and the serious uneven distribution of section quality.

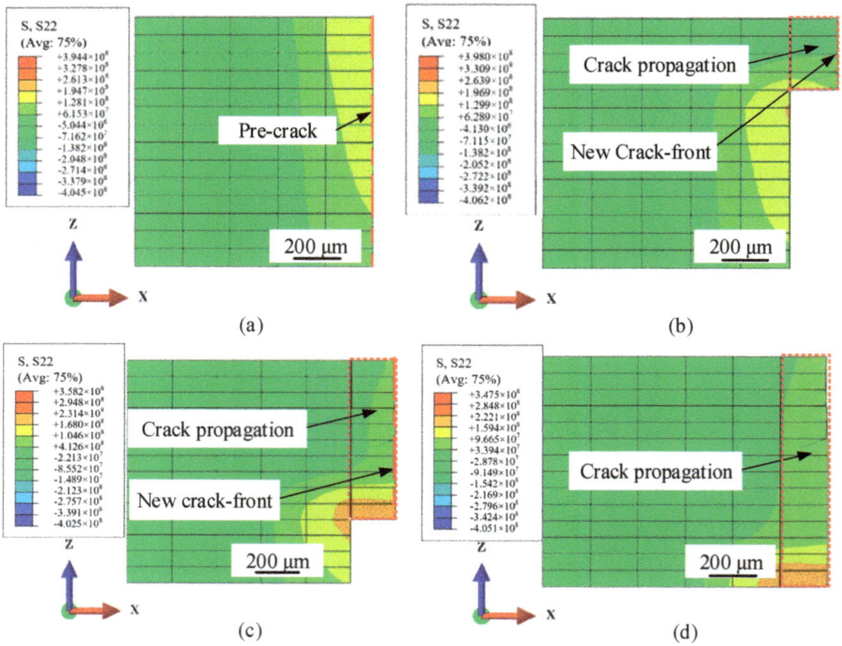

Figure 21. Crack initiation mode of Al_2O_3 ceramic under single-surface heat source at (**a**) 1.6922 s, (**b**) 1.7252 s, (**c**) 1.7269 s and (**d**) 1.7599 s.

Furthermore, it can be derived that the crack initiated at the maximum tensile stress zone. This distribution of transverse tensile stress along the thickness direction in dual-sided cutting mode is consistent with the distribution of the good section quality at the middle depth and poor section quality near the surface in the experiment. Because the more uniform stress distribution in the middle depth led to better quality in the middle depth, and the larger stress gradient near the upper and lower surfaces led to poor section quality.

5. Conclusions

In this work, the microwave-induced dual-sided thermal stress method was proposed to resolve the limited machining capacity and poor fracture quality problem produced by the single-sided heat mode in splitting opaque, brittle material using the thermal-controlled fracture method. The major conclusions from this study are summarized as follows.

The microwave-induced dual-sided thermal-cracking method was first proposed and realized. The experimental results indicate that it could improve the machining capacity by more than double compared with the single-sided mode.

The dual-sided thermal stress method provides better fracture quality and more uniform section morphology compared with the single-sided mode.

A thermal-controlled fracture model based on dual-sided thermal stress was established to calculate the temperature and the thermal stress distribution and simulate the crack propagation behavior. The simulation results had good agreement with the experiment result. The modeling was used to advance our understanding of the improvement mechanism of the machining capacity and fracture quality for this method.

The depth of the transverse tensile stress along the thickness direction of the workpiece was proven to have a positive correlation with the machining capacity by contrasting the dual-sided and single-sided thermal stress cutting method.

The reason for the better fracture quality and more uniform section morphology in the dual-sided thermal stress cutting mode is that the middle sites in the thickness direction of the workpiece could have an approximate equality transverse tensile stress zone accounting for more than 70%, and this state could be maintained in more than 80% of the cutting periods.

This study provides an innovative and feasible method for cutting opaque, brittle materials with promising fracture quality and machining capacity for industrial application. An important research direction in the future is the study of the influence of the matching of dual heat sources on stress distribution at crack fronts and its effect on machining quality.

Author Contributions: Conceptualization, Y.W. and X.C.; methodology, H.W. and Z.H.; software, X.C. and Z.H.; validation, X.C., Y.W., Z.H. and H.W.; formal analysis, X.C.; investigation, X.C.; resources, Y.W.; data curation, Z.H.; writing—original draft preparation, X.C.; writing—review and editing, X.C.; visualization, X.C.; supervision, X.C.; project administration, Y.W.; funding acquisition, X.C. All authors have read and agreed to the published version of the manuscript.

Funding: This work was supported by the Research Initiation Foundation of Anhui Polytechnic University (No. S022021002); the scientific research project of Anhui Polytechnic University (No. KZ42022065) and the Key Project of Natural Science of Anhui Provincial Department of Education (No. KJ2021A0487).

Institutional Review Board Statement: Not applicable.

Informed Consent Statement: Not applicable.

Data Availability Statement: Not applicable.

Conflicts of Interest: The authors declare no conflict of interest.

References

1. Rakshit, R.; Das, A. A review on cutting of industrial ceramic materials. *Precis. Eng.* **2019**, *59*, 90–109. [CrossRef]
2. Lumley, R.M. Controlled Separation of Brittle Materials Using a Laser. *Am. Ceram. Soc. Bull.* **1968**, *48*, 850–854.
3. Yang, L.J.; Wang, Y.; Tian, Z.G.; Cai, N. YAG laser cutting soda-lime glass with controlled fracture and volumetric heat absorption. *Int. J. Mach. Tool Manuf.* **2010**, *50*, 849–859. [CrossRef]
4. Haupt, O.; Siegel, F.; Schoonderbeek, A.; Richter, L.; Kling, R.; Ostendorf, A. Laser dicing of silicon: Comparison of ablation mechanisms with a novel technology of thermally induced stress. *J. Laser Micro Nanoeng.* **2008**, *3*, 135–140. [CrossRef]
5. Yang, L.; Ding, Y.; Cheng, B.; Mohammed, A.; Wang, Y. Numerical simulation and experimental research on reduction of taper and HAZ during laser drilling using moving focal point. *Int. J. Mach. Tool Manuf.* **2017**, *91*, 1171–1180. [CrossRef]
6. Li, C.; Piao, Y.; Meng, B.; Hu, Y.; Li, L.; Zhang, F. Phase transition and plastic deformation mechanisms induced by self-rotating grinding of GaN single crystals. *Int. J. Mach. Tool Manuf.* **2022**, *172*, 103827. [CrossRef]
7. Li, C.; Wu, Y.; Li, X.; Ma, L.; Zhang, F.; Huang, H. Deformation characteristics and surface generation modelling of crack-free grinding of GGG single crystals. *J. Mater. Process. Technol.* **2020**, *279*, 116577. [CrossRef]
8. Ueda, T.; Yamada, K.; Oiso, K.; Hosokawa, A. Thermal Stress Cleaving of Brittle Materials by Laser Beam. *Cirp. Ann.-Manuf. Technol.* **2002**, *51*, 149–152. [CrossRef]
9. Zhao, C.; Zhang, H.; Wang, Y. Semiconductor laser asymmetry cutting glass with laser induced thermal-crack propagation. *Opt. Laser Eng.* **2014**, *63*, 43–52. [CrossRef]
10. Cai, Y.; Yang, L.; Zhang, H.; Wang, Y. Laser cutting silicon-glass double layer wafer with laser induced thermal-crack propagation. *Opt. Laser Eng.* **2016**, *82*, 173–185. [CrossRef]
11. Kalyanasundaram, D.; Schmidt, A.; Molian, P.; Shrotriya, P. Hybrid CO_2 laser/waterjet machining of polycrystalline diamond substrate: Material separation through transformation induced controlled fracture. *J. Manuf. Sci. Eng.* **2014**, *136*, 1–10. [CrossRef]
12. Furumoto, T.; Hashimoto, Y.; Ogi, H.; Kawabe, T.; Yamaguchi, M.; Koyano, T.; Hosokawa, A. CO_2 laser cleavage of chemically strengthened glass. *J. Mater. Process. Technol.* **2021**, *289*, 116961. [CrossRef]
13. Liu, P.; Deng, L.; Zhang, F.; Chen, H.; Duan, J.; Zeng, X. Study on high-efficiency separation of laminated glass by skillfully combining laser-induced thermal-crack propagation and laser thermal melting. *Appl. Phys. A* **2020**, *126*, 1–13. [CrossRef]
14. Xu, J.; Hu, H.; Zhuang, C.; Ma, G.; Han, J.; Lei, Y. Controllable laser thermal cleavage of sapphire wafers. *Opt. Laser Eng.* **2018**, *102*, 26–33. [CrossRef]
15. Cheng, X.; Yang, L.; Wang, M.; Cai, Y.; Wang, Y.; Ren, Z. Laser beam induced thermal-crack propagation for asymmetric linear cutting of silicon wafer. *Opt. Laser Technol.* **2019**, *120*, 105765. [CrossRef]
16. Cheng, X.; Yang, L.; Wang, M.; Cai, Y.; Wang, Y.; Ren, Z. The unbiased propagation mechanism in laser cutting silicon wafer with laser induced thermal-crack propagation. *Appl. Phys. A* **2019**, *125*, 1–11. [CrossRef]
17. Brugan, P.; Cai, G.; Akarapu, R.; Segall, A.E. Controlled-fracture of prescored alumina ceramics using simultaneous CO_2 lasers. *J. Laser Appl.* **2006**, *18*, 236–241. [CrossRef]

18. Saman, A.M.; Furumoto, T. Evaluation of Separating Process for Different Materials by Thermal Stress Cleaving Technique. *Int. J. Precis. Eng.* **2018**, *7*, 158–161. [CrossRef]
19. Tsai, C.; Chen, H. Laser cutting of thick ceramic substrates by controlled fracture technique. *J. Mater. Process. Technol.* **2003**, *136*, 166–173. [CrossRef]
20. Wang, H.; Zhang, H.; Wang, Y. Splitting of glass and SiC ceramic sheets using controlled fracture technique with elliptic microwave spot. *Ceram. Int.* **2016**, *43*, 1669–1676. [CrossRef]
21. Cheng, X.; Zhao, C.; Wang, H.; Wang, Y.; Wang, Z. Mechanism of irregular crack-propagation in thermal controlled fracture of ceramics induced by microwave. *Mech. Ind.* **2020**, *21*, 610. [CrossRef]
22. Wang, H.; Zhang, H.; Wang, Y.; Wang, M. Thermal controlled fracture of Al2O3 substrate by inducing microwave discharge in graphite coat. *Ceram. Int.* **2019**, *45*, 6149–6159. [CrossRef]
23. Cai, Y.; Wang, M.; Zhang, H.; Yang, L.; Fu, X.; Wang, Y. Laser cutting sandwich structure glass–silicon–glass wafer with laser induced thermal—Crack propagation. *Opt. Laser Technol.* **2017**, *93*, 49–59. [CrossRef]
24. Yin, Y.; Gao, Y.; Yang, C. Sawing characteristics of diamond wire cutting sapphire crystal based on tool life cycle. *Ceram. Int.* **2021**, *47*, 26627–26634. [CrossRef]

crystals

MDPI

Article

Surface Morphology Evaluation and Material Removal Mechanism Analysis by Single Abrasive Scratching of RB-SiC Ceramics

Zhangping You [1,2], Haiyang Yuan [1,2,*], Xiaoping Ye [1,2] and Liwu Shi [3]

1 School of Engineering, Lishui University, Lishui 323000, China; youzping@163.com (Z.Y.); anst_yxp@163.com (X.Y.)
2 Key Laboratory of Digital Design and Intelligent Manufacture in Culture & Creativity Product of Zhejiang Province, Lishui University, Lishui 323000, China
3 College of Mechanical Engineering, Zhejiang University of Technology, Hangzhou 310023, China; liwushi2022@163.com
* Correspondence: yuanhaiyang1156@163.com

Abstract: Reaction-Bonded Silicon Carbide (RB-SiC) ceramics possessing excellent mechanical and chemical properties, whose surface integrities have an essential effect on their performance and service life, have been widely used as substrates in the core parts of aerospace, optics and semiconductors industries. The single abrasive scratching test is considered as the effective way to provide the fundamental material removal mechanisms in the abrasive lapping and polishing of RB-SiC ceramics for the best surface finish. In this study, a novel single abrasive scratching test with an increasing scratching depth has been properly designed to represent the real abrasive lapping and polishing process and employed to experimentally investigate the surface integrity regarding different scratching speeds. Three typical and different material removal stages, including the ductile mode, ductile–brittle transition mode and brittle mode, can be clearly distinguished and it is found that in the ductile material removal stage by increasing the scratching speed would inhibit the plastic deformation and improve its surface integrity. It is also found that in the ductile–brittle transition and brittle material removal stages, to increase the scratching speed would inhibit the plastic deformation due to the fast scratching speed that limits the time of plastic deformation on the target, but it also results in the increased length of lateral cracks with the increased scratching speed which can reflect that the size of brittle chips, like brittle fractures and large grain fragmentations, increases as the scratching speed increases. It can provide the references for the optimization of the abrasive lapping and polishing of RB-SiC ceramics with high efficiency and surface quality.

Keywords: surface morphology; material removal mechanism; reaction-bonded silicon carbide ceramics; abrasive scratching

Citation: You, Z.; Yuan, H.; Ye, X.; Shi, L. Surface Morphology Evaluation and Material Removal Mechanism Analysis by Single Abrasive Scratching of RB-SiC Ceramics. *Crystals* **2022**, *12*, 879. https://doi.org/10.3390/cryst12070879

Academic Editors: Chen Li, Chongjun Wu, Binbin Meng and Shanshan Chen

Received: 8 May 2022
Accepted: 20 May 2022
Published: 21 June 2022

1. Introduction

Due to the distinct mechanical and chemical properties of high erosion and wear resistance, high thermal conductivity, high chemical inertness and low thermal expansion, Reaction-Bonded Silicon Carbide (RB-SiC) ceramics have become the primary structure and substrate materials necessary for the aerospace, optics and semiconductors industries [1–3]. However, the inherent characteristics of high hardness and low fracture toughness bring large challenges in the machining of RB-SiC ceramics, and the surface integrity has an essential influence on their performance and service life in practice. Abrasive machining, as one of the non-traditional machining technologies, is a process to remove material by means of micro-ploughing, micro-cutting, micro-fatigue and micro-cracking, and has been extensively used to machine almost any material, particularly hard brittle materials [4–7]. Currently, the flat substrates of hard brittle materials, like RB-SiC ceramics, are usually prepared by using the abrasive lapping and polishing processes [8,9], and to ensure the machining efficiency and quality, the optimization of processing parameters is necessary by conducting and repeating associated experiments many times, which is to some extent time-consuming and expensive. The single abrasive scratching test is considered one of the effective ways to investigate the fundamental of material removal mechanisms and provides good references for the abrasive machining process [10–13].

For hard brittle materials like RB-SiC ceramics, brittle material removal mode usually dominates the erosion process in terms of cracks and fractures, while by properly controlling the machining process the ductile material removal mode can be also found in removing hard brittle materials in terms of micro-cutting and micro-ploughing [14,15]. Rao et al. [16] employed the Vickers indenter to scratch the RB-SiC ceramics at elevated temperatures and found that the material deformation and adhesive behavior enhanced the ductile material removal and the coefficient of friction at elevated temperatures. Klecka and Subhash [17] investigated the scratch-induced damage in alumina ceramics by considering the different grain sizes and found that the intergranular fracture and grain dislodgement significantly contributed to the lateral crack propagation and resultant material removal from the target surface. Moreover, the multi-scratching tests were conducted by Yang et al. to reveal stress interaction and crack propagation behavior of glass ceramics and it was found that the material removal mechanism was distinctly related to the value and orientation of stress [18].

Most of the current research work on exploring the material removal mechanisms during the scratching process are either employing the nanoindentation or relatively low scratching speed at nanoscale, but in the abrasive lapping and polishing process the speed of abrasive particle scratching on the target surface is relatively large and the scratching depth is varied due to the different pressures on the target surface. Thus, in this study, a novel single abrasive scratching test on RB-SiC ceramics with an increasingly scratching depth will be designed to experimentally investigate the surface integrity regarding different scratching speeds with the assistance of high-tech measuring instruments, and it aims to provide the references for optimizing the abrasive lapping and polishing of RB-SiC ceramics with high efficiency and surface quality.

2. Experimental Work

2.1. Preparation of RB-SiC Ceramic Sample

The RB-SiC ceramic sample with dimensions of 10 mm × 10 mm × 3 mm is selected as the target and the major material properties of RB-SiC ceramics are given in Table 1. In order to guarantee the scratching stability of these samples in the experiment it is necessary to make the inlaid treatment for them, as shown in Figure 1, where the sample is embedded into the cylinder inlaid materials that is fixed by the gripping tong during the scratching process. After the inlaid treatment of the sample, the surface of the sample was first lapped parallel to the surface of the cylinder inlaid materials with P2400 sandpaper, and then it was polished by using the diamond slurry with concentration of 10% by mass and diameters of 2.5 μm and 1 μm, respectively, to realize the surface roughness of about 200 nm, which facilitates for further observation and analysis.

Table 1. Material properties of RB-SiC ceramics.

Density (g/cm^3)	Elastic Modulus (GPa)	Fracture Toughness (MPa·m$^{1/2}$)	Mohs Hardness	Bending Strength (MPa)
3.08	430	3.5	9.5	490

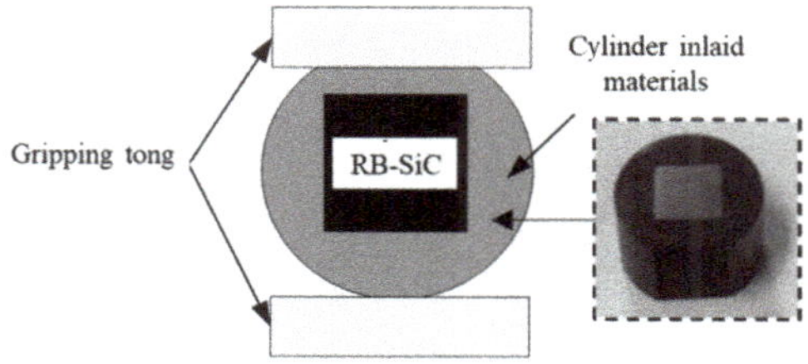

Figure 1. Inlaid treatment of the RB-SiC ceramic sample.

2.2. Experimental Setup

The experimental setup for the single abrasive scratching of RB-SiC ceramics is presented in Figure 2, where the specialized design cutting tool was connected to the spindle of the VMX42SRTi (Hurco Companies Inc., Indianapolis, IN, USA) CNC machine with the maximum spindle speed of 12,000 r/min and the cylinder inlaid sample was put just under the PCD cutter and fixed by the gripping tong. Then, the micrometer was employed to adjust the gripping tong to make it parallel to the working platform for ensuring the accuracy of the experiment.

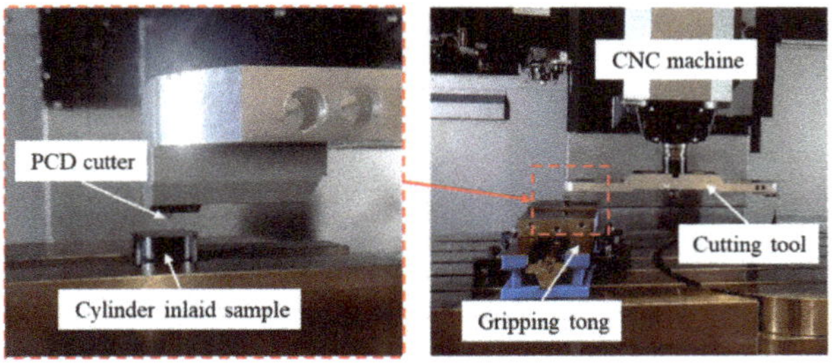

Figure 2. Experimental setup.

To be specific, as shown in Figure 3, the diameter of the specialized design cutting tool was 250 mm, which was large enough to realize the approximately straight scratching line on the target surface, and the triangle single blade PCD cutter (TPGH110302, Kyocera, Kyoto, Japan) with fillet radius of 200 μm was fixed in the bottom of the one side of the cutting tool, while on the other side, the same balancing weight was added in order to keep the stability during the rotating process with the relatively high speed. In the experiment, the single abrasive scratching test with an increasing scratching depth (d_c) from 0 to 30 μm was designed to facilitate the observation of ductile, ductile–brittle transition and brittle material removal modes on the target surface in a single scratching test. Moreover, due to the large mass and rotational inertia of the cutting tool and the consideration of the experimental safety, three levels of scratching speed (v_s) were selected at 1 m/s, 5 m/s and 10 m/s, respectively, by properly controlling the rotation speed of the spindle and the feeding speed in the CNC machine. Each test was repeated 10 times and the average data were taken. After each test, the sample was treated by putting it into the ultrasonic cleaner with the alcohol for 15 min and with the assistance of Zeiss SIGMA VP FE-SEM and Keyence VHX-7000 3D Microscope the surface morphology of the scratched sample can be observed for further comparison and analysis.

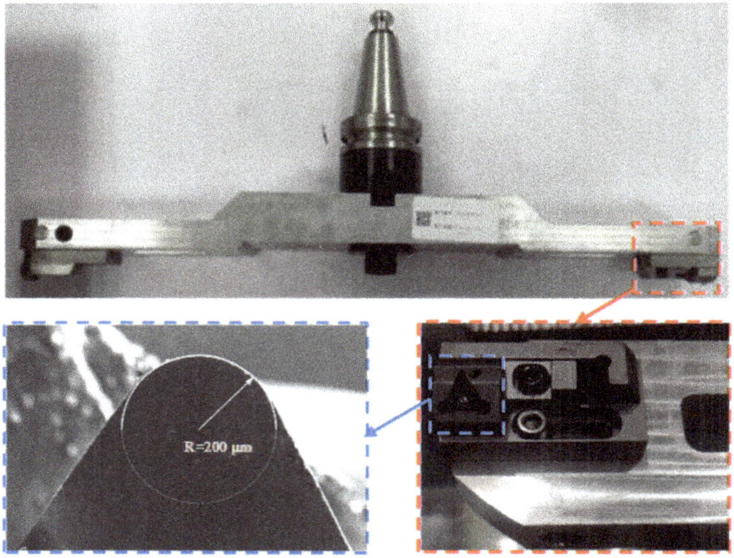

Figure 3. Specialized design cutting tool.

3. Results and Discussion

3.1. Overall Observation of the Surface Morphology

The overall surface morphology of the original RB-SiC ceramic sample after the precision abrasive polishing process can be seen in Figure 4, where the SiC and Si grains can be clearly found on the sample surface (see Figure 4a) and a good surface finish with the roughness of about 200 nm can be measured from Figure 4b, so that by using this kind of sample it facilitates the further distinguishing of the scratched sample surface with different material removal mechanisms. Due to the high hardness and brittleness of RB-SiC ceramic, it presents the typical brittle dominated material removal mode, such as the transgranular and intergranular cracks, as compared with the metals during the machining process [19,20]. However, by properly controlling the machining parameters the ductile material removal from RB-SiC ceramics with good surface finish can also be realized and the single abrasive scratching test is considered as one of the effective ways to investigate this phenomenon and to provide good references for the machining process.

Figure 4. Surface morphology of the original sample. (**a**) 2D view; (**b**) 3D view.

Figure 5 shows the typical surface morphology of the gradually increasing scratched RB-SiC ceramic sample from d_c = 0 to d_c = 30 μm at v_s = 10 m/s and three material removal stages can be distinguished with respect to different underlying material removal mechanisms. In the ductile material removal stage, it can be seen from Figure 5 that a shallow scratched groove is left on the target surface and the surface of the groove seems to be smooth. Like the material removal from metals, RB-SiC ceramics are removed by plastic flow in this ductile material removal stage. To be specific, some materials can generate continuous thin chips in front of the rake face of the cutter and others can accumulate on both sides of the scratched groove to form plastic uplift such that there is almost no damage on the target surface [21,22]. After entering into the ductile–brittle transition material removal stage, the brittle material removal mode, like cracks, begins to dominate the material removal process, which results in the decrease of the smoothness on the scratched target surface. With the further increase of the scratching depth, it goes into the brittle material removal stage, where the width of the scratched groove significantly increases and there are obvious brittle fractures at the edge of the scratched groove caused by the grain spalling and fragmentation. Thus, in order to further analyze the material removal mechanisms in different stages, the surface integrity under different scratching speeds has been observed and compared, as explained in Section 3.2.

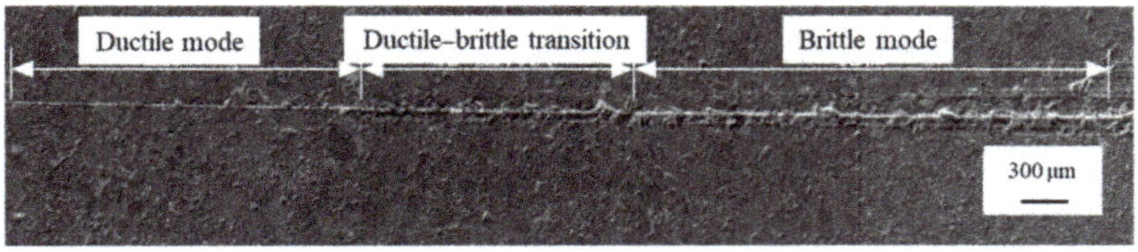

Figure 5. Overall surface morphology of the scratched RB-SiC ceramic sample at v_s = 10 m/s.

3.2. Evaluation of the Surface Intergrity at Different Material Removal Stages

The surface integrity at different material removal stages with corresponding scratching speeds can be evaluated by distinguishing the surface morphologies of the scratched RB-SiC ceramic samples, as shown in Figure 6.

In the ductile material removal (DR) stage, as can be seen from Figure 6(a-i,b-i,c-i) the interaction between the cutter and target surface is mainly caused by plastic deformation with plastic flow and it is because the scratching depth is less than the critical depth of the ductile–brittle transition of RB-SiC ceramic so that the ductile material removal mode dominates the material removal process in the DR stage. It is also found, by comparing these figures, that the surface integrity has been improved with the increase of scratching speed from 1 m/s to 10 m/s in the DR stage. When the scratching speed is 1 m/s, the plastic uplift on both sides of the scratched groove is obvious, while the plastic uplift on both sides of the scratches at v_s = 5 m/s and v_s = 10 m/s is relatively slight. It is attributed to the fact that with the increase of the scratching speed the initiation of cracks can be effectively inhibited due to the limited generation of plastic deformation, hence resulting in the narrower scratched groove with a shallower depth.

In the ductile–brittle transition material removal (D-B-TR) stage, it can be seen from Figure 6(a-ii,b-ii,c-ii) that the plastic flow, brittle crack and fracture can be clearly observed on the scratched target surfaces; in this stage the scratching depth is just near the critical depth of the brittle–ductile transition of the RB-SiC ceramic, where both ductile and brittle material removal modes contribute to the material removal process. By observing the surface morphologies under different scratching speeds, it can be found in the D-B-TR stage that when the scratching speed is 1 m/s there is obvious plastic flow on the left side of the scratched groove and the brittle cracks occur on the right side of the scratched groove (see Figure 6(a-ii)). When the scratching speed increases to 5 m/s, the plastic uplift caused by the plastic flow and obvious cracks could be observed from Figure 6(b-ii) on the target surface. When the scratching speed further increases to 10 m/s, due to the existence of micropores on the target surface of the RB-SiC ceramic, the brittle fracture generated by the initiation, propagation and intersection of cracks can be found on the scratched surface and the radial crack with a short length can be observed from Figure 6(c-ii) as well.

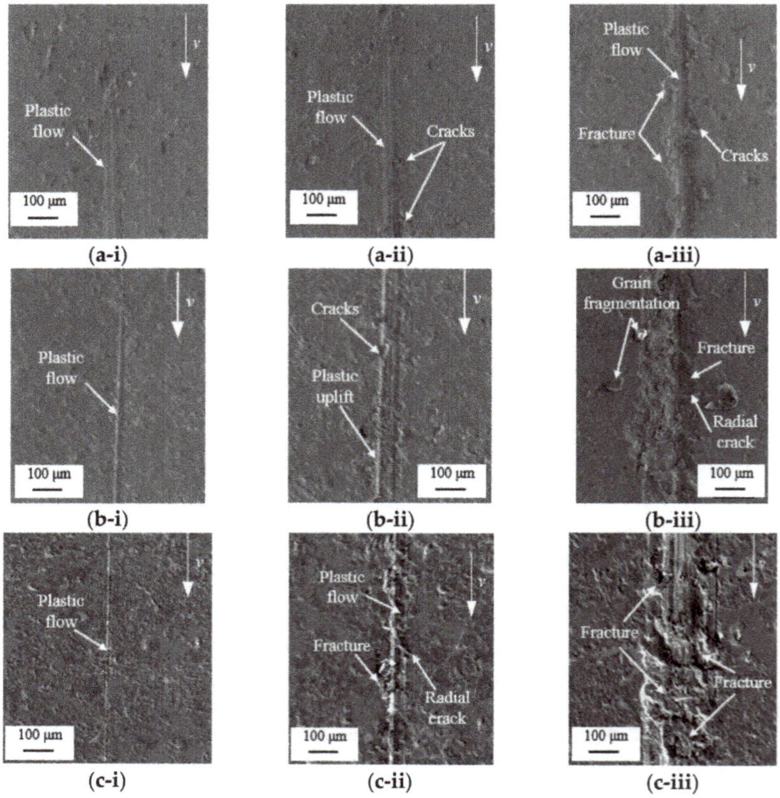

Figure 6. Surface integrity of the scratched RB-SiC ceramic sample at different material removal stages with corresponding scratching speeds. (**a-i**) DR stage, v_s = 1 m/s; (**a-ii**) D-B-TR stage, v_s = 1 m/s; (**a-iii**) BR stage, v_s = 1 m/s; (**b-i**) DR stage, v_s = 5 m/s; (**b-ii**) D-B-TR stage, v_s = 5 m/s; (**b-iii**) BR stage, v_s = 5 m/s; (**c-i**) DR stage, v_s = 10 m/s; (**c-ii**) D-B-TR stage, v_s = 10 m/s; (**c-iii**) BR stage, v_s = 10 m/s.

In the brittle material removal (BR) stage, the overall scratching depth is larger than the critical depth of the ductile–brittle transition of the RB-SiC ceramic, and the brittle fracture mainly contributes to the material removal in this stage, as can be observed from Figure 6(a-iii,b-iii,c-iii). Generally, in the BR stage the median crack can be formed at the bottom of the plastic zone and the radial crack can be generated on the target surface during the scratching process, while the lateral crack is usually formed after the scratching process due to the unloading residual stress, so that the initiation, propagation and intersection of median, radial and lateral cracks could result in the large brittle fracture on the target surface [23,24]. To be specific, when the scratching speed is 1 m/s, obvious cracks and fracture can be observed from Figure 6(a-iii) on the edge of the scratched groove. As the scratching speed increases to 5 m/s, the grain fragmentations caused by the intersections of transgranular and intergranular cracks are found along the scratching groove with some radial cracks as shown in Figure 6(b-iii). With the further increase of the scratching speed to 10 m/s, the resultant increased length of lateral cracks causes more brittle fractures that significantly dominate the material removal process, as can be found both on the bottom and the edge of the scratched groove in Figure 6(c-iii), as similar findings in Ref. [25].

To sum up, in the DR stage, with an increase of the scratching speed it would reduce the generation of plastic flow under the scratched groove and thus inhibit the plastic deformation and improve its surface integrity. In contrast, in the D-B-TR and BR stages, although to increase the scratching speed would inhibit the plastic deformation, it can also result in the increased length of lateral cracks in the abrasive target contact zone, which may facilitate the initiation, propagation and intersection of cracks to form the brittle fractures and the large grain fragmentations on the target surface.

4. Conclusions

The single abrasive scratching test can provide the fundamental material removal mechanisms in the abrasive lapping and polishing of RB-SiC ceramics, which can be used to optimize the associated processing parameters for the best surface finish. In this paper, a novel single abrasive scratching test with an increasing scratching depth from 0 to 30 μm has been properly designed to represent the real abrasive lapping and polishing process and has been employed to experimentally investigate the surface integrity regarding different scratching speeds. Three typical and different material removal stages, including the ductile mode, ductile–brittle transition mode and brittle mode, can be clearly distinguished by observing the surface morphology after the single abrasive scratching test with the assistance of high-tech measuring instruments. To be specific, it is found that in the ductile material removal stage, by increasing the scratching speed it would inhibit the plastic deformation and improve its surface integrity. It is also found that in the ductile–brittle transition and brittle material removal stages, although to increase the scratching speed would inhibit the plastic deformation, it can also result in the increment of the length of lateral cracks in the abrasive target contact zone that can induce the initiation, propagation and intersection of cracks to form the typical brittle fractures and large grain fragmentations on the target surface. However, it is also noticed that the large grain fragmentations and brittle fractures would result in materials being removed from the target surface, which could have an effect on the subsequent scratching process; this effect is worth investigating in the future work. Therefore, the findings in this paper can provide the references for the abrasive lapping and polishing of RB-SiC ceramics that by properly controlling the abrasive lapping and polishing pressure and speed it can realize the brittle material removal with high efficiency and ductile material removal with high surface quality.

Author Contributions: Conceptualization, L.S.; data curation, Z.Y.; investigation, Z.Y. and X.Y.; supervision, H.Y.; writing—original draft, Z.Y.; writing—review and editing, H.Y. All authors have read and agreed to the published version of the manuscript.

Funding: This research was supported by Zhejiang Provincial Natural Science Foundation of China under Grant Nos. LY20E050002, LZ21F020003, Basic Public Welfare Research Project of Zhejiang Province under Grant No. LGG18E050001.

Institutional Review Board Statement: Not applicable.

Informed Consent Statement: Not applicable.

Data Availability Statement: The data presented in this study are available on request from the corresponding author.

Conflicts of Interest: The authors declare no conflict of interest.

References

1. Li, C.; Zhang, F.; Meng, B.; Liu, L.; Rao, X. Material removal mechanism and grinding force modelling of ultrasonic vibration assisted grinding for SiC ceramics. *Ceram. Int.* **2017**, *43*, 2981–2993. [CrossRef]
2. Grinchuk, P.S.; Kiyashko, M.V.; Abuhimd, H.M.; Alshahrani, M.S.; Solovei, D.V.; Stepkin, M.O.; Akulich, A.V.; Shashkov, M.D.; Kuznetsova, T.A.; Danilova-Tretiak, S.M.; et al. Advanced technology for fabrication of reaction-bonded SiC with controlled composition and properties. *J. Eur. Ceram. Soc.* **2021**, *41*, 5813–5824. [CrossRef]
3. Li, C.; Piao, Y.; Meng, B.; Hu, Y.; Li, L.; Zhang, F. Phase transition and plastic deformation mechanisms induced by self-rotating grinding of GaN single crystals. *Int. J. Mach. Tools Manuf.* **2022**, *172*, 103827. [CrossRef]

4. Cheng, Z.; Qin, S.; Fang, Z. Numerical modeling and experimental study on the material removal process using ultrasonic vibration-assisted abrasive water jet. *Front. Mater.* **2022**, *9*, 895271. [CrossRef]
5. Xie, Y.; Sheng, Y.; Qiu, M.; Gui, F. An adaptive decoding biased random key genetic algorithm for cloud workflow scheduling. *Eng. Appl. Artif. Intell.* **2022**, *112*, 104879. [CrossRef]
6. Shi, H.; Liu, G.; Yang, G.; Bi, Q.; Zhao, Y.; Wang, B.; Sun, X.; Liu, X.; Qi, H.; Xu, W.; et al. Analytical modelling of edge chipping in scratch of soda-lime glass considering strain-rate hardening effect. *Ceram. Int.* **2021**, *47*, 26552–26566. [CrossRef]
7. Qi, H.; Qin, S.; Cheng, Z.; Teng, Q.; Hong, T.; Xie, Y. Towards understanding performance enhancing mechanism of micro-holes on K9 glasses using ultrasonic vibration-assisted abrasive slurry jet. *J. Manuf. Process.* **2021**, *64*, 585–593. [CrossRef]
8. Li, H.N.; Yang, Y.; Zhao, Y.J.; Zhang, Z.; Zhu, W.; Wang, W.; Qi, H. On the periodicity of fixed-abrasive planetary lapping based on a generic model. *J. Manuf. Process.* **2019**, *44*, 271–287. [CrossRef]
9. Li, C.; Piao, Y.; Hu, Y.; Wei, Z.; Li, L.; Zhang, F. Modelling and experimental investigation of temperature field during fly-cutting of KDP crystals. *Int. J. Mech. Sci.* **2021**, *210*, 106751. [CrossRef]
10. Hu, W.; Teng, Q.; Hong, T.; Saetang, V.; Qi, H. Stress field modeling of single-abrasive scratching of BK7 glass for surface integrity evaluation. *Ceram. Int.* **2022**, *48*, 12819–12828. [CrossRef]
11. Qi, H.; Shi, L.; Teng, Q.; Hong, T.; Tangwarodomnukun, V.; Liu, G.; Li, H.N. Subsurface damage evaluation in the single abrasive scratching of BK7 glass by considering coupling effect of strain rate and temperature. *Ceram. Int.* **2022**, *48*, 8661–8670. [CrossRef]
12. Li, C.; Wu, Y.; Li, X.; Ma, L.; Zhang, F.; Huang, H. Deformation characteristics and surface generation modelling of crack-free grinding of GGG single crystals. *J. Mater. Process. Technol.* **2020**, *279*, 116577. [CrossRef]
13. Li, C.; Piao, Y.; Meng, B.; Zhang, Y.; Li, L.; Zhang, F. Anisotropy dependence of material removal and deformation mechanisms during nanoscratch of gallium nitride single crystals on (0001) plane. *Appl. Surf. Sci.* **2022**, *578*, 152028. [CrossRef]
14. Zum Gahr, K.H. Wear by hard particles. *Tribol. Int.* **1998**, *31*, 587–596. [CrossRef]
15. Xu, L.; Wang, L.; Chen, H.; Wang, X.; Chen, F.; Lyu, B.; Hang, W.; Zhao, W.; Yuan, J. Effects of pH values and H_2O_2 concentrations on the chemical enhanced shear dilatancy polishing of Tungsten. *Micromachines* **2022**, *13*, 762. [CrossRef]
16. Rao, X.; Zhang, F.; Luo, X.; Ding, F.; Cai, Y.; Sun, J.; Liu, H. Material removal mode and friction behaviour of RB-SiC ceramics during scratching at elevated temperatures. *J. Eur. Ceram. Soc.* **2019**, *39*, 3534–3545. [CrossRef]
17. Klecka, M.; Subhash, G. Grain size dependence of scratch-induced damage in alumina ceramics. *Wear* **2008**, *265*, 612–619. [CrossRef]
18. Yang, X.; Qiu, Z.; Wang, Y. Stress interaction and crack propagation behavior of glass ceramics under multi-scratches. *J. Non-Cryst. Solids* **2019**, *523*, 119600. [CrossRef]
19. Gastaldi, D.; Vena, P.; Contro, R. Hybrid microstructural finite element modeling for intergranular fracture in ceramic composites and coated systems. *Comp. Mater. Sci.* **2008**, *44*, 26–31. [CrossRef]
20. Li, Z.; Zhang, F.; Luo, X.; Cai, Y. Fundamental understanding of the deformation mechanism and corresponding behavior of RB-SiC ceramics subjected to nano-scratch in ambient temperature. *Appl. Surf. Sci.* **2019**, *469*, 674–683. [CrossRef]
21. Ji, R.; Zhang, L.; Zhang, L.; Li, Y.; Lu, S.; Fu, Y. Processing method for metallic substrate using the liquid metal lapping-polishing plate. *Front. Mater.* **2022**, *9*, 896346. [CrossRef]
22. Zhang, L.; Zheng, B.; Xie, Y.; Ji, R.; Li, Y.; Mao, W. Control mechanism of particle flow in the weak liquid metal flow field on non-uniform curvature surface based on lippmann model. *Front. Mater.* **2022**, *9*, 895263. [CrossRef]
23. Qi, H.; Fan, J.; Wang, J.; Li, H. Impact erosion by high velocity micro-particles on a quartz crystal. *Tribol. Int.* **2015**, *82*, 200–210. [CrossRef]
24. Chen, H.Y.; Xu, Q.; Wang, J.H.; Li, P.; Yuan, J.L.; Lyu, B.H.; Wang, J.H.; Tokunagad, K.; Yao, G.; Luo, L.M.; et al. Effect of surface quality on hydrogen/helium irradiation behavior in tungsten. *Nucl. Eng. Technol.* 2022; *in press*. [CrossRef]
25. Lin, B.; Li, S.; Cao, Z.; Zhang, Y.; Jiang, X. Theoretical modeling and experimental analysis of single-grain scratching mechanism of fused quartz glass. *J. Mater. Process. Technol.* **2021**, *293*, 117090. [CrossRef]

crystals

MDPI

Article

Determination of Heat Transfer Coefficient by Inverse Analyzing for Selective Laser Melting (SLM) of AlSi10Mg

Chongjun Wu [1], Weichun Xu [2], Shanshan Wan [3], Chao Luo [4], Zhijian Lin [4] and Xiaohui Jiang [3,*]

1 College of Mechanical Engineering, Donghua University, Shanghai 201620, China
2 Shanghai Academy of Space Flight Technology, Shanghai 201109, China
3 School of Mechanical Engineering, University of Shanghai for Science and Technology, Shanghai 200093, China
4 Aplos Machines Manufacturing (Shang Hai) Co., Ltd., Shanghai 201306, China
* Correspondence: jiangxh@usst.edu.cn

Abstract: Heat treatment can improve performance and control quality in the additive manufacturing process. In the numerical simulation of heat treatment, the accuracy of the heat transfer coefficient will have a significant impact on the accuracy of the simulated temperature field. At present, The inverse analysis method is the most common and effective method to determine the heat transfer coefficient. Taking the actual temperature curve as the input condition, the heat transfer coefficient values of the heating, quenching, and air cooling components in the heat treatment process are successfully obtained. Based on the obtained heat transfer coefficient, a mathematical model of the heat transfer coefficient change with temperature during heat treatment is established. The heat transfer coefficient obtained by the inverse analysis method is then applied to the simulation of heat treatment, and more accurate simulation results are obtained. It is proven in this work that the inverse analysis method can improve the accuracy of the simulation model in the heat treatment process of AlSi10Mg.

Keywords: heat transfer coefficient; heat treatment; SLM; DEFORM; inverse analysis

Citation: Wu, C.; Xu, W.; Wan, S.; Luo, C.; Lin, Z.; Jiang, X. Determination of Heat Transfer Coefficient by Inverse Analyzing for Selective Laser Melting (SLM) of AlSi10Mg. *Crystals* **2022**, *12*, 1309. https://doi.org/10.3390/cryst12091309

Academic Editors: Indrajit Charit and Shouxun Ji

Received: 1 September 2022
Accepted: 9 September 2022
Published: 16 September 2022

1. Introduction

For the manufacturing process of parts, the traditional material removal [1–3] has been extended to printing parts through additive manufacturing. The combination of additive and subtractive processing technology helps to produce more multifunctional complex structures [4–6]. The aluminum alloy AlSi10Mg is a typical alloy with significantly improved strength and hardness, which is suitable for thin-walled, complex geometrical parts in automotive, aerospace and aerospace industrial-grade prototypes and production components [7,8]. AlSi10Mg is widely used in additive manufacturing. The effect of heat treatment on improving additive manufacturing process can be analyzed by heat treatment of parts made of this material.

There are many factors affecting the heat transfer coefficient [9–11], and it is difficult to accurately solve the interfacial heat transfer coefficient by using the conventional surface temperature and temperature gradient methods. The inverse analysis method is the most common and effective method to solve the heat transfer coefficient. First of all, temperature data [12–14] in the heat treatment process were obtained through experimental measurement, and the data were imported into DEFORM. The optimal program in DEFORM compared simulated time and temperature data with experimental time and temperature data, and optimized the operation until the optimal value was calculated.

Shokoufeh et al. [15] used CFD simulation to explore the heat transfer during the heat up portion of the curing cycle in an autoclave to improve the production rate without compromising quality. Su et al. [16] investigated the heat transfer performance of electrostatic spraying used in machining using the newly developed device by transient heat transfer tests. Kim et al. [17] calculated the interfacial heat transfer coefficient between the die and the workpiece, using the inverse analysis method to measure the temperature of the die in the hot stamping process. Gianfranco et al. [18] took the temperature data of the process collected by thermocouple as the input value, and simulated the whole sand mold casting process with the commercial software package MAGMASOFT using the finite difference method. By increasing the number of temperature control points, the interfacial heat transfer coefficient was obtained. Using the temperature data of sand and metal collected by thermocouple, Sun Mikhail et al. [19] presented a technique for process-induced residual strain modeling for thermoset composite material parts and used the technique to clarify the distribution of the heat transfer coefficient on the surface of the part and mold using the CFD method. And distribution of heat transfer coefficients were obtained in ANSYS CFX under the appropriate process conditions. Modeling simplifications were proposed by Ramos et al. [20] for an efficient numerical discretization of infill structures. With the prospect of choosing correct thermal boundary conditions expressing the natural convection between printed material and its environment, values for the convective heat transfer coefficient and ambient temperature were calibrated through numerical data fitting of experimental thermal measurements. Kadam et al. [21] proposed an inverse method based on the transient temperature of the back surface using the solution of the three-dimensional inverse heat conduction problem to estimate the transient temperature of the collision side and then evaluate the heat transfer coefficient. Kang et al. [22] proposed an empirical formula for laminar natural convection of an outer finned tube heat exchanger with a wide range of structural parameters and calculated the heat transfer coefficient by using the transient temperature response of the heat exchanger. Piotr [23] simplified the three-dimensional transient heat conduction problem to an axisymmetric heat conduction problem and used the inverse method to calculate the surface heat transfer coefficient. Farzad et al. [24] used three-dimensional elliptic mesh generation technology to mesh irregular bodies. A new and highly effective sensitivity analysis program was introduced to the gradient-based optimization method to calculate the sensitivity coefficient, and a more accurate thermal conductivity, heat transfer coefficient, and heat flux density could be obtained by this solution.

With the development of intelligent algorithms, some new inverse analysis methods have emerged in recent years. Parida et al. [25] used Green's function method considering transient convective boundary conditions and transient radiant heat loss to estimate the total heat transfer coefficient and adiabatic heat transfer during the jet impact heat transfer process by inputting the transient temperature of the non-collision surface obtained by the experiment. Zhang et al. [26] introduced the complex variable differentiation method (CVDM) in the commercial finite element software Abaqus, developed a complex variable finite element model by using the user element subroutine (UEL), and accurately calculated the Levenberg-Marquardt algorithm to solve the key parameter sensitivity matrix coefficient of the inverse problem. This technique provided a general method for solving the inverse problem of three-dimensional transient nonlinear heat conduction in irregular and complex structures. Ming et al. [27] established a three-way equivalent heat conduction model including heat conduction and heat convection in the cutting process based on the structural characteristics of the honey comb core. After using the inverse calculation process of the Fourier number characterization model to improve the stability and accuracy of the inverse calculation, the model was used to study the influence of alloy materials and process parameters on the heat transfer coefficient of the casting-mold interface during the die casting process. The functional relationship between the casting-mold interface heat transfer coefficient and the solidification fraction and solidification rate of the die-casting process was also established. At the same time, it was found that, among the die-casting process parameters, the initial surface temperature of the mold had the greatest influence on the heat transfer coefficient [28–30]. Chen et al. [31] proposed to take the screw as a one-dimensional rod heat transfer system to solve the thermal error of the screw by establishing a thermal characteristic equation. In Ho et al.'s [32] investigation, two commercial-scale porous lattice heat exchangers of different lattice unit cell sizes were fabricated by SLM. Experiments were performed in a wind tunnel to characterize the thermal-hydraulic performances of the heat exchangers. Jiang et al. [33] explored the effects of heat treatment on microstructure and residual stress of SLM AlSi10Mg alloy. Wang et al. [34] reported a study of condensation heat transfer and pressure drop of R-134a inside four enhanced tubes and one plain tube fabricated by SLM. The effects of fin height, refrigerant flow direction and mass flux on the heat transfer coefficient and pressure drop were studied. Martin et al. [35] presented an experimental study of the mechanical strength and numerical analysis of the thermal behaviour during SLM fabrication of ALSi10Mg block support structures.

In Huiping et al.'s [36] investigation, A new method to calculate the temperature-dependent surface heat transfer coefficient during quenching process is presented, which applies finite-element method (FEM), advance–retreat method and golden section method to the inverse heat conduction problem, and can calculate the surface heat transfer coefficient according to the temperature curve gained by experiment. In Apmann et al.'s [37] research, the influence of the connector between the two microchannels was studied. By studying the effect of Reynolds number and the introduction of nanoparticles into the base liquid on the heat transfer coefficient of the connector, it was shown that these two factors played an important role in the influence of the connector on the heat transfer coefficient. Khooshehchin et al. [38] studied the effects of vertical and horizontal surface roughness on bubble dynamics and heat transfer coefficient during pool boiling. The experimental results showed that with the increase of surface roughness, the nucleation position increases, leading to the enhancement of heat transfer. The experimental data were verified by a conventional model. This paper aims to import the temperature data measured in the experiment into DEFORM and calculate the heat transfer coefficient in the heat treatment process through its inverse heat conduction module. As professional finite element software for volume forming, DEFORM-3D can not only simulate the thermodynamic coupling calculation of the workpiece heat treatment process, but also directly solve the heat transfer coefficient between the workpiece and the medium with its inherent Inverse Heat Transfer Wizard model [39]. Using the DEFORM Inverse Heat Transfer model to solve the surface heat transfer coefficient involves the selection and determination of parameters such as temperature control point and the initial value of the surface heat transfer coefficient. The reasonable selection of these parameters will have a significant influence on the calculated value of the surface heat transfer coefficient and the predicted temperature field [40]. In this paper, the temperature curve in the heating and cooling process in the heat treatment process is obtained through experiments, and the finite element model is established to solve the heat transfer coefficient [41]. By adjusting the initial value of the heat transfer coefficient, the simulated temperature curve corresponds closely to the experimental results to obtain an accurate heat transfer coefficient value, and the heat transfer coefficient results obtained by the solution are fitted to establish the mathematical relationship between the heat transfer coefficient and the temperature.

2. Experimental Procedure

The AlSi10Mg alloy molded by selective laser melting (SLM) used in this experiment was manufactured by the Space M200 manufacturing system (Figure 1). In the SLM process, the thermocouples are installed on the substrate for measurement, that is, "The preparation of substrate" shown in Figure 1. This research uses thermocouples to measure temperature data during subsequent heat treatment of molded samples. The print size was 250 mm × 250 mm × 250 mm, and the sample size was 40 mm × 30 mm ×80 mm. The SLM molding process parameters used in this project were laser power of P = 400 W, laser scanning speed of v = 200 mm/s, a scanning layer thickness of 50 μm, scanning time interval of 150 μm, and a preheating temperature of 80 °C.

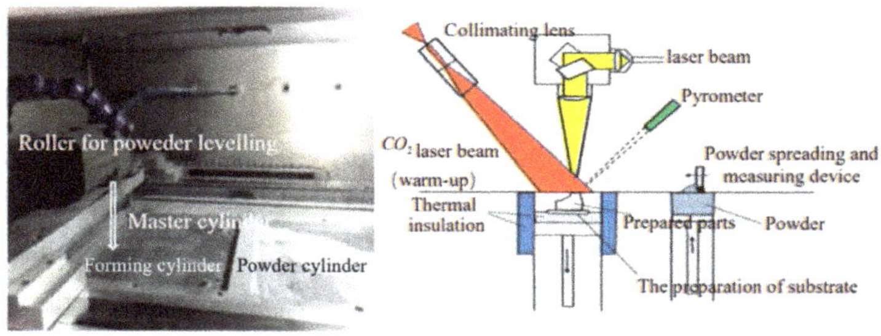

Figure 1. Space M200 manufacturing system.

A JXR1200-30 muffle furnace was used in the heat treatment process.

As shown in Figure 2a, shows the SLM printing equipment.The working size was 300 mm × 150 mm × 150 mm, the highest operating temperature was 1200 °C, and a high quality resistance wire containing molybdenum was used as the heating element. The Proportional Integral Derivative (PID) control mode and a K type thermocouple were employed, and the constant temperature accuracy was ±1 °C. In this experiment, the size of the AlSi10Mg sample was 8 mm × 25 mm × 80 mm(Figure 2b). When the measurement depth of the thermocouple is 4 mm, it can measure the temperature data in the center area of the sample. In order to ensure that the results of different regions are representative, four points are selected in different locations for analysis, which can be used for comparison and verification of simulation results. Therefore, four holes were punched in the workpiece to a depth of 4 mm, and a thermocouple was installed to measure the real-time temperature (Figure 2c). As shown in Figure 2c, the thermocouple is inserted into a small hole on the surface of the sample to measure and obtain the temperature data during the heat treatment process in real time. In order to reduce the experimental error and ensure the reliability of the temperature data, three groups of the same temperature measurement experiments were done, and the temperature data were not different. The average temperature was imported into DEFORM as an input parameter to solve the heat transfer coefficient of each stage. As shown in Figure 2d, shows the comparison curve of the measured temperature and the simulated temperature.

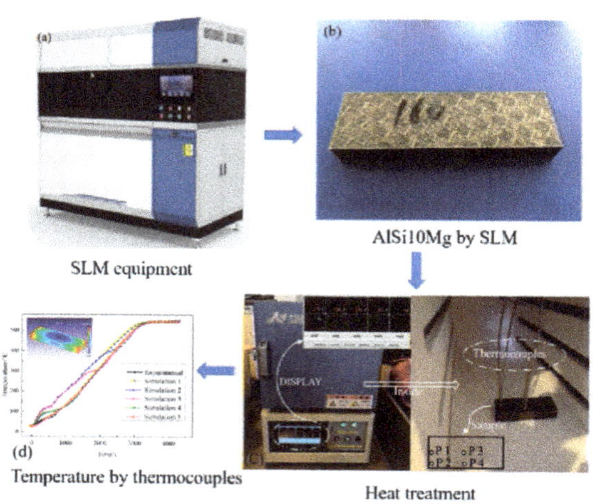

Figure 2. The process of heat treatment temperature measurement.(**a**) The SLM printing equipment. (**b**) The size of the AlSi10Mg sample. (**c**) The thermocouple is inserted into a small hole on the surface of the sample to measure and obtain the temperature data during the heat treatment process in real time. (**d**) The comparison curve of the measured temperature and the simulated temperature.

As shown in Figure 2, based on the above experimental conditions, time-temperature data of each stage in the heat treatment process can be measured, which will be used as input parameter when solving the heat transfer coefficient of each stage with DEFORM.

3. Modeling Procedure

3.1. Mathematical Model

In the process of heat treatment, heat conduction is the main mechanism inside the workpiece. The three-dimensional heat conduction differential equation is:

$$\rho c\frac{\partial T}{\partial t}=\frac{\partial}{\partial x}\left(\lambda\frac{\partial T}{\partial x}\right)+\frac{\partial}{\partial y}\left(\lambda\frac{\partial T}{\partial y}\right)+\frac{\partial}{\partial z}\left(\lambda\frac{\partial T}{\partial z}\right)+\dot{Q} \tag{1}$$

where ρ is density, kg/m^3; c is the specific heat capacity, J/(kg °C); T is temperature, °C; t is time, s; $\dot{Q}$ is the heat generated by the internal heat source, W/m^3; λ is the thermal conductivity coefficient, W/(m °C).

Initial conditions are the starting point of calculation. To solve the above equation, initial conditions must be given. Quenching starts from room temperature furnace heating, at this time:

$$T|_{x=0}=T_0 \tag{2}$$

where T_0 is the known temperature and is a constant.

The boundary conditions of heat transfer generally fall into the following three categories:

(1) When the boundary function is known, it is the first kind of the boundary condition:

$$T|_s=T_w(x,y,z,t) \tag{3}$$

where, s is the boundary range, x, y, z are coordinate values, and t is time.

(2) When the surface heat flux q_w of the object is known, it is the second kind of the boundary condition:

$$-k\frac{\partial T}{\partial n}\bigg|_s=q_w(z,r,t) \tag{4}$$

where, q_w is the heat flux of the workpiece surface.

(3) The third kind of boundary condition, also known as Newton boundary conditions, are given for the convective heat transfer coefficient on the boundary surface:

$$-k\frac{\partial T}{\partial n}\bigg|_s=h_k\left[T_w-T_f\right] \tag{5}$$

where, h_k is the heat transfer coefficient, T_w is the workpiece temperature, and T_f is the medium temperature.

Inverse heat transfer Wizard in DEFORM was used to solve the inverse heat conduction problem, and the algorithm was based on Beck's regularization law. In the solution process, the heat conduction inverse algorithm is determined according to the nonlinear estimation method of Beck et al. [40,41], as shown in Figure 3a,b. The heat flux varying with time can be discretized into a number of heat flux values with time intervals of $\Delta\theta(q_i=q_1,q_2\cdots q_M\cdots q_N)$. In order to improve the stability of the inverse algorithm, the concept of future time step is introduced when solving the heat flow at a certain moment; that is, the heat flow values within the following time interval are assumed to be equal. Thus:

$$q_M=q_{M+1}=q_{M+2}=\ldots=q_{M+R} \tag{6}$$

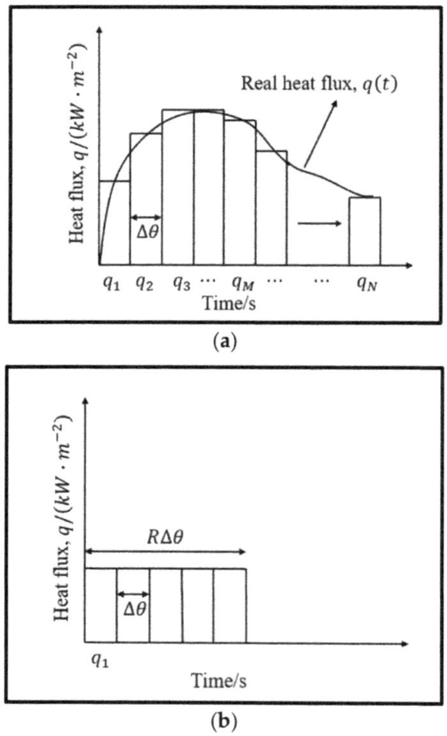

Figure 3. Heat flux solved by discrete solution. (**a**) The actual heat flux $q_{(t)}$ is dispersed into a series of values. (**b**) The heat flux is assumed to be constant in the $R\ \Delta\theta$ time.

Each heat flow is obtained through the iterative calculation of the criterion of the minimum error value of temperature measurement and calculation, which can be expressed by the following formula:

$$F(q_M) = \sum_{i=1}^{N} \sum_{j=0}^{r-1} \left(T^{m}_{i,\ M+j} - T^{c}_{i,M+j} \right)^2 \tag{7}$$

where $T^{m}_{i,\ M+j}$ is the actual measured temperature at the measuring point i in the time period of $M+j$; $T^{c}_{i,M+j}$ is the temperature calculated and solved at the measuring point i in the time period of $M+j$; N is the total number of temperature points measured; r is the total number of the future period of the measurement point.

Minimization is the main objective in the following calculation. To solve the extreme value of $F(q_M)$, the derivative of the above expression is taken and set as equal to 0, obtaining:

$$\sum_{i=1}^{N} \sum_{j=0}^{r-1} \left(T^{m}_{i,\ M+j} - T^{c}_{i,M+j} \right) \frac{\partial T^{c}_{i,M+j}}{\partial q_M} = 0 \tag{8}$$

In order to solve $T^{c}_{i,M+j}$ and $\frac{\partial T^{c}_{i,M+j}}{\partial q_M}$, Taylor series expansion is applied to $T^{c}_{i,M+j}$ at $q_M = q^{*}_{M}$:

$$T^{c}_{i,M+j}(q_M) = T^{c}_{i,M+j}(q^{*}_{M}) + \left(\frac{\partial T^{c}_{i,M+j}}{\partial q_M} \right) (q_M - q^{*}_{M}) \tag{9}$$

Among them, there are: $\Delta q_M = q_M - q_M^*$, so:

$$T_{i,M+j}^c(q_M) = T_{i,M+j}^c(q_M^*) + \left(\frac{\partial T_{i,M+j}^c}{\partial q_M}\right)\Delta q_M \tag{10}$$

The equations can be obtained simultaneously as:

$$\Delta q_M = \frac{\sum_{i=1}^{N}\sum_{j=0}^{r-1}\left(T_{i,\ M+j}^{m} - T_{i,M+j}^{c}\right)\frac{\partial T_{i,M+j}^c}{\partial q_M}}{\sum_{i=1}^{N}\sum_{j=0}^{r-1}\left(\frac{\partial T_{i,M+j}^c}{\partial q_M}\right)^2} \tag{11}$$

where q_M and Δq_M are the interface heat flow value and interface heat flow modification value in the current calculation period, respectively.

Function F is an object function that is computed iteratively. With mathematical derivation, the interface heat flow can be solved by the following two expressions:

$$\Delta q_M = \frac{\sum_{j=1}^{J}\sum_{i=1}^{R}\left(T_{j,M+i} - C_{j,M+i}\right)\phi_{j,M+i}}{\sum_{j=1}^{J}\sum_{i=1}^{R}\phi_{j,M+i}^2} \tag{12}$$

$$q_{M+1} = q_M + \Delta q_M \tag{13}$$

where $\phi_{j,M+i}$ is the sensitivity coefficient, which is defined as the response of temperature at a certain position in the casting or mold with the change of unit heat flow, and is expressed as:

$$\phi_{j,M+i} = \frac{\partial T_{j,M+i}}{\partial q_{j,M+i}} \tag{14}$$

In the iterative calculation of heat flow, q_M is modified continuously with the above formula. When the difference between the measured temperature and the calculated temperature meets the given convergence error, the currently calculated heat flow q_M can be obtained. The process is repeated until the heat flow q at all times is calculated.

When the heat flow q at all times is solved, the surface heat transfer coefficient can be obtained by the following equation:

$$H = q/(T_w - T_c) \tag{15}$$

where H is the surface heat transfer coefficient/N/(mm s °C); T_w is the workpiece surface temperature/°C; T_c is the medium temperature/°C.

The DEFORM reverse heat transfer model first assumes the initial value of the surface heat transfer coefficient and calculates the internal temperature value through the thermal conductivity differential equation. It then continuously revises the set value according to the difference between the calculated value and the experimentally measured value and, finally, makes the calculated value approach the measured value. The accuracy of the calculated surface heat transfer coefficient can be determined by comparing the predicted temperature value with the measured one.

3.2. Finite Element Model

An AlSi10Mg alloy sample printed by selective laser melting (SLM) is investigated in this work. A rectangular parallelepiped sample with a length of 80 mm, width of 25 mm, and height of 8 mm is selected. This alloy has the advantages of high strength, good thermal performance, and light weight. AlSi10Mg powder is spherical; its chemical composition is shown in Table 1 and is within the size of 20~63 um, prepared by the atomization method.

Table 1. Chemical composition of AlSi10Mg alloy powder (wt.%).

Al	Si	Mg	Fe	N	O	Ti
Bal	9.0–11	0.25–0.45	<0.25	<0.2	<0.2	<0.15
Zn	**Mn**	**Ni**	**Cu**	**Pb**	**Sn**	
<0.1	<0.1	<0.05	<0.05	<0.02	<0.02	

Due to the vigorous development of metal 3D printing technology in recent years, the sample of AlSi10Mg alloy formed by SLM is a relatively new material. Thus no corresponding material is available in the material libraries of the simulation software, and we must establish a new material model here.

In the process of material transformation changing with temperature, the thermophysical parameters change non-linearly. Tables 2 and 3 shows the thermophysical parameters and the mechanical property parameters of AlSi10Mg. Physical parameters of materials are crucial to the prediction accuracy of the model. The physical parameters of AlSi10Mg alloy referenced by the model in this subject are as follows.

Table 2. Thermophysical parameters of AlSi10Mg alloy.

Temperature, $T/°C$	20	100	200	300	400
Thermal conductivity, $K_s\,(\frac{W}{m}°C)$	147	155	159	159	155
Specific heat capacity, $C\,\left(\frac{J}{kg}°C\right)$	739	755	797	838	922
Density, $\rho(kg/m^3)$			2650		

Table 3. Mechanical property parameters of AlSi10Mg alloy material.

Temperature, $T/°C$	20	100	200	300	400
Modulus of elasticity, EE GPa	69	67	62	53	41
The yield strength, σ_s(MPa)	195	150	105	70	30
Coefficient of thermal expansion, $\theta\left(\frac{10^{-6}}{°C}\right)$	21.7	22.5	23.5	23.3	25.5
Poisson's ratio, (μ)			0.33		

4. Results and Discussion

The inverse problem of heat conduction is a relatively important inverse problem. It has important guiding significance to improve the accuracy of simulation model, determine and predict the temperature field, and further improve the efficiency of heat treatment. In this paper, through constructing the simulation model of the AlSi10Mg alloy sample, the inverse calculation of heat conduction is carried out in the Inverse Heat Transfer Wizard module of DEFORM. As shown in Figure 4, the specific operation is to heat-treat a workpiece, and a thermocouple is used to collect the time-temperature data corresponding to the workpiece during the heat treatment process. The parameters of the simulation part are shown in Table 4. On the one hand, the future time steps will affect the accuracy of the calculation result, on the other hand, it will affect the convergence of the result, that is, it will affect the calculation time.

Table 4. The parameters of the simulation part.

Number of Elements	Control Points	Time per Step	Step Increm-Ent to Save	Relative Improve-Ment Less Than (%)	Maximum Iterations	Maxi-Mum Simula-Tions	Objective Function Less Than	Decision Vector Change Less Than
2000	3	0.01	10	2	500	5000	1	1×10^{-6}

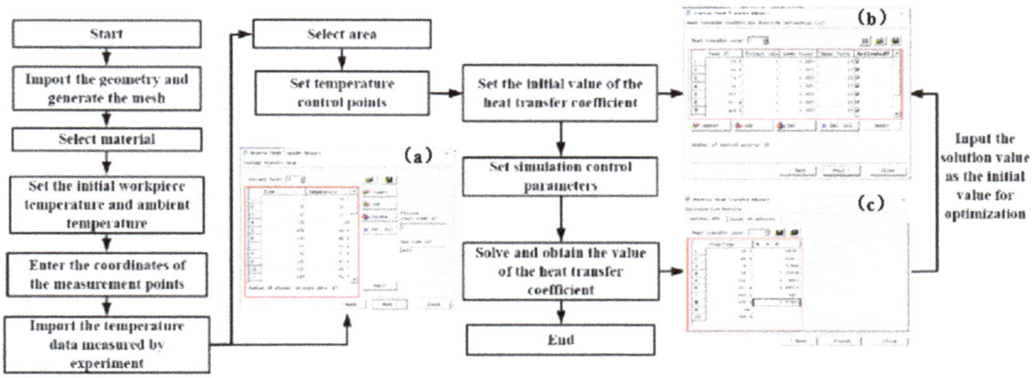

Figure 4. The process of solving the heat transfer coefficient in DEFORM.

It is assumed that the initial heat transfer coefficient is constant 1, the initial temperature of the workpiece is 20°, and the temperature distribution is uniform. The simulated ambient temperature is set as the highest temperature obtained during the experiment. Based on the initial assumption of the heat transfer coefficient, the heat transfer coefficient under different temperature conditions is solved. Finally, the simulated time and temperature data are compared with the experimental time and temperature data, and the solution value is input as the initial value for the optimization operation until the optimal value is calculated.

The stages of heat treatment are shown in Figure 5. In the heat treatment process, assuming that the bottom surface of the sample is fully constrained, the inverse calculation of the heat transfer coefficient value of the AlSi10Mg alloy sample can be roughly divided into the following three cases. The first one is the heat transfer coefficient value of the AlSi10Mg alloy during the heating process. The second is the heat transfer coefficient during quenching. The third is the value of the heat transfer coefficient in the process of air cooling.

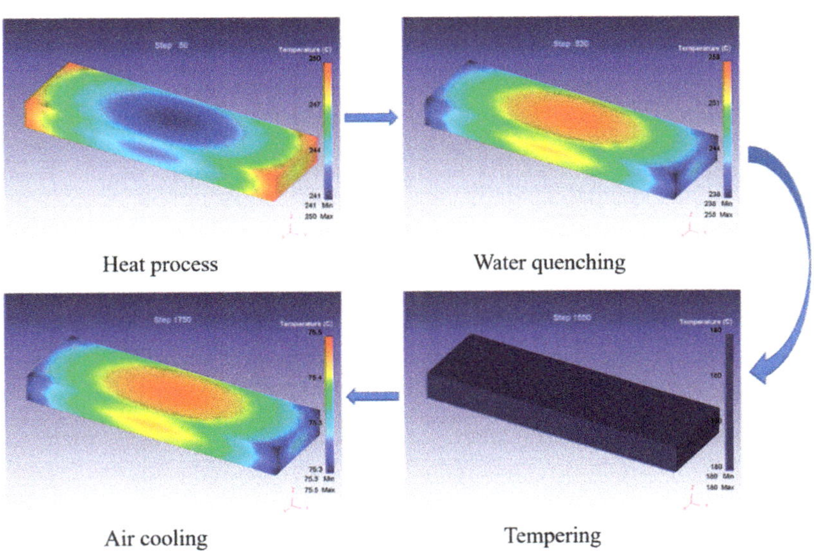

Figure 5. Stages of heat treatment.

4.1. Detection and Analysis of Heat Transfer Coefficient during Heating

During the heating process, the temperature of the AlSi10Mg alloy sample rises with the temperature in the furnace, and the temperature values at all points change in a relatively consistent way. The temperature data of one point is imported into the Inverse Heat Transfer Wizard module of DEFORM for simulation, and the simulation results are shown in Figure 6.

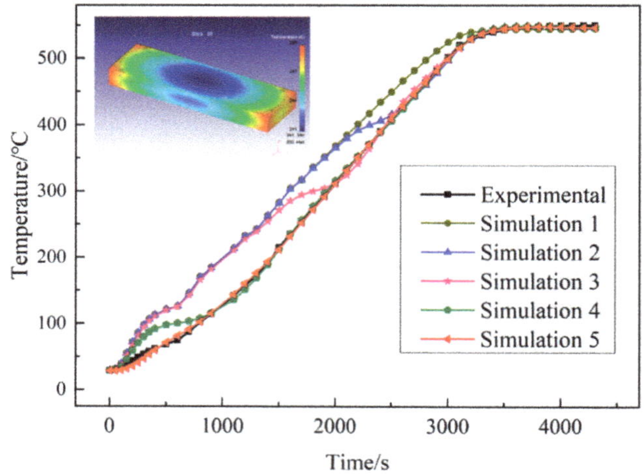

Figure 6. Temperature curve in the heating process.

Figure 7 below shows the simulation results of five inverse calculations. It can be seen from the figure that although the simulated temperature value obtained from the first simulation result differs greatly from the experimental value, the temperature variation trend is completely consistent, indicating that the heat conduction inverse calculation module of DEFORM has reliability. In the second and third simulations, the simulation temperature of the high temperature part is relatively close to the experimental value, but there is still a big difference at low temperature.

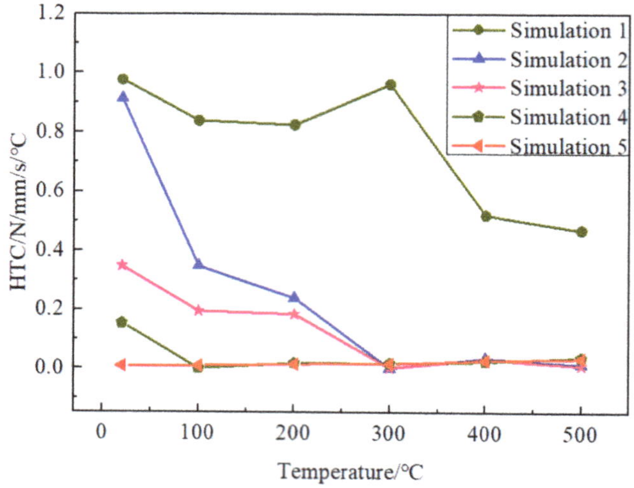

Figure 7. The heat transfer coefficient curve in the heating process.

The closer the initial value of heat transfer coefficient is to the actual situation, the shorter the calculation time and the more accurate the simulation result is. First of all, set the heat transfer coefficient of the initial value is 1, solving the heat transfer coefficient to get a new set of values, to solve the resulting value used in the simulation of the heat treatment process, comparing the simulation results and experimental results, and then use this value as the initial value to solve the next set of heat transfer coefficient, and used to simulate the heat treatment process, until the simulation results was coincident with the experimental results. By adjusting the heat transfer coefficient at low temperature, after several adjustments of the simulation, the simulation temperature curve gradually fits with the experimental value, the final curve is almost exactly the same, and the fitting effect is good.

Table 5 shows the value of the heat transfer coefficient obtained by the simulation solution. The T here refers to the temperature of the point measured by the thermocouple. The fitting results in Origin are as shown in Figure 8. The following equation is the mathematical model obtained by fitting:

$$H = 0.95 + 2.97 \times 10^{-4}T + 1 \times 10^{-5}T^2 \quad (16)$$

where H is the value of the heat transfer coefficient, and the unit is N/mm·s·°C × 10^{-2}; T is the temperature, and the unit is °C; R^2 is 0.96.

Table 5. The temperature data of the heating progress.

Time/s	Experimental	Simulation 1	Simulation 2	Simulation 3	Simulation 4	Simulation 5
1	29.2	28.9961	28.9962	28.9978	28.9988	28.9999
50	30.1	30.8804	30.8605	30.4314	29.9407	29.1261
100	33.1	39.2818	39.1822	37.2974	34.6504	29.9117
150	38	55.4681	55.1671	51.7825	45.5795	32.1536
200	43.8	71.9046	71.4698	67.9396	58.6956	36.1405
250	48.8	85.9944	85.4938	82.2656	70.3767	41.4023
300	53.4	97.681	97.1544	94.295	79.7749	47.4175
350	57.8	106.393	105.972	103.709	86.8414	53.7204
400	61.8	112.949	112.636	110.907	91.8843	59.9506
500	68.7	121.182	120.98	119.908	97.7388	71.4138
600	74.7	126.145	126.026	125.385	100.381	81.2586
700	87.3	145.922	145.331	142.338	103.2	91.0137
800	102.2	170.45	169.73	166.104	108.093	102.4
900	114.1	184.481	184.086	182.064	115.227	115.351
1100	142	214.063	213.599	210.909	136.221	144.012
1200	158.1	232.317	231.738	227.384	151.1	159.98
1300	171.7	242.423	242.13	239.072	168.628	176.39
1400	190.6	263.318	262.636	254.466	188.727	193.41
1500	214.7	283.373	282.751	271.081	211.848	212.16
1600	235.1	304.382	303.697	285.267	235.064	232.251
1700	252.7	317.692	317.118	294.642	256.976	252.496
1800	273.6	336.086	334.808	299.937	277.56	272.462
1900	292.5	351.39	349.847	304.769	297.45	292.42
2000	312.2	368.234	365.431	312.512	316.487	312.094
2100	331.6	384.614	379.915	324.467	335.035	331.884
2200	351.4	401.144	392.131	341.607	353.16	351.586
2300	370.5	417.059	399.861	363.754	370.914	371.027
2400	390.1	434.291	406.531	389.339	388.563	390.36
2500	408.2	449.59	416.156	414.394	406.084	409.341
2600	427.2	466.5	428.51	435.182	424.268	427.89
2700	445.3	481.756	443.328	453.561	443.048	446.114
2800	464.4	497.148	459.974	470.534	461.961	463.942
2900	483.1	511.61	478.084	486.896	480.97	481.63

Table 5. *Cont.*

Time/s	Experimental	Simulation 1	Simulation 2	Simulation 3	Simulation 4	Simulation 5
3000	501.8	524.546	496.99	502.794	499.804	499.101
3100	520	534.75	516.088	518.288	518.252	516.28
3200	528.5	539.852	530.897	530.215	532.182	529.585
3300	535.7	541.867	539.516	537.492	540.162	537.681
3400	539.8	542.971	544.474	542.138	544.754	542.695
3500	543.3	543.58	546.988	544.905	547.094	545.545
3600	545.4	543.975	548.188	546.552	548.219	547.129
3700	547	544.223	548.571	547.407	548.578	547.862
3800	548	544.4	548.587	547.819	548.59	548.146
3900	548.7	544.524	548.353	547.903	548.359	548.112
4000	549.3	544.613	547.958	547.756	547.969	547.864
4100	549.6	544.675	547.494	547.469	547.51	547.5
4200	549.8	544.721	547.016	547.111	547.034	547.085
4300	549.8	544.751	546.527	546.702	546.548	546.636

The temperature data at a certain time obtained by the experiment can be input into the above formula to obtain the corresponding heat transfer coefficient value, which can be used in the subsequent heat treatment process simulation

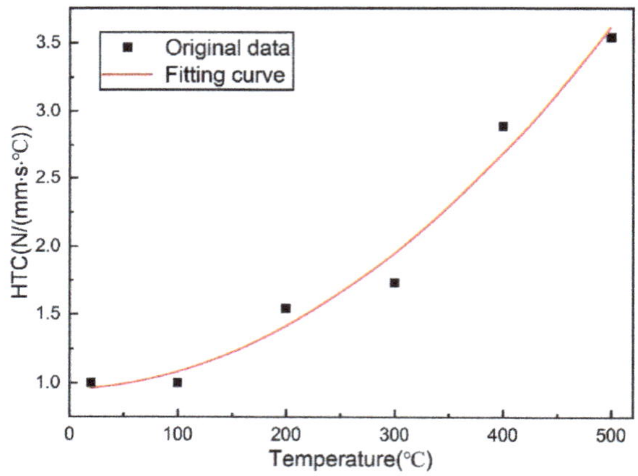

Figure 8. Fitting curve of heat transfer coefficient in the heating process.

4.2. Detection and Analysis of Heat Transfer Coefficient during Quenching

Due to the good thermal conductivity of aluminum alloy, the cooling rate is large, and the temperature drops quickly during quenching. The temperature drops at each point of the sample are not consistent, and the surface and boundary drop the fastest.

Three points of symmetry on the surface of AlSi10Mg alloy are taken with the depth of 2 mm: point 1 (12.5, 40, 6), point 2 (23, 40, 6) and point 3 (12.5, 77, 6).

As shown in Figure 9, the black upper triangle symbol is the experimental test temperature. During quenching, the temperature drops sharply. Due to the sharp drop in temperature during quenching, the temperature data measured have a sharp decline trend. Among them, in the beginning, the temperature of points 2 and 3 drops with the center point 1. After 3–4 s, the temperature decreases slowly and gradually lowers to room temperature over time.

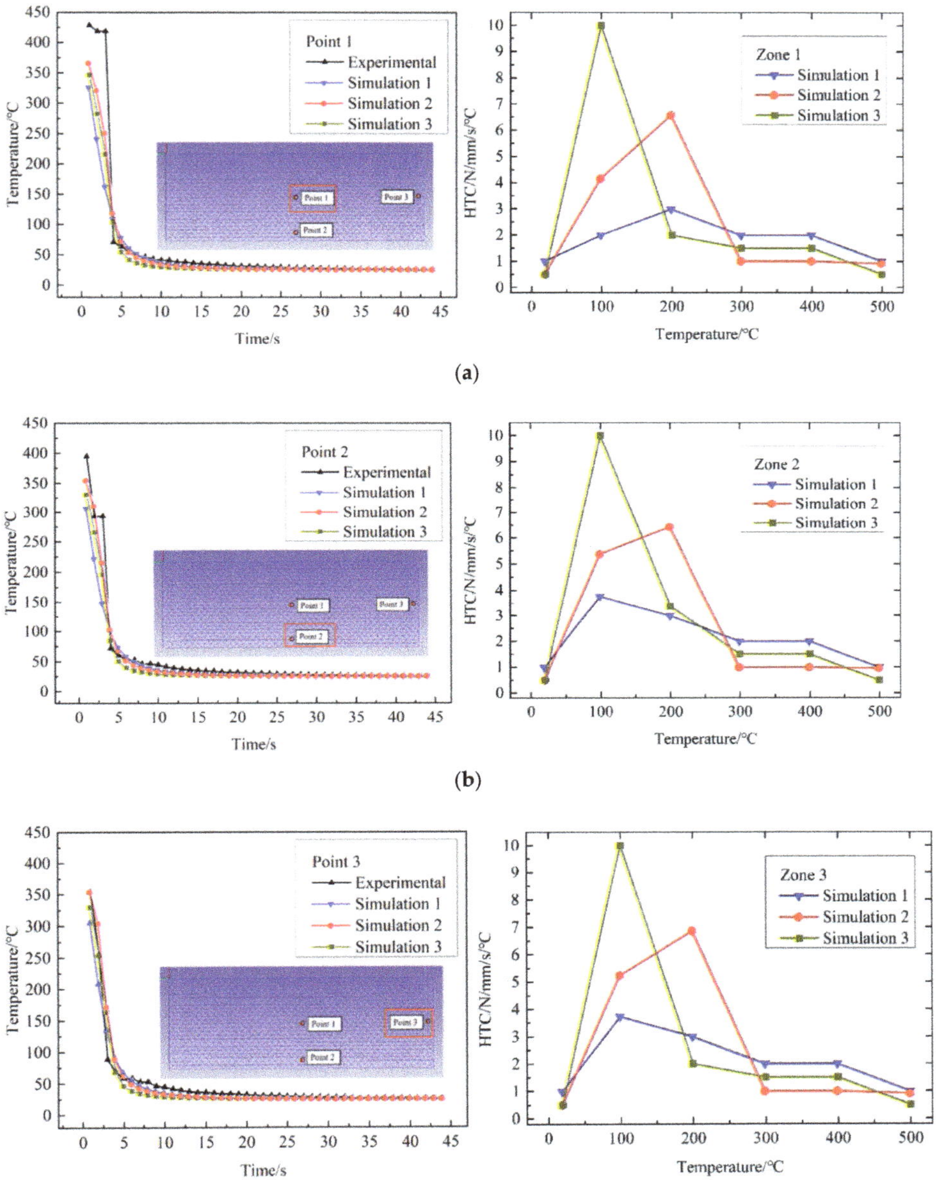

Figure 9. Temperature and heat transfer coefficient curves during quenching. (**a**) Comparison between the experimental temperature and the simulation temperature of point 1. (**b**) Comparison between the experimental temperature and the simulation temperature of point 2. (**c**) Comparison between the experimental temperature and the simulation temperature of point 3.

The selected temperature range ranges from room temperature to the highest temperature used in the heat treatment process. The trend of the simulated temperature data curve is basically the same as that of the experimental temperature data curve, so it can be concluded that the data is selected under the condition that the experimental error allows. Temperature changes at three points in the simulation over time are shown in Figure 9. The decreasing trend of each point is basically consistent with the measured value in the experiment. Compared with the first and third simulations, the simulation value of the second simulation is closer to the experimental value. The results of the three simulations are consistent with the experimental values. However, due to the rapid temperature decline during quenching, the number of measured temperature values is small. As a result, the simulation accuracy is slightly lower than that in the heating process.

Table 6 shows the value of the heat transfer coefficient obtained by the simulation solution. The fitting results in Origin are as shown in Figure 10. The following equation is the mathematical model obtained by fitting:

Table 6. Heat transfer coefficient values obtained during the fifth simulation.

Temperature/°C	20	100	200	300	400	500
Heat transfer coefficient/ N/mm·s·°C	1	1	1.54	1.73	2.89	3.54

The temperature data at a certain time obtained by the experiment can be input into the above formula to obtain the corresponding heat transfer coefficient value, which can be used in the subsequent heat treatment process simulation

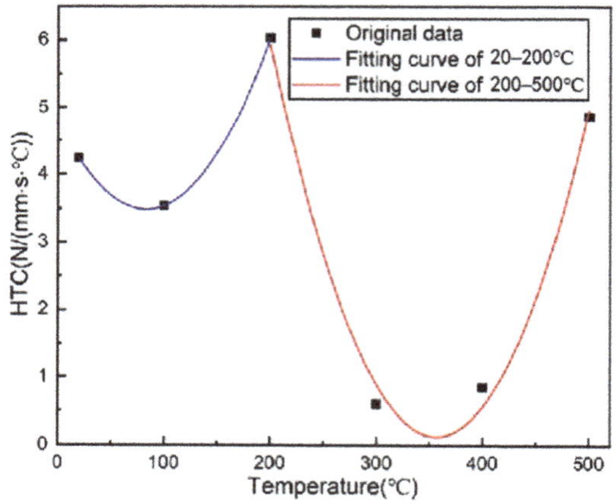

Figure 10. Fitting curve of heat transfer coefficient during quenching.

When the temperature is 20–200 °C:

$$H = 0.95 + 2.97 \times 10^{-4}T + 1 \times 10^{-5}T^2 \quad (17)$$

When the temperature is 200–500 °C:

$$H = 0.95 + 2.97 \times 10^{-4}T + 1 \times 10^{-5}T^2 \quad (18)$$

where H is the value of heat transfer coefficient, the unit is N/mm·s·°C $\times 10^{-2}$; T is the temperature, the unit is °C; R^2 is 1 and 0.98, respectively.

4.3. Detection and Analysis of Heat Transfer Coefficient in Air Cooling Process

The indoor temperature is 26 °C during air cooling. The temperature of the surface and edge of the AlSi10Mg alloy sample drops rapidly, but the difference is not significant. The temperature at the center of the sample is taken as the experimental value, and the simulated temperature is obtained by inserting the Inverse Heat Transfer Wizard module, as shown in Figure 11. In the process of temperature drop, the simulation temperature is lower than the experimental value between 42 and 30 °C, while the simulation temperature is slightly higher than the experimental value between 30 °C and room temperature, but the difference is not more than 5 °C.

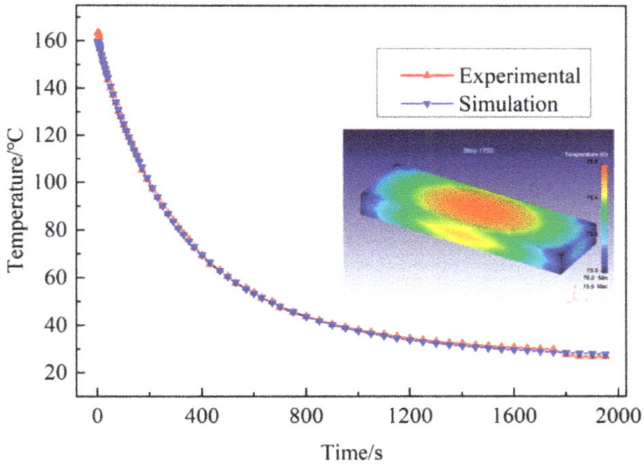

Figure 11. Comparison between the simulation and experiment of the temperature drop process during air cooling.

In the process of air cooling, the temperature basically remains unchanged, that is, the ambient temperature of the sample remains unchanged. The simulated values of heat transfer coefficients in each region are shown in Table 7. The simulated heat transfer coefficient values during air cooling are shown in Table 8.

Table 7. Heat transfer coefficient values obtained in the third simulation.

Temperature/°C	20	100	200	300	400	500
Heat transfer coefficient/ N/mm·s·°C	4.24	3.54	6.03	0.60	0.85	4.87

Table 8. Simulated heat transfer coefficient values during air cooling.

Temperature/°C	20	100	200
Heat transfer coefficient/ N/mm·s·°C	1	1	1

4.4. Simulation of the Entire Heat Treatment Process

The entire heat treatment process of quenching and tempering is simulated in the Heat treatment Wizard module of DEFORM. The comparison between simulated temperature and experimental temperature is shown in Figure 12. As can be seen from the figure, during the whole heat treatment process and air cooling process, the relative error control of experimental temperature and simulation temperature is within 2%. In the water quenching process, only when the time is between 7000–9000 s, the simulated temperature rise is slightly less than the experimental value, and the relative error between the experimental temperature and the simulation temperature is large. The relative error of the other time periods is controlled within 1%. However, throughout the heat treatment process, the simulation temperature curve has the same experimental temperature curve trend and the coincidence degree is high, indicating that the heat transfer coefficient value obtained by the inverse analysis method and the heat treatment simulation are accurate and reliable.

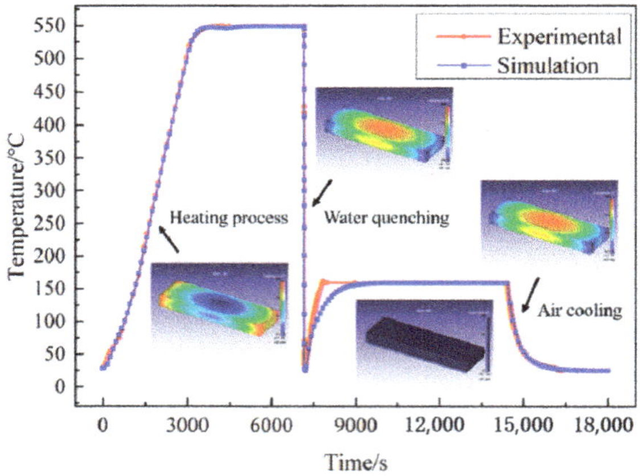

Figure 12. Comparison of the simulation and experimental temperatures in the heat treatment process.

Figure 13 shows the microstructure of samples before and after heat treatment. Figure 13a is the micrograph of the SLM molded parts after polishing and corrosion. In the figure, there are not only a certain number of pores, but also the laser scanning path in the SLM molding process can be clearly seen. Figure 13b shows the microstructure of the workpiece after polishing and corrosion after heat treatment. It can be seen from the figure that after heat treatment, not only the pores are finer and uniform, but the grains are fine, and the precipitated Si tends to be distributed along the grains.

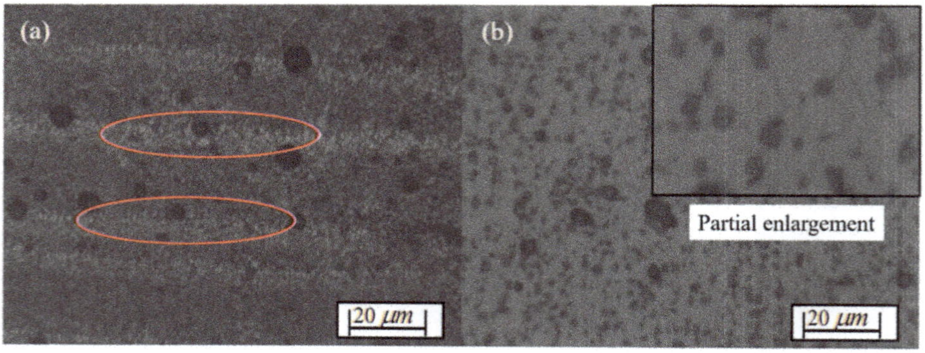

Figure 13. Microstructure of samples before and after heat treatment. (**a**) Micrograph of SLM molding parts. (**b**) micrograph of workpiece after heat treatment.

5. Conclusions

For samples prepared with SLM, heat treatment process is needed to improve the performance of the samples. It is very important to obtain accurate heat transfer coefficient for simulating heat treatment process. Combined with the temperature data in the experiment, the heat transfer coefficient of the material under different conditions was calculated by inverse calculation in the reverse heat conduction module in DEFORM. The main conclusions are as follows:

1. Based on the nonlinear evaluation method, the inverse analysis model of heat transfer coefficient in the heat treatment process was established. Taking the actual temperature curve as the input condition, the heat transfer coefficient values of heating, quenching and air cooling parts in the heat treatment process were obtained successfully.

2. In the tempering process, when the temperature is from 100 to 160 °C, the simulated temperature rise is slightly smaller than the experimental value. In the process of temperature drop, the simulation temperature is lower than the experimental value between 42 and 30 °C, while the simulation temperature is slightly higher than the experimental value between 30 °C and room temperature, but the difference is not more than 5 °C.
3. The mathematical model of heat transfer coefficient changing with temperature during heat treatment was established.
4. The heat transfer coefficient obtained by the inverse analysis method was used to simulate the heat treatment process, and the obtained simulation temperature curve had a high coincidence degree with the experimental temperature curve.

Author Contributions: Conceptualization, X.J.; Data curation, C.W.; Investigation, W.X., C.L. and Z.L.; Writing—original draft, S.W. All authors have read and agreed to the published version of the manuscript.

Funding: This project is supported by the Natural Science Foundation of China (No.52005098) and the Shanghai Natural Science Foundation (Grant No.20ZR1438000, 22ZR1402400). The authors wish to record their gratitude for their generous supports.

Data Availability Statement: The data and materials that support the findings of this study are available from the corresponding author upon reasonable request.

Acknowledgments: In this section, you can acknowledge any support given which is not covered by the author contribution or funding sections. This may include administrative and technical support, or donations in kind (e.g., materials used for experiments).

Conflicts of Interest: The authors declare no conflict of interest.

References

1. Luo, H.; Wang, Y.; Zhang, P. Study on Surface Quality of 7A09 Aluminum Alloy Milling Based on Single Factor Method. *Surf. Technol.* **2020**, *49*, 327–333.
2. Li, C.; Piao, Y.C.; Meng, B.B.; Hu, Y.; Li, L.; Zhang, F. Phase transition and plastic deformation mechanisms induced by self-rotating grinding of GaN single crystals. *Int. J. Mach. Tools Manuf.* **2022**, *172*, 103827. [CrossRef]
3. Wu, C.J.; Li, B.Z.; Liu, Y.; Liang, S.Y. Surface roughness modeling for grinding of Silicon Carbide ceramics considering co-existing of brittleness and ductility. *Int. J. Mech. Sci.* **2017**, *133*, 167–177. [CrossRef]
4. Zhang, Y.; Bai, Q.; He, X.; Qing, L.; Yue, S.; Liu, X. Wear analysis of CVD diamond tool in micro-milling AA356 aluminum alloy. *Diam. Abras. Eng.* **2021**, *41*, 44–50. [CrossRef]
5. Xing, M.; Lei, X.; Guan, X.; Song, Y.; Zhou, Y. Research on Construction of Super hydro phobic Surface of Aluminum Alloy and Its Stability and Self-cleaning Performance. *Surf. Technol.* **2021**, *50*, 152–161.
6. Yan, Y.; Wang, J.; Geng, Y.; Zhang, G. Material Removal Mechanism of Multi-Layer Metal-Film Nanomilling. *CIRP Ann.—Manuf. Technol.* **2022**, *71*, 61–64. [CrossRef]
7. Wu, C.J.; Pang, J.Z.; Li, B.Z.; Liang, S.Y. High Speed Grinding of HIP-SiC Ceramics on Transformation of Microscopic Features. *Int. J. Adv. Manuf. Technol.* **2019**, *102*, 1913–1921. [CrossRef]
8. Ding, Z.S.; Sun, J.; Guo, W.C.; Jiang, X.; Wu, C.; Liang, S.Y. Thermal Analysis of 3J33 Grinding Under Minimum Quantity Lubrication Condition. *Int. J. Precis. Eng. Manuf.-Green Technol.* **2021**, *91*, 1–19. [CrossRef]
9. Xie, Y.; Wang, C.; Zhang, K.; Liang, C.; Liang, C.; Zhou, C.; Lin, D.; Chen, Z. Optimizing Laser Cladding on Aluminum Alloy Surface with Numerical Simulation and Rare Earth Modification. *Surf. Technol.* **2020**, *49*, 144–155.
10. Wang, J.Q.; Yan, Y.D.; Li, Z.H.; Geng, Y.Q. Towards understanding the machining mechanism of the atomic force microscopy tip-based nanomilling process. *Int. J. Mach. Tools Manuf.* **2021**, *162*, 103701. [CrossRef]
11. Zhu, Y.G.; Wang, Y.G.; Niu, S.W.; Xie, Y.J.; Lei, Y.Y. Effect of environmental friendly complexing agent and oxidant on CMP of aluminium alloy under low pressure. *Diam. Abras. Eng.* **2020**, *40*, 74–78. [CrossRef]
12. Guo, W.C.; Wu, C.J.; Ding, Z.S.; Zhou, Q. Prediction of surface roughness based on a hybrid feature selection method and long short-term memory network in grinding. *Int. J. Adv. Manuf. Technol.* **2021**, *112*, 1853–2871. [CrossRef]
13. Li, C.; Piao, Y.; Meng, B.; Zhang, Y.; Li, L.; Zhang, F. Anisotropy dependence of material removal and deformation mechanisms during nanoscratch of gallium nitride single crystals on (0001) plane. *Appl. Surf. Sci.* **2022**, *578*, 152028. [CrossRef]
14. Sun, Y.; Jin, L.; Gong, Y.D.; Wen, X.L.; Yin, G.; Wen, Q.; Tang, B. Experimental evaluation of surface generation and force time-varying characteristics of curvilinear grooved micro end mills fabricated by EDM. *J. Manuf. Processes* **2022**, *73*, 799–814. [CrossRef]

15. Shokoufeh, M.; Krishna, K.; Charbel, A.K.; Jeffrey, R.; Eric, P. Heat transfer simulation and improvement of autoclave loading in composites manufacturing. *Int. J. Adv. Manuf. Technol.* **2021**, *112*, 2989–3000. [CrossRef]
16. Su, Y.; Jiang, H.; Liu, Z.Q. An experimental investigation on heat transfer performance of electrostatic spraying used in machining. *Int. J. Adv. Manuf. Technol.* **2021**, *112*, 1285–1294. [CrossRef]
17. Kim, D.; Kim, H.; Lee, S. Life Estimation of Hot Press Forming Die by Using Interface Heat Transfer Coefficient Obtained from Inverse Analysis. *Int. J. Automot. Technol.* **2015**, *16*, 285–292. [CrossRef]
18. Gianfranco, P.; Vito, P.; Antonio, P. Determination of Interfacial Heat Transfer Coefficients in a Sand Mold Casting Process Using an Optimized Inverse Analysis. *Appl. Therm. Eng.* **2015**, *78*, 682–694. [CrossRef]
19. Mikhail, K.; Mikhail, K.; Iuliia, T.; Nikolai, E.S.; Alexey, B. Method for residual strain modeling taking into account mold and distribution of heat transfer coefficients for thermoset composite material parts. *Int. J. Adv. Manuf. Technol.* **2021**, *117*, 2429–2443. [CrossRef]
20. Ramos, N.; Mittermeier Christoph Kiendl, J. Experimental and numerical investigations on heat transfer in fused filament fabrication 3D-printed specimens. *Int. J. Adv. Manuf. Technol.* **2021**, *118*, 1367–1381. [CrossRef]
21. Kadam, A.R.; Hindasageri, V.; Kumar, G.N. Inverse Estimation of Heat Transfer Coefficient and Reference Temperature in Jet Impingement. *ASME J. Heat Transf.* **2020**, *142*, 092302. [CrossRef]
22. Kang, H.C.; Chang, S.M. The Correlation of Heat Transfer Coefficients for the Laminar Natural Convection in a Circular Finned-Tube Heat Exchanger. *ASME J. Heat Transf.* **2018**, *140*, 031801. [CrossRef]
23. Piotr, D. Simplification of 3D Transient Heat Conduction by Reducing It to an Axisymmetric Heat Conduction Problem and a New Inverse Method of the Problem Solution. *Int. J. Heat Mass Transf.* **2019**, *143*, 118492. [CrossRef]
24. Farzad, M.; Mathieu, S. Estimation of Thermal Conductivity, Heat Transfer Coefficient, and Heat Flux Using a Three Dimensional Inverse Analysis. *Int. J. Therm. Sci.* **2016**, *99*, 258–270. [CrossRef]
25. Parida, R.K.; Madav, V.; Hindasageri, V. Analytical Solution to Transient Inverse Heat Conduction Problem Using Green's Function. *J. Therm. Anal. Calorim.* **2020**, *141*, 2391–2404. [CrossRef]
26. Zhang, B.W.; Mei, J.; Cui, M. A General Approach for Solving Three-Dimensional Transient Nonlinear Inverse Heat Conduction Problems in Irregular Complex Structures. *Int. J. Heat Mass Transf.* **2019**, *140*, 909–917. [CrossRef]
27. Ming, W.W.; Yu, W.W.; Qiu, K.X.; An, Q.L.; Chen, M. Modelling of the temperature distribution based on equivalent heat transfer theory and anisotropic characteristics of honeycomb core during milling of aluminum honeycomb core. *Int. J. Adv. Manuf. Technol.* **2021**, *115*, 2097–2110. [CrossRef]
28. Guo, Z.P.; Xiong, S.M.; Cho, S.H. Heat Transfer between Casting and Die during High Pressure Die Casting Process of AM50 Alloy-Modeling and Experimental Results. *J. Mater. Sci. Technol.* **2008**, *1*, 131–135. [CrossRef]
29. Guo, Z.P.; Xiong, S.M.; Liu, B.C. Effect of Process Parameters, Casting Thickness, and Alloys on the Interfacial Heat-Transfer Coefficient in the High-Pressure Die-Casting Process. *Metall. Mater. Trans. A Phys. Metall. Mater. Sci.* **2009**, *39*, 2896–2905. [CrossRef]
30. Guo, Z.P.; Xiong, S.M.; Liu, B.C. Determination of the Heat Transfer Coefficient at Metal-Die Interface of High Pressure Die Casting Process of AM50 Alloy. *Int. J. Heat Mass Transf.* **2009**, *51*, 6032–6038. [CrossRef]
31. Chen, Y.; Chen, J.H.; Xu, G.D. Screw thermal characteristic analysis and error prediction considering the two-dimensional heat transfer structure. *Int. J. Adv. Manuf. Technol.* **2021**, *115*, 2433–2448. [CrossRef]
32. Ho, J.Y.; Leong, K.C.; Wong, T.N. Experimental and numerical investigation of forced convection heat transfer in porous lattice structures produced by selective laser melting. *Int. J. Therm. Sci.* **2018**, *137*, 276–287. [CrossRef]
33. Jiang, X.H.; Xiong, W.J.; Wang, Y.F. Heat treatment effects on microstructure-residual stress for selective laser melting AlSi10Mg. *Mater. Sci. Technol.* **2018**, *36*, 168–180. [CrossRef]
34. Wang, X.W.; Ho, J.Y.; Leong, K.C. Condensation heat transfer and pressure drop characteristics of R-134a in horizontal smooth tubes and enhanced tubes fabricated by selective laser melting. *Int. J. Heat Mass Transf.* **2018**, *126*, 949–962. [CrossRef]
35. Martin, L.; Tobias, M.; Avik, S. Mechanical and thermal characterisation of AlSi10Mg SLM block support structures. *Mater. Des.* **2019**, *183*, 108138. [CrossRef]
36. Huiping, L.; Guoqun, Z.; Shanting, N.; Yiguo, L. Inverse heat conduction analysis of quenching process using finite-element and optimization method. *Finite Elem. Anal. Des.* **2006**, *42*, 1087–1096. [CrossRef]
37. Kevin, A.; Ryan, F.; Branden, S.; Sawyer, G.; Jake, W.; Saeid, V. Nanofluid Heat Transfer: Enhancement of the Heat Transfer Coefficient inside Microchannels. *Nanomaterials* **2022**, *12*, 615. [CrossRef]
38. Khooshehchin, M.; Fathi, S.; Mohammadidoust, A.; Salimi, F.; Ovaysi, S. Experimental study of the effects of horizontal and vertical roughnesses of heater surface on bubble dynamic and heat transfer coefficient in pool boiling. *Heat Mass Transf.* **2022**, *58*, 1319–1338. [CrossRef]
39. James, V.B. Transient Sensitivity Coefficients for the Thermal Contact Conductance. *Int. J. Heat Mass Transf.* **1967**, *10*, 1615–1617. [CrossRef]
40. Beck, J.V.; Litkouhi, B.; Clair, C.R. Efficient Sequential Solution of the Nonlinear Inverse Heat Conduction Problem. *Numer. Heat Transf.* **1982**, *5*, 275–286. [CrossRef]
41. Beck, J.V. Combined Parameter and Function Estimation in Heat Transfer with Application to Contact Conductance. *ASME J. Heat Transf.* **1988**, *110*, 1046–1058. [CrossRef]

crystals

MDPI

Article

Machining Performance Analysis of Rotary Ultrasonic-Assisted Drilling of SiC_f/SiC Composites

Jingyuan He [1,2], Honghua Su [1,*], Ning Qian [1,2,*] and Pengfei Xu [1]

[1] National Key Laboratory of Science and Technology on Helicopter Transmission, Nanjing University of Aeronautics and Astronautics, Nanjing 210016, China
[2] JITRI Institute of Precision Manufacturing, Nanjing 211806, China
* Correspondence: shh@nuaa.edu.cn (H.S.); n.qian@nuaa.edu.cn (N.Q.); Tel.: +86-(0)25-84890644 (N.Q.)

Abstract: An SiC_f/SiC composite has the following excellent properties: high strength, low specific gravity, and high temperature resistance, which has great prospects in the combustion chamber of rockets or aero engines. Hole-making in SiC_f/SiC parts is an important processing method. Generally, water-based or oil-based coolants are avoided, so dry drilling is the primary hole-making approach for SiC_f/SiC. However, the abrasion resistance and high hardness of SiC_f/SiC often lead to fast tool wear as well as serious damage to the fiber and matrix during dry drilling. This study proposes an innovative strategy for hole-making in SiC_f/SiC parts—rotary ultrasonic-assisted drilling (RUAD) using an orderly arranged brazed diamond core drill. The influence of tool life and wear on drilling accuracy is analyzed. Additionally, the impacts of the process parameters of conventional drilling (CD) and RUAD on drilling force, torque, the surface roughness of the hole wall, and the exit tearing factor are investigated. The results show that the orderly arranged brazed diamond core drill exhibits longer tool life and higher accuracy in hole-making. Meanwhile, compared with CD, RUAD with the proposed core drill effectively improves the drilling quality and efficiency, and reduces the force and torque of drilling. The range of process parameters for dry drilling is broadened.

Keywords: SiC_f/SiC; rotary ultrasonic-assisted drilling; diamond; core-drill; lifetime

Citation: He, J.; Su, H.; Qian, N.; Xu, P. Machining Performance Analysis of Rotary Ultrasonic-Assisted Drilling of SiC_f/SiC Composites. *Crystals* **2022**, *12*, 1658. https://doi.org/10.3390/cryst12111658

Academic Editors: Chen Li, Chongjun Wu, Binbin Meng and Shanshan Chen

Received: 25 October 2022
Accepted: 11 November 2022
Published: 17 November 2022

1. Introduction

Silicon-carbide-fiber-reinforced-silicon-carbide (SiC_f/SiC), one of the ceramic matrix composites, is a newly developed high-temperature-resistant lightweight material. SiC_f/SiC has the following excellent properties: high temperature and wear resistance, high strength, and high hardness, and is thus a promising material for highly heat-resistant structural parts and long-life functional parts, e.g., combustor or blades in the LEAP-X1C or GE9X aero-engines [1–3]. These aero-engine components contain multiple holes for various purposes, such as assembling, connecting, and cooling. At the same time, SiC_f/SiC is challenging to drill [4]. First, the tools wear prematurely, so the normal drilling operation is hard to maintain. Second, conventional drilling (CD) of SiC_f/SiC results in high roughness of hole walls and severe tearing of hole exits because of the large drilling force. Third, SiC_f/SiC composites must be dry-drilled because conventional coolants may adversely affect the fiber/matrix interface. Therefore, it is essential to research suitable tools and strategies for drilling SiC_f/SiC composites.

Similar to other ceramic matrix composites (CMCs), SiC_f/SiC parts are prepared by near-net-shape processes [5]. However, most parts still need precision machining to meet the requirements for precision and surface quality [6]. Hole-making on the SiC_f/SiC parts occupies a large proportion [3]. Conventional drilling causes severe tool wear and has poor quality, low efficiency, and high cost [7]. Non-conventional drilling methods mainly include the following: laser drilling [8], abrasive waterjet drilling [9], electric discharge drilling (EDD) [10], and ultrasonic vibration-assisted drilling [11,12]; each method has its advantages and disadvantages. Laser drilling is non-contact, eliminating issues such as defects and frequent tool changes caused by tool wear during machining [13]. However, the disadvantages of laser drilling include low efficiency and difficulties in maintaining high accuracy, particularly when making holes with relatively large depths or diameters [14]. In addition, laser processing creates a heat-affected zone within the irradiated area. It may worsen the material's mechanical properties in that zone, thus deteriorating the lifetime and performance of the part [15]. Abrasive waterjet drilling has fast processing speed and high efficiency, whereas its disadvantages are poor surface quality, low accuracy, and severe edge tearing in hole-making [16]. EDD can machine various shapes through diverse electrode shapes and perform deep cutting; however, it also creates a heat-affected zone in the area affected by the electric arc on the surface/subsurface of a material. Moreover, ceramic matrix composites are mostly electrical insulators (or semiconductors), and the machining efficiency of EDD is very low, which limits the further development of this technology [17]. Ultrasonic-assisted drilling (UAD) reduces drilling forces, improves machined surface quality, alleviates tool wear, and extends tool life. Therefore, it has attracted considerable attention in hard and brittle material processing [18]. Hocheng et al. [19] conducted a comparative processing test on CD and UAD of C/SiC. The results showed that the tool wear was minor, and the material removal rate and drilling quality of UAD were better than those of CD. Feng et al. [20] conducted RUAD of C/SiC using a diamond core drill with various grain sizes. They found that the hole exit tearing was positively correlated with the thrust force, and RUAD could reduce hole exit tear defects by 60%. Wang et al. [21,22] performed multiple rotary ultrasonic machining (RUM) tests on 2D-C/SiC with an 8 mm diameter diamond core drill and showed that RUM could ameliorate the hole surface quality. Ding et al. [23] machined SiC ceramic using ultrasonic-assisted grinding (UAG) and concluded that UAG ultimately decreased the surface/subsurface breakage and surface roughness by reducing grinding force and energy. Therefore, UAD is a suitable machining method for the high-quality drilling of ceramic matrix composites.

However, most of the published studies on the RUM of CMC used coolants. At the same time, SiC_f/SiC composites generally need to be machined under dry conditions to avoid contamination of the interior material by the coolant. Moreover, tool wear is faster during dry drilling, and hole quality is worse than the drilling with a coolant [24]. The major solution to this problem is to use diamond tools for drilling. Commonly used diamond tools are prepared by electroplating, sintering, and brazing. Electroplated diamond tools have a relatively simple preparation process that is mature and stable, and thus have a low price. However, because the abrasive grains and the electroplating layer are combined through simple mechanical embedding, the holding strength between the abrasive grains and the electroplating matrix is low, resulting in the easy fall-off of the abrasive grains and a quick loss of the processing ability during the machining of difficult-to-cut material [25]. Sintered diamond tools have high grain holding strength, but most grains are buried in the metal binding agent with a low protrusion height, so the cutting ability is relatively low [26]. Brazed diamond tools have a high grain protrusion height and high matrix-grain holding strength, capable of maintaining good machining performance under high drilling forces and temperatures [27]. In addition, the orderly arranged brazed diamond tools have adequate space for storing chips, and high tool sharpness results in excellent dry machining performance [28]. However, there is little research on improving tool life and drilling quality of SiC/SiC composites by using the orderly arranged brazed diamond core drill and RUAD with the dry drilling method.

It is still in the preliminary research stage that dry drilling of SiC_f/SiC composites is carried out with orderly arranged brazed diamond tools. Studies on the machining performance of such tools for drilling SiC_f/SiC composites are insufficient. Therefore, this study investigates dry rotary ultrasonic-assisted drilling (RUAD) of SiC_f/SiC composites using an orderly arranged brazed diamond core drill. This study aims to provide fundamental guidance and basic process data on the efficient processing method and high-quality dry drilling of SiC_f/SiC composites.

2. Materials and Methods

2.1. Workpiece

In this study, the workpiece is a 2D SiC_f/SiC composite plate, and the thickness is 4 mm. The material comprises SiC fiber, BN interface layer, and SiC matrix; it also contains randomly distributed pores. The porosity inside the material is 20%. The SiC fiber's type is T300, with a diameter of 5–7 μm. The chemical vapor infiltration (CVI) method is used for material preparation [29]. First, SiC fiber yarns are made into a fiber preform according to a 2D weaving method, as shown in Figure 1a; second, the fiber preform is put into the CVI reactor. The gaseous precursor enters the reactor according to a certain proportion and penetrates the void of the fiber preform by diffusion; a chemical reaction occurs on the surface of the SiC fiber and deposits in situ to form a BN interface layer and SiC matrix. In general, in order to reduce the CMCs machining challenges (fast tool wear and poor machining quality), the SiC_f/SiC composites in machining are not final-state materials, which means the materials are not completely densified [30]. Afterward, the densification is again conducted to increase the properties of the SiC_f/SiC composite, such as density, hardness, strength, etc. Additionally, after the densification, whether further machining is needed or not depends on specific quality requirements.

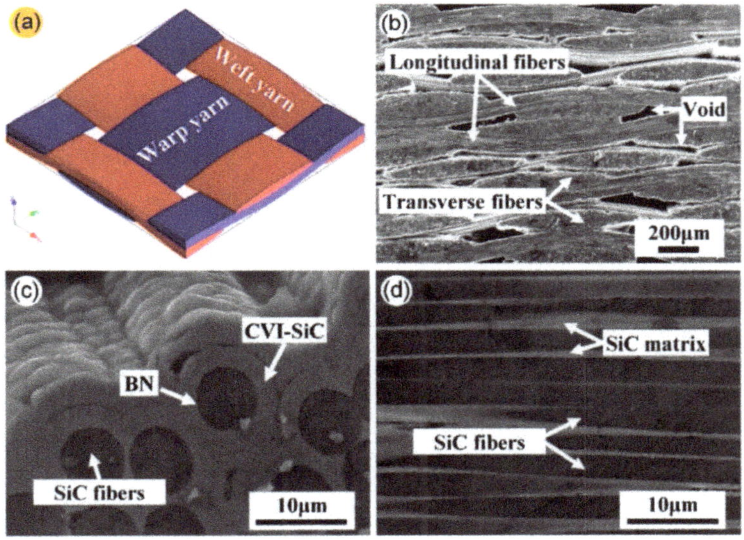

Figure 1. The structure of 2D SiC_f/SiC composites: (**a**) model of SiC fiber preform, (**b**) micromorphology by SEM, (**c**) micro-structure along transverse fibers, (**d**) micro-structure along longitudinal fibers.

We observe the micromorphology on the side of the polished SiC_f/SiC composite. The SiC fibers of each layer are arranged in transverse and longitudinal directions, and the fibers in different directions are woven in a plain weave pattern, as shown in Figure 1a. Pores of various sizes are distributed between plain weave layers (Figure 1b), and the presence of pores is expected to affect the drilling force during mechanical drilling. Figure 1c,d presents the micromorphology of SiC fibers in transverse and longitudinal directions, where the SiC fibers are tightly wrapped by BN interfacial layer and SiC matrix. Compared with SiC ceramics, this woven structure endows SiC_f/SiC composites with higher fracture toughness and better impact resistance under extreme conditions. The mechanical properties of SiC_f/SiC composites and SiC ceramics are shown in Table 1 [31,32].

Table 1. Mechanical properties of the SiC and SiC_f/SiC workpiece [31,32].

Material	Density (g/cm^3)	Elastic Modulus (GPa)	Hardness (GPa)	Fracture Toughness ($MPa \cdot m^{1/2}$)	Flexural Strength (MPa)
SiC	3.214	410	25	3	500–600
SiC_f/SiC	2.29	289	20	26.4	570

2.2. Drilling Tool

The drilling tool used in this study is an orderly arranged brazed diamond core drill developed in-house. This tool has high abrasive grain protrusion height, adequate space for drilling chips, and high holding strength between the abrasive grains and the tool matrix; it is suitable for dry-drilling CMCs. Figure 2a shows the overall appearance and the following three sections of the orderly arranged brazed diamond core drill: the working section containing diamond abrasive grains, tool neck, and holding shank. The working section's diameter is 4 mm, and the diamond abrasive grain size is 40/45 mesh (average size in the range of 425–450 μm). The abrasive grains of the working section are orderly arranged into eight 15-grain rows with a length of 7 mm, and there are four symmetrical arc grooves with diamond abrasive grains welded on the tool end (as shown in Figure 2b,c). Hence, this structure can effectively accommodate the abrasive chips generated during dry drilling and discharge them from the drilling area. In addition, new tools are used for each set of tests to avoid the impact of tool wear on the test data in this study.

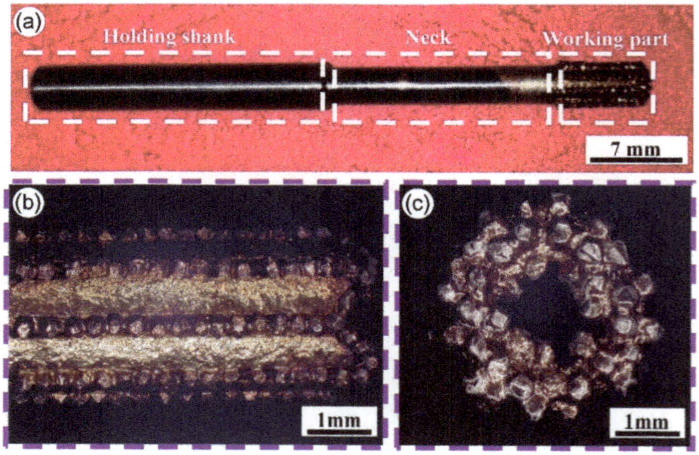

Figure 2. An orderly arranged brazed diamond core drill, (**a**) the composition of tool, (**b**) the side abrasive of tool, (**c**) the top abrasive of tool.

2.3. Experimental Setup

Conventional drilling and RUAD are performed on a 5-axis vertical high-speed machining center (DMG Ultrasonic 20 Linear, Germany), as shown in Figure 3a. The maximum spindle speed of machining center is 42,000 rpm, and the maximum travel of the X, Y, Z axis are 200 mm, 220 mm, and 280 mm respectively. The RUAD is performed using the HSK-E32 ultrasonic tool holder with a 20,000 rpm maximum speed and an in-house ultrasonic system. The workpiece is clamped by a pressure plate on a special fixture for drilling tests, as shown in Figure 3b. All tests are performed under dry drilling conditions, following specific machining parameters shown in Table 2. This paper first tests the drilling performance of an orderly arranged brazed diamond core drill in RUAD (n = 12,000 rpm, v_w = 30 mm/min, f = 23.5 kHz, A = 5 μm), and then, the influence of drilling parameters on drilling force, torque and drilling quality are investigated by single factor test with new tools. Vacuum cleaning removes chips during drilling to avoid contaminating the environment and the machine spindle.

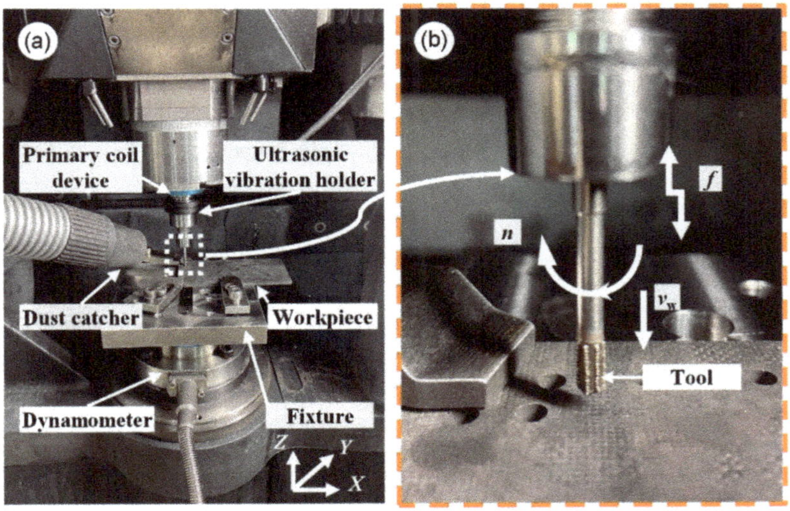

Figure 3. Photo of the machining center: (**a**) overall experimental setup, (**b**) features of RUAD.

Table 2. Experimental parameters of rotary ultrasonic dry drilling.

Parameters	Value
Spindle speed n (r/min)	4000, 8000, 12,000, 16,000, 20,000
Feed rate v_w (mm/min)	10, 20, 30, 40, 50
Ultrasonic frequency f (kHz)	23.5
Amplitude A (μm)	0, 5
Coolant	Dry drilling

2.4. Measurement Method

Before ultrasonic vibration-assisted drilling, the top of the tool's amplitude is measured by an eddy current sensor (DT3301) with a sampling frequency of 100 kHz (Figure 4a). To record the high-frequency alternating voltage signal measured by the eddy current sensor is connected by the oscilloscope as shown in Figure 4b. The relationship between ultrasonic amplitude and voltage can be calculated by Equation (1) [33].

$$A_U = \varphi \times U, \tag{1}$$

where A_U is the top of the tool's ultrasonic amplitude, φ is the amplitude-to-voltage ratio (50 μm/V), and U is the output voltage. When the spindle speed n is 8000 rpm, the resonant frequency is 23.5 kHz, and the ultrasonic amplitude at the core drill tip is 5 μm.

The dynamometer (Kistler 9272 with a 5070 charge amplifier) measures the drilling forces and torques, and then these process signals are collected by a data acquisition card into a computer (Figure 4b). Dynoware, a dynamometer-specific software, saves and processes the collected data. The tearing of the hole exit after drilling is recorded by a 3D video microscope (HIROX KH-7700). The confocal microscope (Sensofar Neo S) measures the hole wall's surface roughness Sa, and the results are obtained by measuring each hole three times.

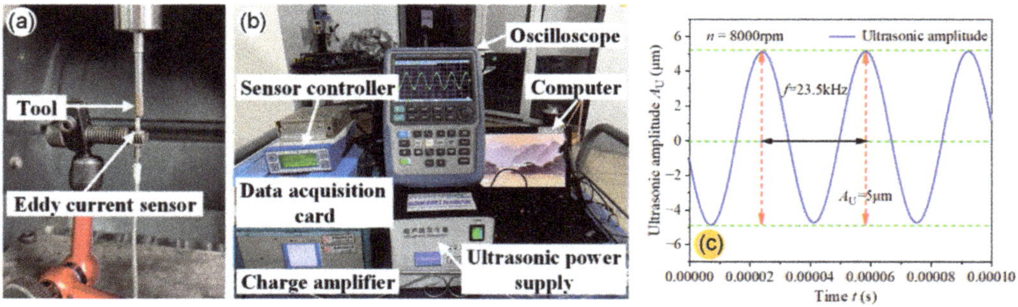

Figure 4. Ultrasonic amplitude testing device, (**a**) eddy current sensor, (**b**) dynamometer and amplitude measurement controller, (**c**) axial ultrasonic amplitude at 8000 rpm.

3. Results and Discussion

3.1. Kinematic Analysis of RUAD Process

The mechanism of RUAD and the abrasive grain motion trajectory are shown in Figure 5. The trajectory of abrasive grain motion during CD consists of rotation along the tool axis and uniform motion in the feed direction. In the RUAD, a frequency of vibration greater than 20 kHz is applied along the Z-axis in addition to CD, as shown in Figure 5a. The motion trajectories of the tool during CD and RUAD of a single diamond grain can be depicted by Equations (2) and (3) [12].

$$\begin{cases} X_{CD}(t) = R_1 \cos\left(\frac{2\pi nt}{60}\right) \\ Y_{CD}(t) = R_1 \sin\left(\frac{2\pi nt}{60}\right) \\ Z_{CD}(t) = v_w t \end{cases}, \quad (2)$$

$$\begin{cases} X_{RUAD}(t) = R_1 \cos\left(\frac{2\pi nt}{60}\right) \\ Y_{RUAD}(t) = R_1 \sin\left(\frac{2\pi nt}{60}\right) \\ Z_{RUAD}(t) = v_w t + A\sin(2\pi f t) \end{cases}, \quad (3)$$

where R_1 is the distance from the vertex of the abrasive grains on the tool to the axis, f is the frequency, A is the amplitude, t is the drilling time, v_w is the feed rate, and n is the spindle speed.

According to Figure 5b,c, the diamond grains at the tooltip are the first to contact the workpiece and remove the largest volume of workpiece material during drilling. The abrasive grains in the front portion of the diamond core drill side grind the hole wall material and eventually form the hole wall morphology. Although they remove the least amount of material, the rear part of the tool's abrasive grains significantly contributes to hole consistency and increase the useful tool life during continuous hole-making.

As shown in Figure 5d, Equations (2) and (3) visualize a single diamond grain's trajectory. The comparison between the single diamond grain's motion trajectory in RUAD and CD shows that the diamond grain's motion path length under the superposition of ultrasonic vibration is longer than that during CD. This indicates a shorter depth of cut for the single abrasive grain during RUAD compared with that in CD [12]. In addition, under specific processing parameters, the abrasive grains at the top of the core drill maintain a "contact-separation" state in the SiC_f/SiC composite. It can change the stress state of the impact point between the abrasive grain and material, thereby increasing the microscopic fragmentation of the abrasive grains and improving the self-sharpening ability [34].

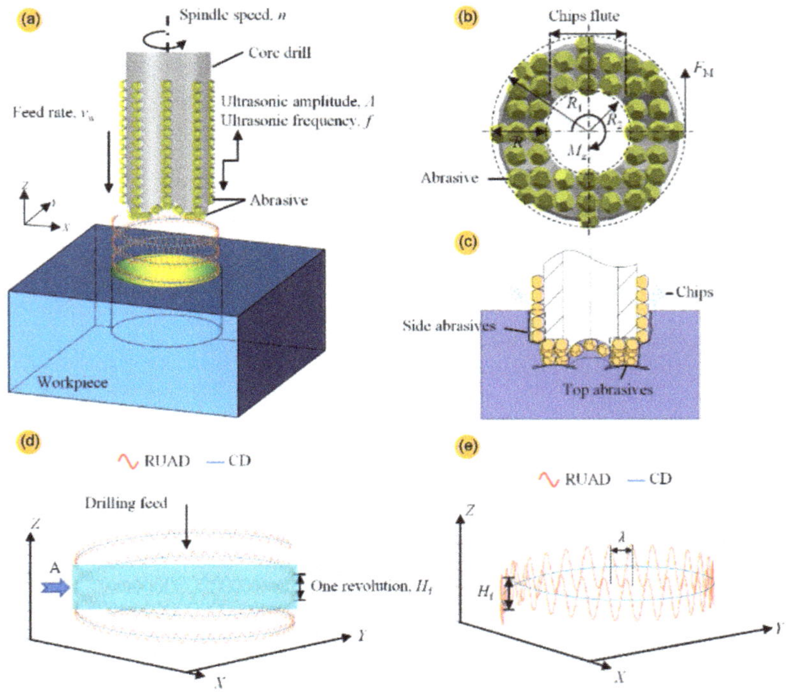

Figure 5. Diagram of the machining process, (**a**) drilling process, (**b**) tip of tool, (**c**) side of the tool, (**d**) the single abrasive grain motion trajectory during RUAD and CD, and (**e**) grain motion trajectory at one revolution.

3.2. Tool Life

While drilling SiC_f/SiC composites, the tool life directly affects the machining efficiency and production cost. This study first conducts tool-life tests on orderly brazed diamond core drills. The experiments are conducted under dry RUAD using the ultrasonic machining settings specified in Table 2, a spindle speed n of 12,000 rpm, and a feed rate v_w of 30 mm/min. During the test, the tool is continuously used to drill until failure.

Figure 6a indicates that the orderly arranged brazed diamond core drill makes 102 holes under dry RUAD conditions. Under the same conditions, the SiC_f/SiC composite is drilled by a commercial electroplated diamond core drill (Berun Japan, Φ 4 mm and 100 mesh). The electroplated core drill can make a hole under the same dry RUAD conditions. In this case, the in-house developed, orderly arranged brazed diamond core drill has a significant advantage over a long lifetime and improves drilling efficiency.

From hole 1 to hole 92, the process is in a stable wear state, during which the hole diameter ranges from 4.01 to 4.04 mm, and the difference from the maximum to minimum diameter is 0.03 mm. From hole 93 to hole 102, the process is in a rapid wear state, and the hole diameter ranges from 3.95 to 3.99 mm, and the maximum and minimum hole diameters are 0.09 mm in the whole state. In this stage, the tool no longer meets the requirements of high precision hole making (4 ± 0.03 mm) and thus is considered to have failed. Figure 6b shows the trend in the drilling force throughout the hole-making process. Before hole 80, the drilling force is below 8 N. In this case, the top and side abrasive grains slowly wear out, the abrasive grains involved in drilling are in the stable wear stage, and the hole diameter remains within a certain range. As the tool kept making more holes, the wear of the top abrasive grains progressed, resulting in a sharp increase in drilling force (a maximum increase of 62.5%), indicating the start of the tool's rapid wear stage. In actual manufacturing, when the tool is in the rapid wear stage, the machining should be stopped. The tool should be replaced in time to reduce the generation of scrap caused by the excessive difference in hole diameters. According to the hole diameters and variation in drilling force between hole 81 and hole 92, the tool is about to enter the rapid wear stage; the drilling force increases sharply, while the hole diameter is still maintained within the deviation of 0.03 mm. This can provide operators with more indicators to avoid scrap generation before the complete failure of the tool.

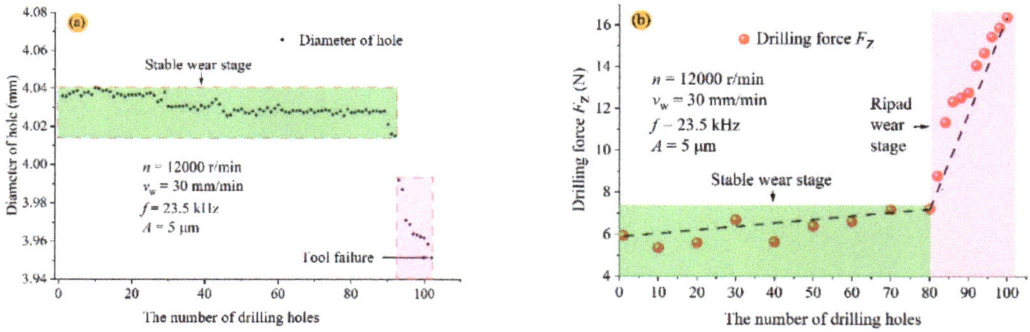

Figure 6. The result of tool life test, (**a**) relationship between the diameter of hole and the number of holes, (**b**) relationship between force and the number of holes.

The orderly arranged brazed diamond core drill combined with the RUAD increases the number of high-quality holes produced in SiC_f/SiC composites under dry conditions. The tool structure ensured stable wear of abrasive grains, and RUAD significantly reduced the axial force, thus extending the core drill life and improving the drilling quality. The following sections assessed the impact of different drilling methods and process parameters on drilling force and hole-making quality in the stable tool wear stage.

3.3. Drilling Force and Torque

The original signals of drilling force and torque obtained in the ultrasonic-assisted dry drilling of SiC_f/SiC composite by the brazed diamond tool reveal that the whole drilling process has gone through the following four stages: AB, BC, CD, and DE (Figure 7). AB is the entrance drilling phase, where the diamond grains at the top of the brazed diamond tool enter the uppermost SiC_f/SiC workpiece at moment A. As more abrasive grains participate in drilling, the drilling force gradually increases. BC is the stable drilling stage, in which the diamond grains at the top and the front side of the core drill are involved in processing; the axial force at this stage is relatively stable. CD is the exit drilling stage, during which the diamond grains drill the bottom of the SiC_f/SiC composite at moment C, and the end abrasive grains drill through the bottom material at moment D. Therefore, the drilling force gradually decreases. The last stage, DE, is the hole-trimming stage. Because the working section of the tool has a stepped structure (Figure 5c), the diamond grains on the side of the tool continue to grind the hole wall at this stage, thus trimming the hole wall, improving the hole quality, and maintaining the final hole diameter. In this stage, the least amount of material is removed, so the drilling force is very small. The mean drilling force and torque of section BC are analyzed in this study.

The patterns of variation in the drilling force and torque, as well as their reduction with spindle speed n and feed rate v_w for CD and RUAD, are shown in Figures 8 and 9. The drilling force and torque of the CD are higher than those of the RUAD, according to the measured specifications. Figure 5e depicts how ultrasonic vibration lengthens the machining trajectories of abrasive grains and decreases the depth of a single-grain cut, reducing force and torque [34]. Additionally, as seen in Figures 8a and 9a, the drilling force and torque increase with rising feed rate v_w while decreasing with rising spindle speed

n. This research examines how rotary ultrasonic-assisted drilling affects the drilling force and torque when the same drilling parameters are used. To simplify the analysis, reduced magnitudes K_F and K_M were used in this study to characterize the effects of RUAD on the drilling force and torque under the same drilling parameters, as expressed in Equations (4) and (5) [35].

$$K_F = (F_{Z-CD} - F_{Z-RUAD})/F_{Z-CD} \times 100\%, \tag{4}$$

$$K_M = (M_{Z-CD} - M_{Z-RUAD})/M_{Z-CD} \times 100\%, \tag{5}$$

where $F_{Z\text{-}CD}$ and $M_{Z\text{-}CD}$ are the drilling force and torque for CD, respectively, whereas $F_{Z\text{-}RUAD}$ and $M_{Z\text{-}RUAD}$ are the drilling force and torque for rotary ultrasonic-assisted drilling, respectively.

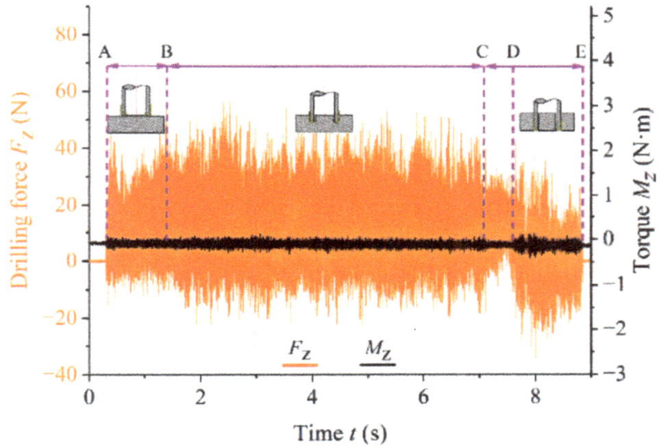

Figure 7. Drilling force and torque signals.

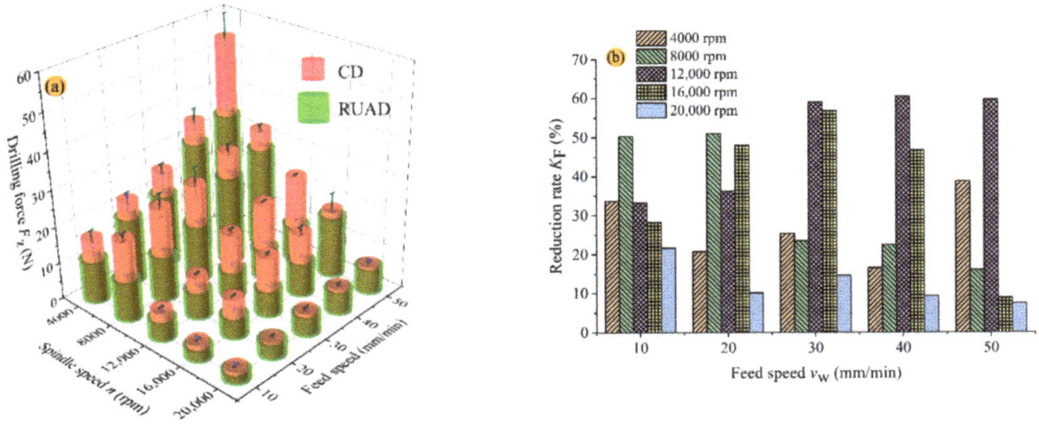

Figure 8. Influence of spindle speeds and feed rates on drilling force, (**a**) variation trend of drilling force, (**b**) reduction rate K_F of drilling force.

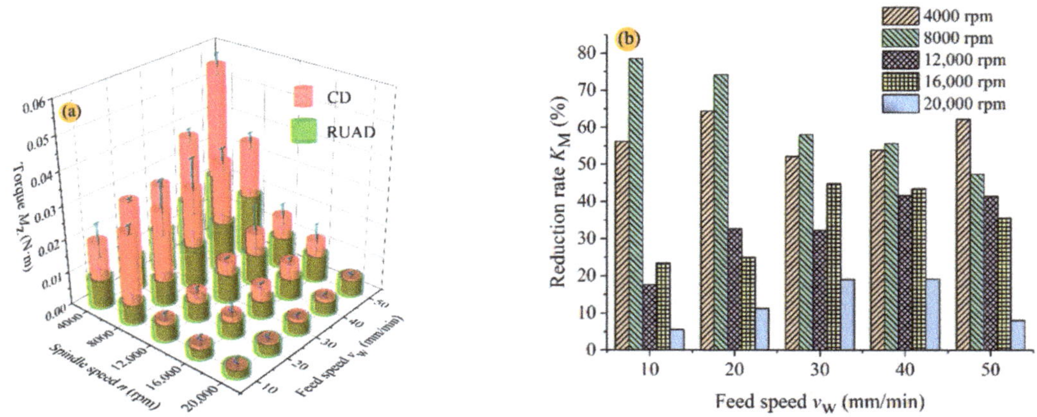

Figure 9. Effect of different machining parameters on torque, (**a**) variation trend of torque, (**b**) variation trend K_M of torque reduction.

Figures 8b and 9b show the impact of drilling parameters on the reduction in drilling force and torque. The reduction in the drilling force, K_F, decreases from 60.6% to 21.7% with the increase in speed spindle n from 4000 to 20,000 rpm and in feed rate v_w from 10 to 50 mm/min; similarly, the torque reduction, K_M, decreases from 78.6% to 5.6%. The results indicate that the effect of speed spindle n on the reduction rate is greater than that of feed rate v_W. For example, at the speed spindle n of 8000 rpm, the feed rate v_w from 10 to 50 mm/min increases the K_F and K_M values from 16% to 51% and from 47.3% to 78.6%, respectively. The K_F and K_M values rise from 14.6% to 59.1% and from 19.1% to 58%, respectively, at the feed rate v_W of 30 mm/min when the spindle speed n is increased from 4000 to 20,000 rpm. The impact of ultrasonic vibration is weakened by increasing spindle speed, gradually diminishing the ultrasonic processing effect to the level of conventional processing. During RUAD, the wavelength λ of ultrasonic vibration can be expressed by Equation (6) [24]

$$\lambda = \frac{v_s}{f} = \frac{\pi n D}{f}, \tag{6}$$

where D is the diameter of the tool, v_s is the linear velocity of the abrasive grains during drilling, and n is the spindle speed. Figure 5a illustrates the diamond abrasive grain's track during RUAD as a sinusoidal curve spinning downward around the spindle axis. Therefore, the number of wavelengths generated by ultrasonic vibration per revolution of diamond abrasive grain in the feeding process (N) can be expressed by Equation (7).

$$N = \frac{\sqrt{H_f{}^2 + L^2}}{\lambda}, \tag{7}$$

where L is the circumference of the diamond grain rotating for one turn, and its value is πD; H_f is the feed per revolution of the diamond grain. As the value of H_f per revolution is very small (0.5–2.5 μm) compared to the circumference L, it can be neglected. Combined with Equation (6), the number of wavelengths N is expressed as follows:

$$N = \frac{f}{n}, \tag{8}$$

According to Equation (8), as the spindle speed adds, both the number of ultrasonic wavelengths per revolution of diamond abrasive grain in the feeding process and the number of hammering on the workpiece by abrasive grains decrease. Therefore, the ultrasonic vibration effect diminishes when the spindle speed increases.

In the reported studies, the core drill's contact area with the SiC_f/SiC workpiece has been lubricated by adding a coolant, thereby reducing the damage caused by friction-induced heat to the tool and hole-making quality [36]. However, the present study focuses on improving the tool-workpiece friction behavior during dry drilling of SiC_f/SiC composites and thus will provide new insights into this extreme machining method. Figure 5b,c illustrate the contact process between the SiC_f/SiC workpiece and the diamond grains at the top of the brazed core tool, where F_M and M_z are the friction and torque on the diamond grains at the top of the tool, which can be expressed by Equations (9) and (10) [37].

$$F_M = F_Z \cdot \mu, \quad (9)$$

$$M_Z = F_M \cdot R, \quad (10)$$

where μ is the equivalent friction coefficient between the SiC_f/SiC workpiece and the diamond grains at the top of the tool, and R is the width of the working surface for the top of the core drill with $R = R_1 - R_2$ (Figure 5b). Hence, Equations (9) and (10) may be combined into Equation (11) to obtain the equivalent friction coefficient μ as follows:

$$\mu = \frac{M_Z}{F_Z \cdot R}, \quad (11)$$

In conventional drilling, the coefficient of friction is 0.21–0.31, and in RUAD, it is 0.13–0.19, as shown in Figure 10. Therefore, RUAD can effectively reduce the coefficient of friction. In dry drilling, a smaller coefficient of friction means less heat generated by friction and thus less tool wear, which is more conducive to maintaining the sharpness of the tool during machining.

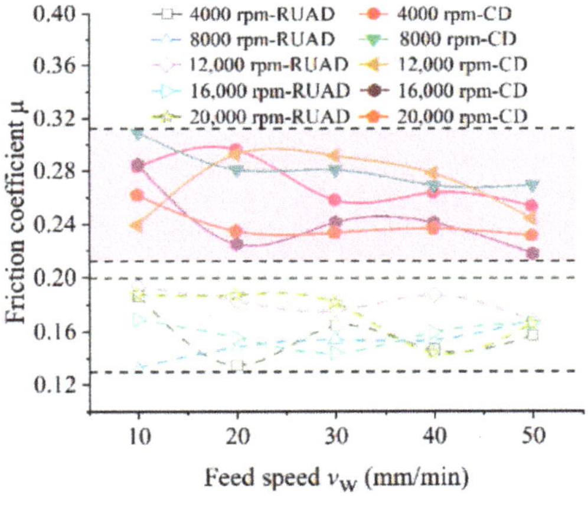

Figure 10. Effect of machining parameters on friction coefficient with and without ultrasonic drilling process.

3.4. Surface Roughness of Hole

Surface roughness is an important indicator reflecting surface integrity, which directly reveals the impact of processing parameters, abrasive cutting trajectory, and other variables on the workpiece's surface quality. As SiC_f/SiC is an anisotropic material, its surface condition is difficult to characterize comprehensively because of the large variations in the linear roughness *Ra* measured in different directions. Therefore, as illustrated in Figure 11, we adopt the surface roughness *Sa* to evaluate the grinding surface quality. Based on the results, the surface roughness *Sa* of holes has been strongly impacted by the drilling parameters. In particular, the quality of the machined surface improves by reducing the feed rate v_w and increasing the spindle speed n. Additionally, the surface roughness of the holes that RUAD produces is superior to that produced by CD when the same parameters are examined (Figure 12a,b). Under all testing parameters, the surface roughness *S*a of all holes is 2.60–13.03 μm for RUAD and 2.91–19.03 μm for CD. During RUAD, ultrasonic vibration changes the state of motion of the abrasive grains, resulting in a longer cutting path than in CD, thus leading to a smaller single-diamond grain chip size. Moreover, RUAD has a larger overlap area of abrasive grains on the edge of the diamond core drill than CD, which enhances the roughness of drilled holes' walls [24]. In addition, under relatively low spindle drilling speeds, in contrast to CD, RUAD can maintain good hole surface roughness when the feed rate increases. This provides a good solution for improving the efficiency of hole-making and expanding the application range of this drilling process.

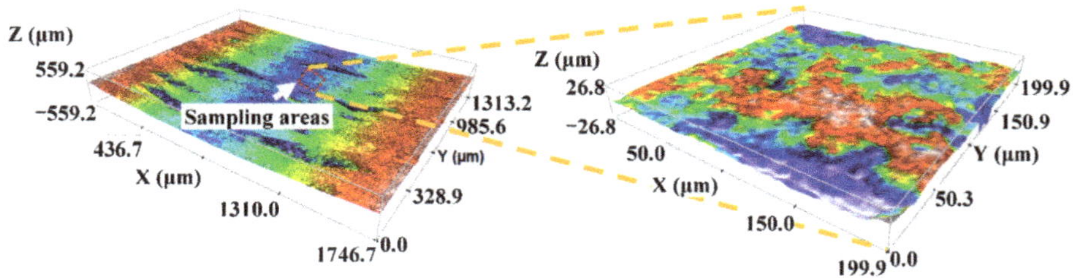

Figure 11. *Sa* sampling method.

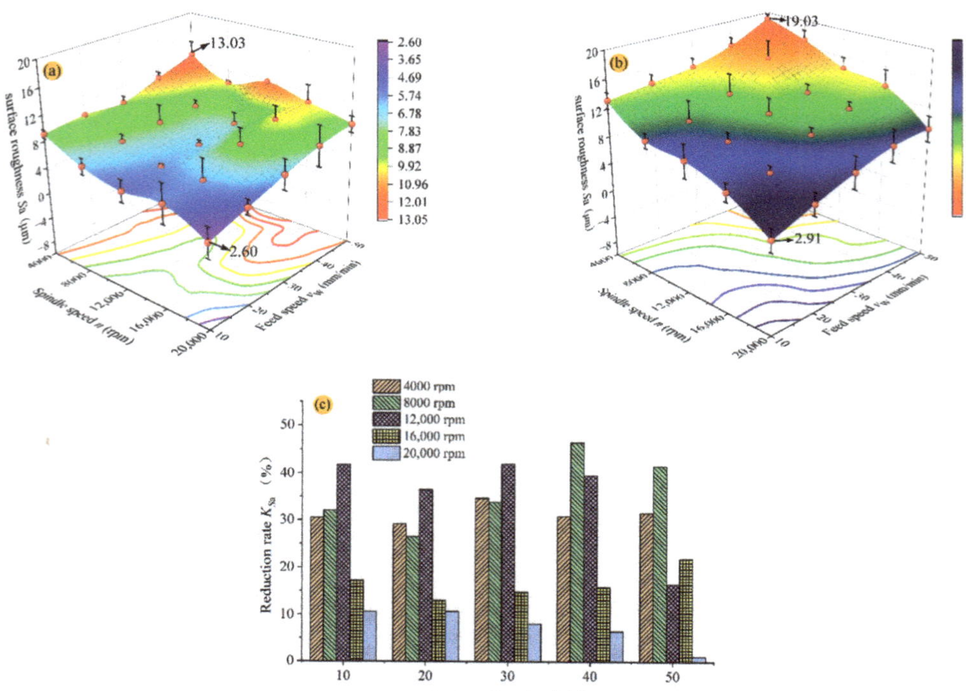

Figure 12. Surface roughness and reduction rate effects of drilling parameters, (**a**) surface roughness variation trend in RUAD, (**b**) surface roughness variation trend in CD, and (**c**) reduction rate.

This study defines K_{Sa} as the magnitude of the reduction in hole wall surface roughness from CD to RUAD performed at the same machining parameters, expressed in Equation (12).

$$K_{Sa} = (S_{a-CD} - S_{a-RUAD})/S_{a-CD} \times 100\%, \quad (12)$$

where $S_{a\text{-}CD}$ is the surface roughness of the hole wall for CD, and $S_{a\text{-}RUAD}$ is the surface roughness of the hole wall for RUAD. K_{Sa} expresses the magnitude of the influence of ultrasonic machining on the quality of the hole wall.

The spindle speed n has a greater influence on the decrease in surface roughness (K_{Sa}) of the hole walls than the feed rate v_w. The impact of RUAD on surface roughness of hole improvement is more pronounced when K_{Sa} is between 30% and 42% and the spindle speed n is between 4000 and 12,000 rpm, as illustrated in Figure 12c. When the feed rate v_w is in the range of 10–50 mm/min and the spindle speed n approaches 20,000 rpm, K_{Sa} drops to 1–10%, suggesting that at high spindle speeds, the influence of RUAD on the surface roughness of drilled holes progressively reduces to a level comparable to CD. To summarize, increasing spindle speed reduces the hole wall roughness Sa during CD and RUAD; however, it also reduces the degree of improvement in surface roughness Sa after RUAD.

3.5. Tearing Factor of the Hole Exit

The tearing factor is frequently employed to quantify drilling damage to composite materials surrounding the entry or exit of the hole [3]. As shown in Figure 13, the tearing factor of the hole exit L_D can be calculated as follows:

$$L_D = \frac{D_m - D_h}{D_h}, \quad (13)$$

where D_m is the maximum diameter of the tearing area at the hole exit, and D_h is the actual drilled hole diameter of the SiC_f/SiC composite.

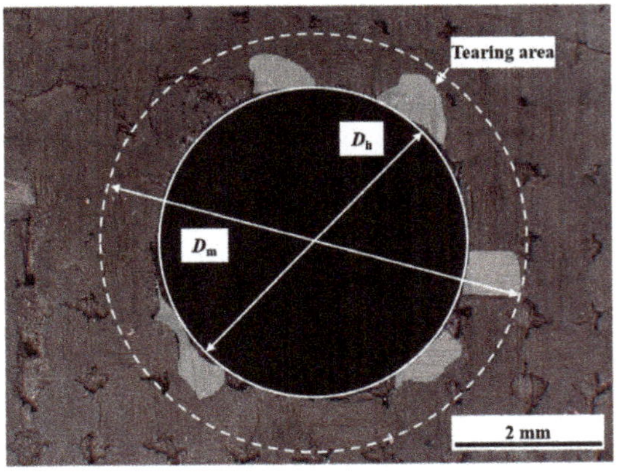

Figure 13. Schematic of the tearing factor.

The hole exit tearing factor is directly related to the drilling parameters (spindle speed, feed rate), and different processing methods (RUAD, CD) of the hole made by the tearing factor show different phenomena. Figure 14 depicts the development of the tearing factor of the hole exit L_D at various spindle speeds and feed rates in RUAD and CD. When the feed rate v_w and the spindle speed n increase from 10 to 50 mm/min and from 4000 to 20,000 rpm, respectively, the tearing factor of the hole exit L_D is better in RUAD than in CD. When the spindle speed n is 4000 rpm and the feed rate v_w is 10 mm/min, the tearing factor of the hole exit L_D is 0.39 in CD and 0.36 in RUAD. In particular, throughout the experimental domain, the tearing factor of the hole exit L_D increases from 0.1 to 0.5 in RUAD and from 0.11 to 0.7 in CD, as shown in Figure 14a,b. The morphology of the hole exit also illustrates that with the increment of spindle speed, the tearing area reduces for both RUAD and CD, while the increased feed rate can deteriorate the hole exit quality, which means an increase in the tearing area, as shown in Figures 15 and 16. In the studied range of the parameters, the exit tearing factor in RUAD varies to a smaller extent compared to CD. Furthermore, RUAD results in smaller exit tear size throughout the entire range of drilling parameters, achieving a maximum reduction of 37.26% (n = 16,000 rpm, v_W = 10 mm/min, A = 5 μm, f = 23.5 kHz) as shown in Figure 14c. The morphology of the hole exit also demonstrates the advantages of ultrasonic vibration in reducing the exit tearing, as shown in Figures 15c and 16c.

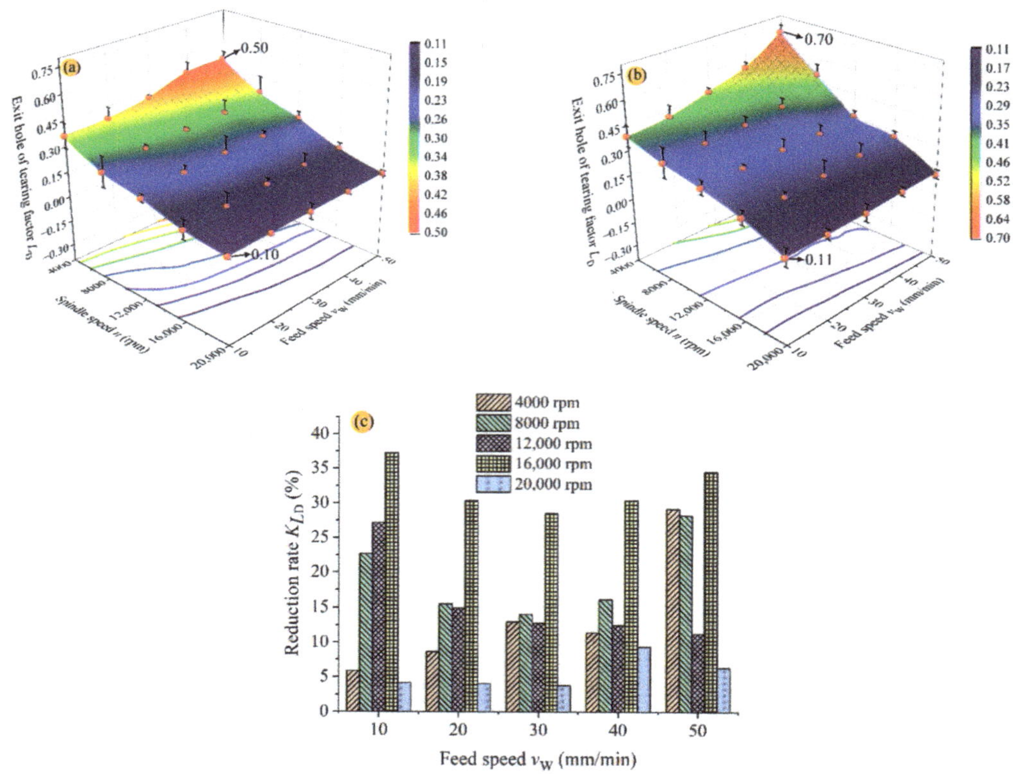

Figure 14. Effect of different parameters on exit tearing factor and its reduction, (**a**) variation trend of exit hole of tearing factor in RUAD, (**b**) variation trend of exit hole of tearing factor in CD, and (**c**) reduction rate.

As both the SiC matrix and the SiC fibers in SiC_f/SiC composites are exceedingly hard and brittle, the exit tear during drilling is primarily caused by cracks generated under the drilling force and their extension to the hole exit. This paper describes the formation process of exit tearing in detail, as shown in Figure 17. The tear damage formation during the drilling of SiC_f/SiC is similar to that during the drilling of the C/SiC composite, as reported by Feng et al. [21]. It mainly includes fiber-matrix interface debonding, fiber bending, and fiber breaking. When the diamond core drill reaches the bottom of the material, cracks occur at the fiber-matrix contact as a result of the drilling force, as shown in Figure 17a. The crack expands along the fiber-matrix interface until debonding. The hole exit is composed of fewer composite layers and thus has low support strength. The fiber is bent and deformed as the core drill proceeds toward the hole exit, further expanding the interface's debonding length, as illustrated in Figure 17b. With further feeding of the core drill, the fibers continue to bend until breaking when the compressive stress from the drilling force exceeds the flexural strength of the fibers, eventually causing tearing at the hole exit, as shown in Figure 17c. The chip morphology of the SiC_f/SiC composite after drilling through the material and the fiber fracture morphology at the edge of the chip reflect the state of the hole exit tear, as illustrated in Figure 17d.

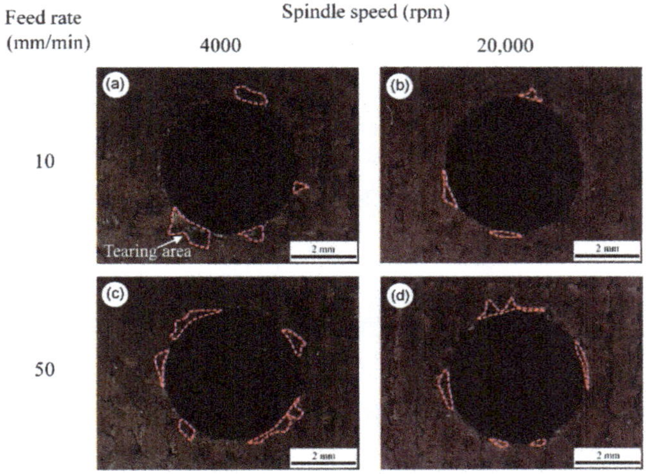

Figure 15. The morphology of the hole exit tearing areas in RUAD with (**a**) n = 4000 rpm, v_W = 10 mm/min; (**b**) n = 20,000 rpm, v_W = 10 mm/min; (**c**) n = 4000 rpm, v_W = 50 mm/min; (**d**) n = 20,000 rpm, v_W = 50 mm/min.

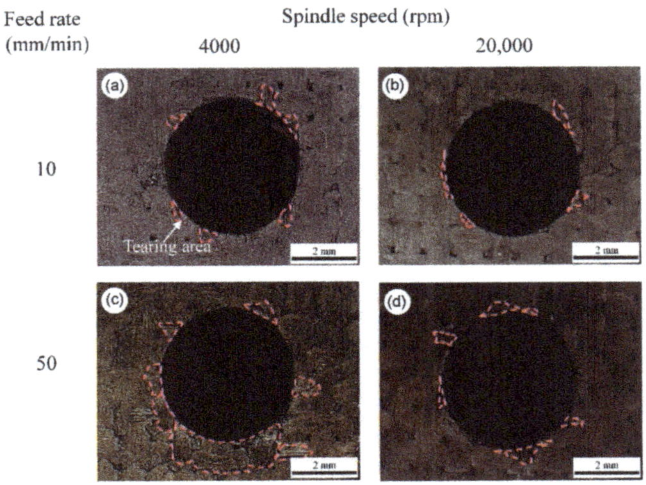

Figure 16. The morphology of the hole exit tearing areas in CD with (**a**) n = 4000 rpm, v_W = 10 mm/min; (**b**) n = 20,000 rpm, v_W = 10 mm/min; (**c**) n = 4000 rpm, v_W = 50 mm/min; (**d**) n = 20,000 rpm, v_W = 50 mm/min.

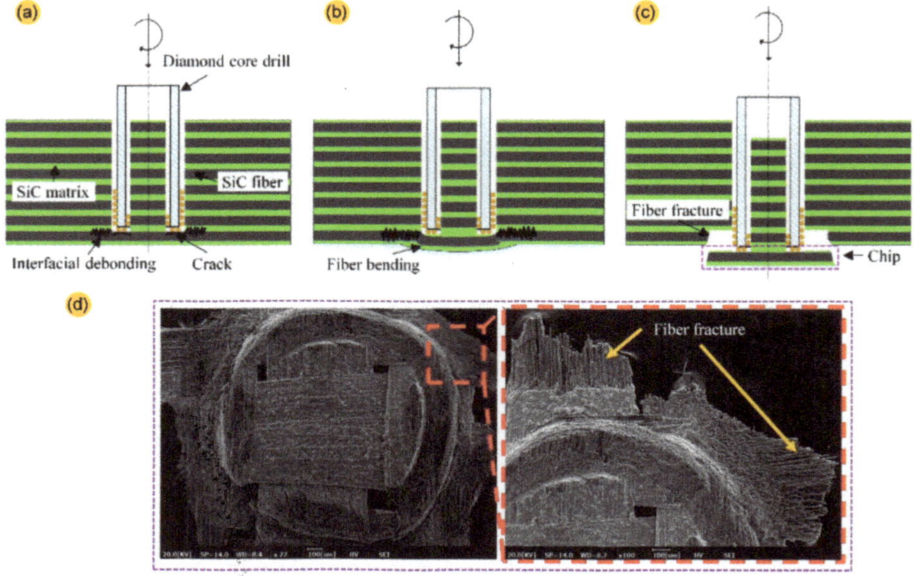

Figure 17. Hole exit tearing damage formation process, (**a**) fiber-matrix interface debonding, (**b**) fiber bending, (**c**) fiber breaking, and (**d**) chip morphology.

The drilling force F_Z is the key element influencing the degree of hole exit tearing. According to Figure 18, the drilling force and hole exit tearing are positively associated, meaning that the higher the drilling force, the bigger the hole exit tearing. This is similar to the results of Feng [20] and Chen [38]. As a result, RUAD effectively minimizes the drilling force, lowering the hole exit tearing factor.

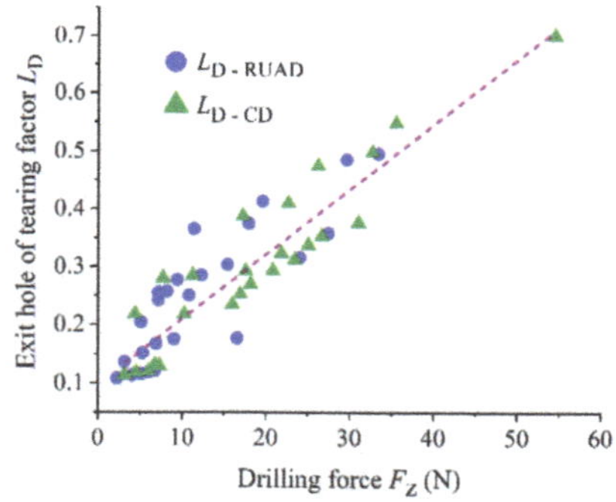

Figure 18. Drilling force versus hole exit tearing factor.

4. Conclusions

This study proposes an innovative strategy for dry drilling of small holes in SiC_f/SiC composites—dry rotary ultrasonic-assisted drilling (RUAD) using an orderly arranged brazed diamond core drill. We investigate the effect of tool life and wear on drilling accuracy during dry machining and the effects of processing parameters on drilling force, torque, surface roughness *Sa* of holes, and exit tearing. We also validated the benefits of RUAD with orderly arranged brazed diamond core drills for machining SiC_f/SiC composites. The main conclusions are as follows.

(1) When the feed rate is 30 mm/min and the spindle speed is 12,000 rpm, the studied core drill made 102 holes during RUAD until failure. The diameters of the drilled holes ranged from 3.95 to 4.04 mm, whereas before hole 92, the maximum error in hole diameter was 0.03 mm, indicating a relatively high drilling accuracy for this process.

(2) RUAD and conventional drilling (CD) are compared using multiple parameters. Ultrasonic vibration changed the contact state and friction behavior between the core drill and SiC_f/SiC workpiece, and the equivalent friction coefficient of dry RUAD is lower by 38% than that of CD. Furthermore, RUAD showed significantly lower drilling force and torque than CD as follows: the drilling force during RUAD is in the range of 2.25–33.40 N with a maximum reduction of 60.6%, and torque is in the range of 0.0024–0.02096 N m with a maximum reduction of 78.6%.

(3) Compared with CD, RUAD exhibits a smaller drilling force and a larger trajectory overlap region for the abrasive grains on the tool, which effectively reduces the surface roughness of holes and the exit tearing factor. At the tested parameters, the hole wall surface roughness *Sa* is 2.60–13.03 μm for RUAD and 2.91–19.03 μm for CD, RUAD can reduce the *Sa* up to 46.4%. The exit tearing factor L_D is 0.1–0.5 for RUAD with a maximum reduction of 37.3%, while the tearing factor is 0.11–0.7 for CD.

(4) RUAD maintains high hole-making quality and improves the hole-making efficiency even at relatively high feed rates and low spindle speeds, thus extending the process domain for efficient and high-quality hole-making in SiC_f/SiC composites.

Author Contributions: Conceptualization, J.H.; methodology, J.H. and N.Q.; software, P.X.; validation, J.H. and P.X.; formal analysis, J.H., H.S. and N.Q.; investigation, J.H.; resources, H.S. and N.Q.; data curation, J.H.; writing—original draft preparation, J.H. and N.Q.; writing—review and editing, H.S.; funding acquisition, H.S. and N.Q. All authors have read and agreed to the published version of the manuscript.

Funding: This research was funded by National Natural Science Foundation of China, (grant number 92060203); China Postdoctoral Science Foundation, (grant number 2021M701696); Outstanding Postdoctoral Program of Jiangsu Province, (grant number 2022ZB204); and Science Center for Gas Turbine Project, (grant number P2022-A-IV-002-001).

Institutional Review Board Statement: Not applicable.

Informed Consent Statement: Not applicable.

Data Availability Statement: Not applicable.

Acknowledgments: The authors would like to thank the CHEN Yurong from JITRI Institute of Precision Manufacturing for his kind assistance with the equipment.

Conflicts of Interest: The authors declare no conflict of interest.

References

1. Annerino, A.; Lawson, M.; Gouma, P.I. Future Insights on High Temperature Ceramics and Composites for Extreme Environments. *Int. J. Ceram. Eng. Sci.* **2022**, *4*, 296–301. [CrossRef]
2. Gao, T.; Zhang, Y.; Li, C.; Wang, Y.; Chen, Y.; An, Q.; Zhang, S.; Li, H.N.; Cao, H.; Ali, H.M.; et al. Fiber-Reinforced Composites in Milling and Grinding: Machining Bottlenecks and Advanced Strategies. *Front. Mech. Eng.* **2022**, *17*, 24. [CrossRef]
3. Xing, Y.; Deng, J.; Zhang, G.; Wu, Z.; Wu, F. Assessment in Drilling of C/C-SiC Composites Using Brazed Diamond Drills. *J. Manuf. Process.* **2017**, *26*, 31–43. [CrossRef]
4. Huang, B.; Wang, W.; Jiang, R.; Xiong, Y.; Liu, C. Experimental Study on Ultrasonic Vibration-Assisted Drilling Micro-Hole of SiCf/SiC Ceramic Matrix Composites. *Int. J. Adv. Manuf. Technol.* **2022**, *120*, 8031–8044. [CrossRef]
5. Xiao, C.; Han, B. Grinding, Machining Morphological Studies on C/SiC Composites. *J. Inst. Eng. India Ser. D* **2018**, *99*, 209–215. [CrossRef]
6. Garcia Luna, G.; Axinte, D.; Novovic, D. Influence of Grit Geometry and Fibre Orientation on the Abrasive Material Removal Mechanisms of SiC/SiC Ceramic Matrix Composites (CMCs). *Int. J. Mach. Tools Manuf.* **2020**, *157*, 103580. [CrossRef]
7. Xu, J.; Li, C.; Chen, M.; Ren, F. A Comparison between Vibration Assisted and Conventional Drilling of CfFR/Ti6Al4V Stacks. *Mater. Manuf. Process.* **2019**, *34*, 1182–1193. [CrossRef]
8. Liu, Y.; Zhang, R.; Li, W.; Wang, J.; Yang, X.; Cheng, L.; Zhang, L. Effect of Machining Parameter on Femtosecond Laser Drilling Processing on SiC/SiC Composites. *Int. J. Adv. Manuf. Technol.* **2018**, *96*, 1795–1811. [CrossRef]

9. Zhang, Y.; Liu, D.; Zhang, W.; Zhu, H.; Huang, C. Hole Characteristics and Surface Damage Formation Mechanisms of C/SiC Composites Machined by Abrasive Waterjet. *Ceram. Int.* **2022**, *48*, 5488–5498. [CrossRef]
10. Sheikh-Ahmad, J.Y. Hole Quality and Damage in Drilling Carbon/Epoxy Composites by Electrical Discharge Machining. *Mater. Manuf. Process.* **2016**, *31*, 941–950. [CrossRef]
11. Li, Z.; Yuan, S.; Ma, J.; Shen, J.; Batako, A.D.L. Cutting Force and Specific Energy for Rotary Ultrasonic Drilling Based on Kinematics Analysis of Vibration Effectiveness. *Chin. J. Aeronaut.* **2021**, *35*, 376–387. [CrossRef]
12. Ding, K.; Fu, Y.; Su, H.; Chen, Y.; Yu, X.; Ding, G. Experimental Studies on Drilling Tool Load and Machining Quality of C/SiC Composites in Rotary Ultrasonic Machining. *J. Mater. Process. Technol.* **2014**, *214*, 2900–2907. [CrossRef]
13. Li, C.; Hu, Y.; Zhang, F.; Geng, Y.; Meng, B. Molecular Dynamics Simulation of Laser Assisted Grinding of GaN Crystals. *Int. J. Mech. Sci.* **2022**, *in press*. [CrossRef]
14. Zhang, R.; Li, W.; Liu, Y.; Wang, C.; Wang, J.; Yang, X.; Cheng, L. Machining Parameter Optimization of C/SiC Composites Using High Power Picosecond Laser. *Appl. Surf. Sci.* **2015**, *330*, 321–331. [CrossRef]
15. Zhai, Z.; Wang, W.; Mei, X.; Li, M.; Cui, J.; Wang, F.; Pan, A. Effect of the Surface Microstructure Ablated by Femtosecond Laser on the Bonding Strength of Ebcs for SiC/SiC Composites. *Opt. Commun.* **2018**, *424*, 137–144. [CrossRef]
16. Thongkaew, K.; Wang, J.; Li, W. An Investigation of the Hole Machining Processes on Woven Carbon-Fiber Reinforced Polymers (CFRPs) Using Abrasive Waterjets. *Mach. Sci. Technol.* **2019**, *23*, 19–38. [CrossRef]
17. Wei, C.; Zhao, L.; Hu, D.; Ni, J. Electrical Discharge Machining of Ceramic Matrix Composites with Ceramic Fiber Reinforcements. *Int. J. Adv. Manuf. Technol.* **2013**, *64*, 187–194. [CrossRef]
18. Sonia, P.; Jain, J.K.; Saxena, K.K. Influence of Ultrasonic Vibration Assistance in Manufacturing Processes: A Review. *Mater. Manuf. Process.* **2021**, *36*, 1451–1475. [CrossRef]
19. Hocheng, H.; Tai, N.H.; Liu, C.S. Assessment of Ultrasonic Drilling of C/SiC Composite Material. *Compos. Part A-Appl. Sci. Manuf.* **2000**, *31*, 133–142. [CrossRef]
20. Feng, P.; Wang, J.; Zhang, J.; Zheng, J. Drilling Induced Tearing Defects in Rotary Ultrasonic Machining of C/SiC Composites. *Ceram. Int.* **2017**, *43*, 791–799. [CrossRef]
21. Wang, J.; Zhang, J.; Feng, P. Effects of Tool Vibration on Fiber Fracture in Rotary Ultrasonic Machining of C/SiC Ceramic Matrix Composites. *Compos. Part B-Eng.* **2017**, *129*, 233–242. [CrossRef]
22. Wang, J.; Zhang, J.; Feng, P.; Guo, P. Experimental and Theoretical Investigation on Critical Cutting Force in Rotary Ultrasonic Drilling of Brittle Materials and Composites. *Int. J. Mech. Sci.* **2018**, *135*, 555–564. [CrossRef]
23. Ding, K.; Li, Q.; Zhang, C. Experimental Studies on Material Removal Mechanisms in Ultrasonic Assisted Grinding of SiC Ceramics with a Defined Grain Distribution Brazed Grinding Wheel. *Int. J. Adv. Manuf. Technol.* **2021**, *116*, 3663–3676. [CrossRef]
24. Wu, W.; Kuzu, A.; Stephenson, D.; Hong, J.; Bakkal, M.; Shih, A. Dry and Minimum Quantity Lubrication High-Throughput Drilling of Compacted Graphite Iron. *Mach. Sci. Technol.* **2018**, *22*, 652–670. [CrossRef]
25. Hwang, T.W.; Evans, C.J.; Malkin, S. High Speed Grinding of Silicon Nitride with Electroplated Diamond Wheels, Part 1: Wear and Wheel Life. *J. Manuf. Sci. Eng.* **2000**, *122*, 42–50. [CrossRef]
26. Yu, Y.Q.; Tie, X.R.; Zhang, G.Q.; Huang, G.Q.; Huang, H.; Xu, X.P. Comparison of Brazed and Sintered Diamond Tools for Grinding of Stone. *Mater. Res. Innov.* **2014**, *18*, S2–S869. [CrossRef]
27. Liu, D.; Long, W.; Wu, M.; Qi, K.; Pu, J. Microstructure Evolution and Lifetime Extension Mechanism of Sn-Added Fe-Based Pre-Alloy Brazing Coating in Diamond Tools. *Coatings* **2019**, *9*, 364. [CrossRef]
28. Chen, Z.; Xiao, B.; Wang, B. Optimum and Arrangement Technology of Abrasive Topography for Brazed Diamond Grinding Disc. *Int. J. Refract. Met. Hard Mat.* **2021**, *95*, 105455. [CrossRef]
29. Wang, P.; Liu, F.; Wang, H.; Li, H.; Gou, Y. A Review of Third Generation SiC Fibers and SiC_f/SiC Composites. *J. Mater. Sci. Technol.* **2019**, *35*, 2743–2750. [CrossRef]
30. Razzell, A.G. Joining and Machining of Ceramic Matrix Composites. In *Comprehensive Composite Materials*; Elsevier: Amsterdam, The Netherlands, 2000; Volume 4, pp. 689–697.
31. Dai, J.; Su, H.; Hu, H.; Yu, T.; Zhou, W.; Ding, W.; Ji, S.; Zheng, Y. The Influence of Grain Geometry and Wear Conditions on the Material Removal Mechanism in Silicon Carbide Grinding with Single Grain. *Ceram. Int.* **2017**, *43*, 11973–11980. [CrossRef]
32. Justine, D.; Eduardo, S.; Nasrin, A.N. Fracture Behaviour of SiC/SiC Ceramic Matrix Composite at Room Temperature. *J. Eur. Ceram. Soc.* **2022**, *42*, 3156–3167.
33. Chen, Y.; Su, H.; Qian, N.; He, J.; Gu, J.; Xu, J.; Ding, K. Ultrasonic Vibration-Assisted Grinding of Silicon Carbide Ceramics Based on Actual Amplitude Measurement: Grinding Force and Surface Quality. *Ceram. Int.* **2021**, *47*, 15433–15441. [CrossRef]
34. Wu, B.; Zhao, B.; Ding, W.; Su, H. Investigation of the Wear Characteristics of Microcrystal Alumina Abrasive Wheels during the Ultrasonic Vibration-Assisted Grinding of Ptmcs. *Wear* **2021**, *477*, 203844. [CrossRef]
35. Ding, K.; Fu, Y.; Su, H.; Cui, F.; Li, Q.; Lei, W.; Xu, H. Study on Surface/Subsurface Breakage in Ultrasonic Assisted Grinding of C/SiC Composites. *Int. J. Adv. Manuf. Technol.* **2017**, *91*, 3095–3105. [CrossRef]
36. Khunt, C.P.; Makhesana, M.A.; Patel, K.M.; Mawandiya, B.K. Performance Assessment of Vegetable Oil-Based Minimum Quantity Lubrication (MQL) in Drilling. *Mater. Today Proc.* **2021**, *44*, 341–345. [CrossRef]

37. Islam, S.; Yuan, S.; Li, Z. Mathematical Modeling and Experimental Studies on Axial Drilling Load for Rotary Ultrasonic Drilling of C/SiC Composites. *Int. J. Adv. Manuf. Technol.* **2020**, *107*, 1309–1326. [CrossRef]
38. Zou, F.; Chen, J.; An, Q.; Cai, X.; Chen, M. Influences of Clearance Angle and Point Angle on Drilling Performance of 2D Cf/SiC Composites Using Polycrystalline Diamond Tools. *Ceram. Int.* **2020**, *46*, 4371–4380. [CrossRef]

crystals

Article

Study on Grinding Force of Two-Dimensional Ultrasonic Vibration Grinding 2.5D-C/SiC Composite Material

Yunguang Zhou *, Chuanchuan Tian, Shiqi Jia, Lianjie Ma *, Guoqiang Yin and Yadong Gong

School of Mechanical Engineering and Automation, Northeastern University, Shenyang 110819, China
* Correspondence: zhouyunguang@neuq.edu.cn (Y.Z.); bcmlj1025@163.com (L.M.)

Abstract: The grinding force is an important index during the grinding process, which affects the surface quality and other aspects after machining. However, the research on the grinding force of ceramic matrix composites assisted by two-dimensional ultrasonic vibration-assisted grinding is very weak. In this paper, the impact of the relationship between the critical cutting depth and the maximum undeformed chip thickness on the removal mode of ceramic matrix composites was analyzed. On this basis, the grinding force model of two-dimensional ultrasonic vibration-assisted grinding were developed for ductile removal and brittle removal, respectively. Finally, the correctness of the model was verified, and the impact of grinding parameters on the grinding force was analyzed. The experimental results show that compared with the conventional grinding force, the two-dimensional ultrasonic vibration assisted grinding force decreases obviously. When the feed rate and grinding depth increase, the grinding force increases. When the grinding velocity and ultrasonic amplitude increase, the grinding force decreases. Compared with the experimental value, the average relative error of normal grinding force is 8.49%, and the average relative error of tangential grinding force is 13.59%. The experimental and theoretical values of the grinding force have a good fitting relationship.

Keywords: grinding; two-dimensional ultrasonic vibration-assisted grinding; grinding force; 2.5D-C/SiC composites material

Citation: Zhou, Y.; Tian, C.; Jia, S.; Ma, L.; Yin, G.; Gong, Y. Study on Grinding Force of Two-Dimensional Ultrasonic Vibration Grinding 2.5D-C/SiC Composite Material. *Crystals* **2023**, *13*, 151. https://doi.org/10.3390/cryst13010151

Academic Editors: Ronald W. Armstrong, Chen Li, Chongjun Wu, Binbin Meng and Shanshan Chen

Received: 7 December 2022
Revised: 28 December 2022
Accepted: 12 January 2023
Published: 15 January 2023

1. Introduction

Ceramic matrix composites (CMCS) are a new type of multiphase material, which are based on conventional ceramic material, adding fiber as reinforcement to the ceramic material, to overcome the shortcomings of the conventional ceramic material, such as high brittleness and high sensitivity. Among them, carbon fiber reinforced silicon carbide-based (C/SiC) composites exhibit high strength, high hardness, low density, excellent wear resistance, good thermal stability and excellent oxidation resistance, and are considered as one of the most promising materials in many high-temperature engineering applications [1]. However, it is still a hard and brittle material in nature. Surface/sub-surface damage, material delamination and other defects will appear during machining, which will affect the reliability and service life of parts and components, and limit the wide application of C/SiC composites. Therefore, high precision machining of C/SiC composites has become one of the hot issues at present.

At present, machining research on hard and brittle materials such as C/SiC composites mainly focuses on grinding [2,3] and special machining methods which combines conventional machining methods with sound, light, electricity and chemical energy. Processing methods include grinding, ultrasonic processing (USM) [4], rotating ultrasonic processing (RUM) [5], abrasive water jet cutting (AWC) [6], electrical discharge machining (EDM) [7] and laser beam machining (LBM) [8]. Ultrasonic vibration-assisted grinding (UVAG) is a kind of ultrasonic machining, which is a precision machining technology combining high-frequency ultrasonic vibration with conventional grinding (CG). Because of the existence of ultrasonic vibration, the movement trajectory of abrasives changes. At the same time, it can effectively reduce the grinding force and heat, and further improve the surface machining quality of material, so it has a wide application prospect in the precision machining field of hard and brittle material, and relevant scholars have conducted in-depth research on this. Ding et al. [9] conducted radial ultrasonic-assisted grinding tests on SiC ceramics, and the results showed that, compared with conventional grinding, radial ultrasonic-assisted grinding could effectively reduce the grinding force ratio, the grinding force and surface roughness. Cao et al. [10] analyzed the brittle-ductile transition behavior of SiC ceramics during axial ultrasonic vibration-assisted internal grinding, and found that the critical chip depth of conventional grinding and ultrasonic vibration assisted grinding are 0.072 μm and 0.093 μm, respectively, which shows that ultrasonic vibration-assisted grinding is easier to realize ductile removal of material and improve surface quality. Chen et al. [11] studied the effect of different fracture mechanisms of carbon fiber on the removal mechanism and surface quality of C_f/SiC composites in the grinding process. The experimental results show that the change in removal mechanism is related to the value of the maximum undeformed chip thickness, and ultrasonic-assisted grinding can effectively reduce this value, promote nano-scale brittle fracture of fiber and improve surface quality. Guo et al. [12] used ultrasonic vibration-assisted grinding to machine linear micro-structured surface. The experimental results show that ultrasonic vibration-assisted grinding can effectively reduce the roughness and ensure the sharpness of microstructure edges, which proves the advantages of ultrasonic vibration-assisted grinding in precision machining of microstructures. To study the removal mechanism during ultrasonic vibration-assisted grinding, Gao et al. [13] performed an ultrasonic vibration-assisted scratch test on SiC ceramics. Compared with the ordinary scratch test, it was found that radial ultrasonic vibration-assisted grinding has obvious advantages in improving critical cutting depth of brittle-ductile transition and material removal rate, but the study of axial ultrasonic vibration-assisted grinding is not detailed. Liang et al. [14] developed a new two-dimensional ultrasonic vibration-assisted grinding processing technology, and verified the feasibility of two-dimensional ultrasonic vibration assisted grinding in machining hard and brittle material through experiments. Yan et al. [15] analyzed the impact of abrasive protrusion height and interference of adjacent abrasive trajectories on material removal mechanism, and established a three-dimensional roughness model for two-dimensional ultrasonic vibration-assisted grinding of zirconia ceramics, which provided a new method for modeling surface roughness of two-dimensional ultrasonic-assisted grinding.

As an important index in the grinding process, grinding force affects grinding temperature, grinding surface quality, grinding wheel wear and so on. For the impact of ultrasonic vibration on the grinding force, the main research focuses on one-dimensional ultrasonic vibration-assisted grinding and develops a variety of prediction models. Sun et al. [16] developed the axial ultrasonic vibration-assisted grinding force model. Compared with the conventional grinding test results, it is found that the normal force and tangential force of axial ultrasonic assisted grinding are reduced by 27.31% and 22.52%, respectively, and the surface roughness value is reduced by 18.0%, which indicates that ultrasonic vibration-assisted grinding can reduce the grinding force and improve the machining surface quality. Yang et al. [17] analyzed the kinematics principle of the tangential ultrasonic vibration-assisted grinding of ZrO_2 ceramics and the contact rate between abrasives and workpieces in the machining process. On this basis, the grinding force prediction model for tangential ultrasonic vibration-assisted grinding of ZrO_2 ceramics was developed, and the impact of grinding parameters and vibration parameters on the grinding force was analyzed. Li et al. [18] carried out a varied-depth nano-scratch test on SiC ceramics, measured the critical chip thickness of brittle-ductile transition of SiC ceramics by AFM and SEM, and developed a grinding force model of one-dimensional axial ultrasonic vibration assisted grinding according to material removal rate. Xiao et al. [19] considered the brittle-ductile transition mechanism in the one-dimensional ultrasonic vibration-assisted grinding of ceramic material, modeled the grinding force under two removal modes, respectively, and developed the final grinding force prediction model on this basis.

However, according to the current research, for hard and brittle materials, the research on the grinding force of UVAG mainly focuses on one-dimensional ultrasonic vibration-assisted grinding, while the research on two-dimensional ultrasonic vibration-assisted grinding (TUVAG) is relatively lacking. At the same time, unlike engineering ceramics such as ZrO_2 and SiC ceramics, C/SiC composites show anisotropy due to the existence of a fiber toughening phase, which makes the removal mechanism of C/SiC composites more complex. Therefore, the research on UVAG of C/SiC composites, especially the research on the grinding force of UVAG is rare, which limits the application of ultrasonic machining technology in C/SiC composites. In this paper, based on the grinding force model of one-dimensional ultrasonic vibration-assisted grinding of ceramic material, the kinematics principle of TUVAG is analyzed, and the different removal stages of material are modeled, respectively. At the same time, the effective abrasives quantity in different removal stages is analyzed, and the grinding force prediction model of TUVAG of 2.5D-C/SiC composites is developed, which is verified by a single factor test.

2. Kinematics Analysis of TUVAG

TUVAG is based on conventional grinding, which can remove material by applying high-frequency vibration in a specific direction and frequency to the workpiece or tool. As shown in Figure 1, due to the existence of ultrasonic vibration, there are three kinds of motions in the machining process: the feeding motion of the grinding tool, the rotating motion of the spindle and ultrasonic vibration. The following assumptions are made to facilitate the analysis:

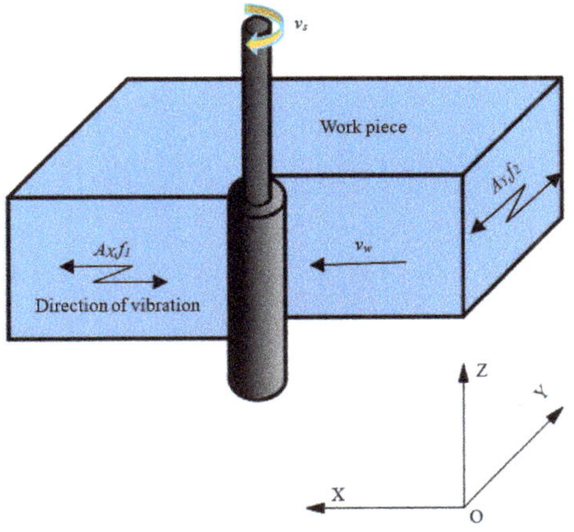

Figure 1. Diagram of TUVAG.

(1) Ultrasonic amplitude and ultrasonic frequency remain unchanged during grinding.

(2) The shape of the abrasives is similar to that of the Vickers indenter, the size is the same, the abrasives are evenly distributed on the grinding tool, and the abrasives will not detach from the grinding tool during grinding.

(3) The shape of the abrasives remains unchanged during grinding.

Based on the above assumptions, as shown in Figure 1, the coordinate system is fixed at the center of the grinding tool, assuming that the abrasives start to contact the workpiece at time 0, and the contact point is P. The kinematic equation of abrasives is shown in Equation (1):

$$S_{TVUAG} = \begin{cases} S_{XTUVAG} \\ S_{YTUVAG} \\ S_{ZTUVAG} \end{cases} = \begin{cases} R\sin(2\pi nt) + v_w t + A_X \sin(2\pi f_1 t + \Phi_1) \\ R - R\cos(2\pi nt) + A_Y \sin(2\pi f_2 + \Phi_2) \\ 0 \end{cases} \tag{1}$$

where R is the radius of the grinding tool, R = 5 mm; v_w is the feeding rate of the grinding tool; n is the spindle speed; A_X and A_Y are ultrasonic amplitudes in X direction and Y direction; f_1 and f_2 are ultrasonic vibration frequencies in X direction and Y direction, $f_1 = f_2$ = 24,600 Hz; Φ_1 and Φ_2 are the initial phases of tangential ultrasonic vibration and radial ultrasonic vibration. It is proved that when the phase difference between two vibration directions is 90°, the effect of vibration combination is the best, and it is a standard elliptical trajectory. Therefore, here $\Phi_1 = 0°$ and $\Phi_2 = 90°$.

The velocity of abrasives during grinding can be obtained by differentiating Equation (1), as shown in Equation (2):

$$V_{TUVAG} = \begin{cases} V_{XTUVAG} \\ V_{YTUVAG} \\ V_{ZTUVAG} \end{cases} = \begin{cases} 2\pi nRcos(2\pi nt) + v_w + 2\pi f_1 A_X cos(2\pi f_1 t) \\ 2\pi nRsin(2\pi nt) + 2\pi f_2 A_Y cos(2\pi f_2 t + \frac{\pi}{2}) \\ 0 \end{cases} \quad (2)$$

Similarly, in conventional grinding, the kinematics and velocity equations of abrasives are shown in Equations (3) and (4):

$$S_{CG} = \begin{cases} S_{XCG} \\ S_{YCG} \\ S_{ZCG} \end{cases} = \begin{cases} Rsin(2\pi nt) + v_w t \\ R - Rcos(2\pi nt) \\ 0 \end{cases} \quad (3)$$

$$V_{CG} = \begin{cases} V_{XCG} \\ V_{YCG} \\ V_{ZCG} \end{cases} = \begin{cases} 2\pi nRcos(2\pi nt) + v_w \\ 2\pi nRsin(2\pi nt) \\ 0 \end{cases} \quad (4)$$

Based on Equations (1)–(4), the motion trajectory and velocity changes of abrasives during CG and TUVAG are obtained, as shown in Figures 2 and 3. It can be seen from Figure 2 that the motion trajectory of a single abrasive in TUVAG is approximately elliptical in a vibration cycle. From the microscopic point of view, in a vibration cycle, the track length of abrasives in TUVAG is longer than that in CG, and under specific grinding parameters, there will be a cycle process of contact-separation-contact between abrasives and workpiece, which makes the workpiece repeatedly ground by abrasives and reduces the residual height of grinding surface. In addition, combined with Figure 3, it can be seen the grinding velocity has been changing in a large range during TUVAG, which shows that due to the existence of ultrasonic vibration, there is obvious impact in grinding, which promotes the removal of material.

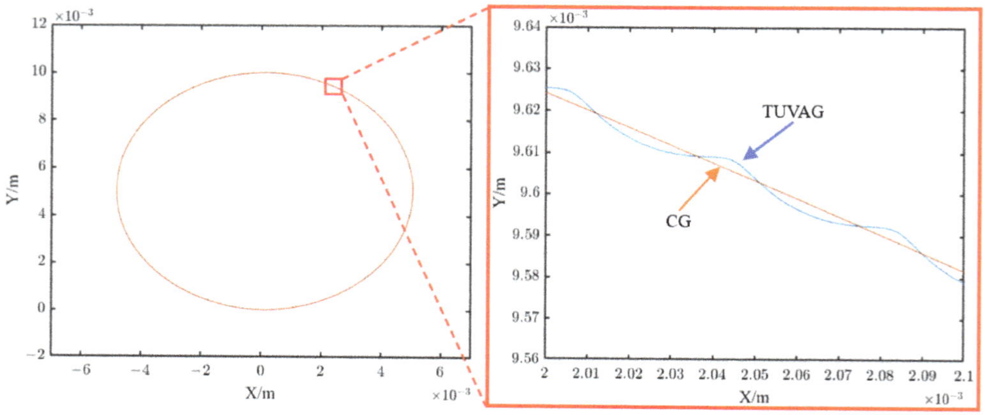

Figure 2. Comparison of abrasive trajectory between TUVAG and CG.

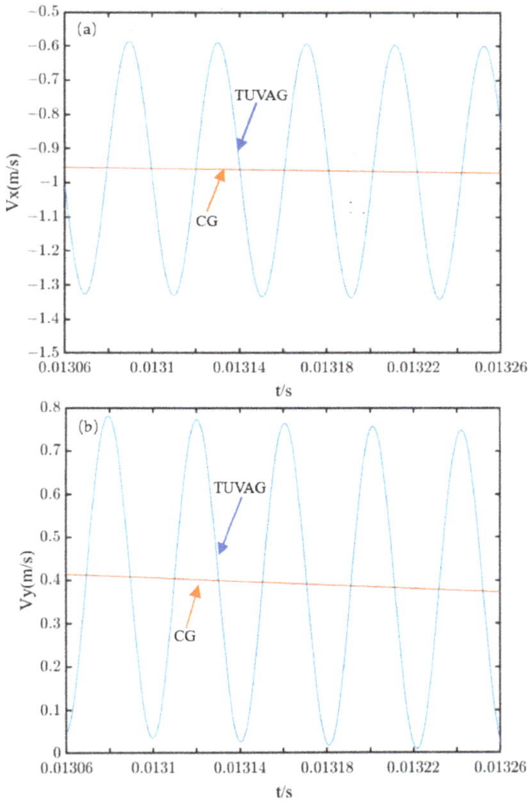

Figure 3. Comparison of abrasive velocity between TUVAG and CG (**a**) X direction (**b**) Y direction.

3. Theoretical Model of Grinding Force in TUVAG

Grinding force, as an important index in grinding process, needs to be studied emphatically. Therefore, the grinding force prediction model of TUVAG of C/SiC composites is developed in this paper. In the process of developing the grinding force model, firstly, the grinding removal mechanism of hard and brittle material such as C/SiC composites is considered, and the models of ductile removal stage and brittle removal stage are established, respectively. Then, considering the number of active abrasives in each stage, a comprehensive grinding force prediction model including the ductile removal stage and brittle removal stage is finally given.

3.1. Grinding Removal Mechanism of C/SiC Composites

Traditionally, the removal mechanism of hard and brittle material during grinding is brittle removal. With the deepening of research, more and more scholars have found that under the specific combination of grinding parameters, hard and brittle materials such as zirconia and monocrystalline silicon will also show ductile removal characteristics. At present, the mainstream view is that in grinding hard and brittle material, the undeformed chip thickness is a process from 0 to the maximum. If the value of the maximum undeformed chip thickness (h_{max}) is less than the value of the critical cutting depth (a_{gc}), the ductile removal of material is dominant, otherwise, the brittle removal of material is dominant [20], as shown in Figure 4. As far as critical cutting depth (a_{gc}) is concerned, it is a value determined by the properties of material themselves. Chen et al. [21] put forward an equation for calculating the critical cutting depth of hard and brittle material, as shown in Equation (5).

$$a_{gc}= cot\theta\sqrt{\frac{2^{\smile}}{\xi}(\frac{K_{IC}}{H_V})^2} \quad (5)$$

where θ is the half apex angle of abrasives; K_{ID} is the dynamic toughness of hard and brittle material. For hard and brittle material, the dynamic toughness is about 30% of the static toughness, that is, $K_{ID} = 0.3K_{IC}$; ξ is the geometric factor of the indenter, and the value is 1.8854 in paper [21]; λ is the comprehensive factor, and its value is (1–1.6) $\times$ 10^4; H_v is the Vickers hardness of the material. It has been pointed out that due to the existence of ultrasonic vibration, the surface hardness of the material decreases by about 30% [22], $H'_v = 0.7H_v$. The critical cutting depth of material in CG and TUVAG can be obtained in Equations (6) and (7):

$$a_{gc}= cot\theta\sqrt{\frac{2^{\smile}}{\xi}(\frac{K_{ID}}{H_v})^2} \quad (6)$$

$$a'_{gc}= cot\theta\sqrt{\frac{2^{\smile}}{\xi}(\frac{K_{ID}}{H'_v})^2} \quad (7)$$

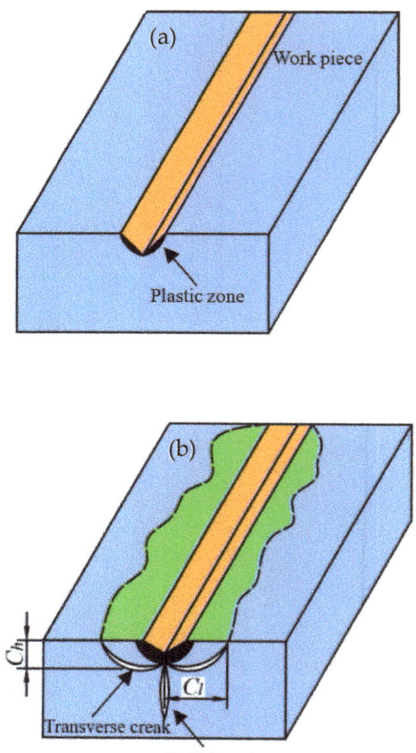

Figure 4. Removal model of hard and brittle material (**a**) Ductile removal model (**b**) Brittle removal model.

Comparing Equations (6) and (7), it can be concluded that the critical cutting depth in TUVAG is larger than that in CG due to the weakening effect of ultrasonic vibration on surface hardness. Under the same combination of machining parameters, the ductile removal of material can be easily realized during TUVAG.

According to the existing research results, the maximum undeformed cutting thickness during CG can be obtained in Equation (8) [23–25]:

$$h_{max} = \left(\frac{3}{Ctan\theta}\right)^{\frac{1}{2}}\left(\frac{v_w}{v_s}\right)^{\frac{1}{2}}\left(\frac{a_p}{2R}\right)^{\frac{1}{4}} \tag{8}$$

where C is the number of active abrasives per unit area; v_w is the feeding rate; v_s is the grinding velocity; a_p is the grinding depth.

To calculate the chip volume conveniently, the chip is treated as a triangular pyramid, and the material removal amount of a single abrasive during CG can be obtained in Equation (9):

$$V_{CG} = \frac{1}{3}h_{max}^2 tan\theta l_{CG} \tag{9}$$

Similarly, during TUVAG, the material removal amount of a single abrasive can be obtained in Equation (10):

$$V_{TUVAG} = \frac{1}{3}h_{max}'^2 tan\theta l_{TUVAG} \tag{10}$$

where, l_{CG} and l_{TUVAG} are the cutting length of a single abrasive during CG and TUVAG, respectively, which can be obtained in Equations (11) and (12).

$$l_{CG} = \int_0^t \sqrt{V_{XCG}^2 + V_{YCG}^2 + V_{ZCG}^2} \tag{11}$$

$$l_{TUVAG} = \int_0^t \sqrt{V_{XTUVAG}^2 + V_{YTUVAG}^2 + V_{ZTUVAG}^2} \tag{12}$$

where t is the cutting time of a single abrasive, and the relationship between t and a_p is shown in Equation (13):

$$a_p = R(1 - cos(2\pi nt) \tag{13}$$

In a complete grinding process of abrasives, the time t used is extremely short. It can be considered that t tends to 0, then $1 - \cos(2\pi nt) \approx (2\pi nt)^2/2$. According to Equation (13), the time used for a complete grinding of abrasives is as shown in Equation (14):

$$t = \frac{1}{\pi n}\sqrt{\frac{a_p}{2R}} \tag{14}$$

From a macro point of view, the final material removal volume of the two processing methods is the same, then $V_{CG} = V_{TUVAG}$, the calculation formula of the maximum undeformed chip thickness of TUVAG can be obtained in Equation (15),

$$h_{max}' = \left(\frac{l_{CG}}{l_{TUVAG}}\right)^{\frac{1}{2}}\left(\frac{3}{Ctan\theta}\right)^{\frac{1}{2}}\left(\frac{v_w}{v_s}\right)^{\frac{1}{2}}\left(\frac{a_p}{2R}\right)^{\frac{1}{4}} \tag{15}$$

By comparing Equations (6)–(8) and (15), it can be concluded that the TUVAG can increase the critical cutting depth of material and reduce the maximum undeformed chip thickness, thus making it easier to realize ductile removal of C/SiC composites.

3.2. Theoretical Model of Grinding Force in Ductile Removal Stage

As shown in Figure 5, when $h_{max} < a_{gc}$, according to the analysis of Section 3.1, the material is in a ductile removal state, and the removal of the material mainly depends on the extrusion and shear between the abrasives and the workpiece, and micro-cracks rarely occur. Referring to the removal mechanism of ductile materials, the removal volume and chip shape is determined by the maximum undeformed chip thickness and the shape of abrasives. As shown in Figure 6, the abrasive is approximately treated as a Vickers indenter. According to the definition of Vickers hardness, the relationship between the instantaneous normal grinding force of a single abrasive and the cutting depth can be obtained in Equation (16):

$$f_{nd} = 2htan\theta\sqrt{(tan^2\theta + 2)}H_v' \quad (16)$$

where h is the instantaneous chip thickness of a single abrasive during grinding.

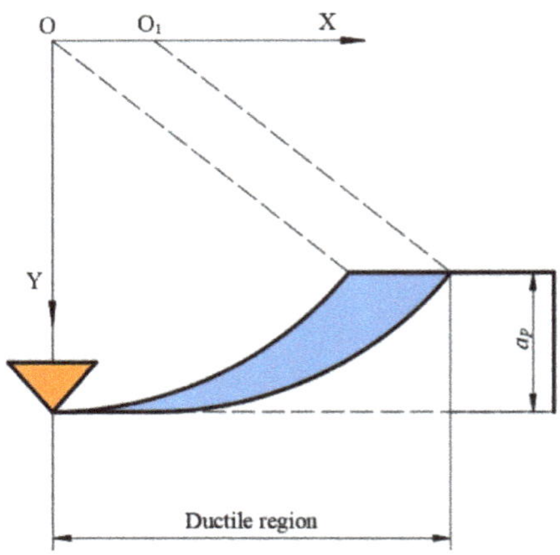

Figure 5. Diagram of chips at the ductile removal stage.

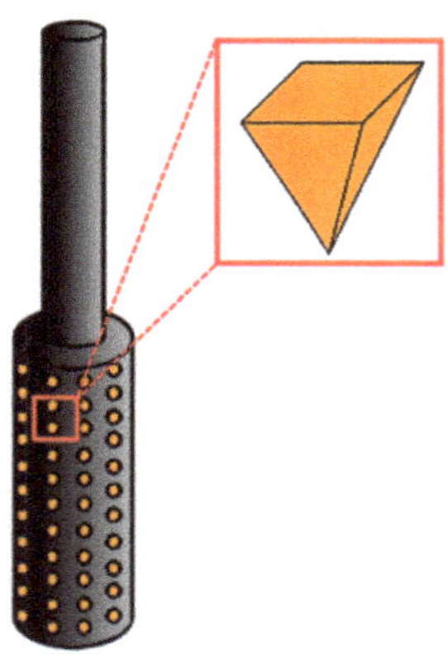

Figure 6. Diagram of grinding tool and abrasive.

For Equation (16), since the grinding depth h changes with the cutting time, the normal grinding force f_{nd} also changes at any time. To obtain the average normal grinding force in the whole ductile removal stage, the average cutting depth of the abrasive during grinding should be obtained. As shown in Figure 7, Xiao et al. [19], to obtain the average cutting depth of abrasives in the ductile removal stage, adopted the equivalent removal volume, that is, the triangular chip is equivalent to the triangular chip so that the removal volume of the two is equal. Based on this idea, the relationship between the average cutting depth and maximum cutting depth of abrasive was solved, as shown in Equation (17):

$$h'_{ave} = \frac{\sqrt{3}}{3}h'_{max} \quad (17)$$

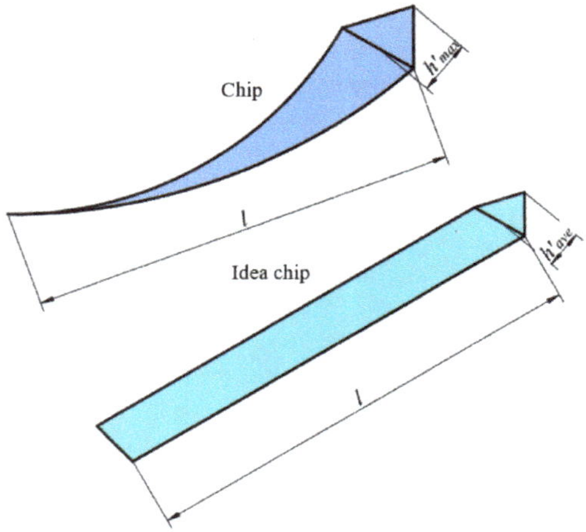

Figure 7. Idealized treatment of chip.

Equation (17) is substituted into Equation (16). When the material is in the condition of ductile removal, the normal grinding force of TUVAG of a single abrasive can be obtained in Equation (18):

$$f_{nd} = \frac{2}{3}h'_{max}tan\theta\sqrt{tan^2+2}H'_v \quad (18)$$

Considering the total number of abrasives N in the contact area between the grinding tool and the workpiece, the average normal grinding force in the ductile removal stage can be obtained in Equation (19):

$$F_{nd}= Nf_{nd} \quad (19)$$

For the value of N, refer to the calculation method proposed by Wang 26:

$$N = (\frac{1}{4}(\frac{C_a}{100})(\frac{6}{\pi D^3}))^{\frac{2}{3}}l_{TUVAG}a_p \quad (20)$$

where Ca is the abrasive concentration, Ca = 100%

3.3. Theoretical Model of Grinding Force in Brittle Removal Stage

As shown in Figure 8, when $h'_{max} > a_{gc}$, the material will undergo a process of ductile removal followed by brittle removal. Given that the normal grinding force at the ductile removal stage is given in Section 3.2, only the normal grinding force at the complete brittle stage are discussed here. With the grinding process, the contact stress between the workpiece and the abrasive gradually increases. According to the fracture mechanics, two main crack systems will be generated inside the material at this time: the radial crack system and the transverse crack system, as shown in Figure 4b. The radial crack is mainly related to the subsurface damage and strength reduction of the material, while the transverse crack is mainly related to the removal of the material and the formation mechanism of the surface. According to existing research results, for Vickers indenter, the depth C_h and length C_l of a transverse crack of hard and brittle materials can be obtained in Equations (21) and (22) [22,26]:

$$C_h= \xi_1(\frac{f_{nb}}{H'_v})^{\frac{1}{2}} \quad (21)$$

$$C_l= \xi_2(\frac{f_{nb}}{K_{ID}})^{\frac{3}{4}} \quad (22)$$

where ξ_1 and ξ_2 are the geometrical factors of the indenter. In fact, due to the existence of the fiber toughening phase, when the micro-cracks propagate in the material and reach the joint surface between the matrix and the fiber, the crack propagation will be temporarily prevented, and the micro-crack propagates again and deflects with the increasing grinding force, which makes the actual size of the micro-crack different from the theoretical size, and the actual calculation is extremely difficult because of the randomness of the micro-crack propagation. We assume that K_1 and K_2 are proportional coefficients, representing the impact of the fiber toughening phase on microcrack propagation. For C/SiC composites, Equations (21) and (22) can be modified as follows:

$$C_h = K_1\xi_1\left(\frac{f_{nb}}{H_v'}\right)^{\frac{1}{2}} \tag{23}$$

$$C_l = K_2\xi_2\left(\frac{f_{nb}}{K_{ID}}\right)^{\frac{3}{4}} \tag{24}$$

where, f_{nb} is the normal grinding force of a single abrasive in the complete brittle removal stage, and H_v' and K_{ID} are the surface hardness and fracture toughness of materials when there are ultrasonic vibration effects.

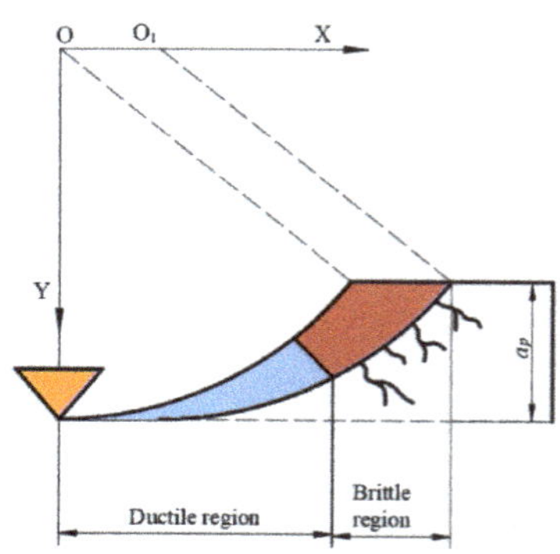

Figure 8. Diagram of chips at the brittle removal stage.

The removal volume of a single abrasive can be obtained in Equation (25)

$$V = 2C_lC_hl_{TUVAG} \tag{25}$$

From the microscopic perspective, the material removal rate (MMR) can be obtained in Equation (26):

$$\text{MMR} = \frac{NV}{t} \tag{26}$$

From the macro perspective, MMR can be obtained in Equation (27):

$$MMR = a_pbv_w \tag{27}$$

where, b is the width of the grinding area. In theory, the values of Equations (26) and (27) should be equal. Equations (23)–(25) should be substituted into Equation (26) and made equal to Equations (26) and (27). The normal grinding force of a single abrasive in the complete brittle removal stage can be obtained in Equation (28).

$$f_{nb} = \left(\frac{a_pbv_wtK_{ID}^{\frac{3}{4}}H_v'^{\frac{1}{2}}}{2K_1K_{2,1,2}l_{TUVAG}N}\right)^{\frac{4}{5}} \tag{28}$$

Since ξ_1, ξ_2, K_1, K_2 are constants related to indenter geometric parameters and material properties, in order to simplify the equation, let $K = \xi_1\xi_2K_1K_2$; Equation (28) can be simplified as:

$$f_{nb}= (\frac{a_p^{\frac{3}{2}} bv_w R^{\frac{1}{2}} K_{ID}^{\frac{3}{4}} H_v'^{\frac{1}{2}}}{\sqrt{2}Kl_{TUVAG}v_s N})^{\frac{4}{5}} \tag{29}$$

In this stage, the average normal grinding force can be obtained in Equation (30):

$$F_{nb}= N_b f_n \tag{30}$$

where, N_b is the number of active abrasives in the brittle removal stage during grinding. Since it is assumed that the size and shape of abrasives are consistent and uniformly distributed on the grinding tool, the active abrasive number N_d involved in ductile region and the active abrasive number N_b involved in the brittle region can be obtained in Equations (31) and (32):

$$N_d = \frac{a'_{gc}}{h'_{max}} N \tag{31}$$

$$N_b = \frac{\left(h'_{max} - a'_{gc}\right)}{h'_{max}} N \tag{32}$$

In Section 3.2, we have obtained the average grinding force in the ductile removal stage, combined with the average grinding force in the complete brittle stage and the number of active abrasives in each stage obtained in Section 3.3, we have obtained the calculation equation of the average grinding force in the whole brittle removal process.

$$F_n = \frac{2}{3} a_{gc}'^2 tan\theta \sqrt{tan^2 + 2} H_v' N_d + N_b (\frac{a_p^{\frac{3}{2}} bv_w R^{\frac{1}{2}} K_{ID}^{\frac{3}{4}} H_v'^{\frac{1}{2}}}{\sqrt{2}Kl_{TUVAG}v_s N})^{\frac{4}{5}} \tag{33}$$

3.4. Theoretical Model of Comprehensive Grinding Force

The removal mechanism of materials, the average normal grinding force at the ductile removal stage and the average normal grinding force at the brittle removal stage have been studied above. Therefore, by synthesizing the above, the grinding force model of TUVAG of C/SiC composites was obtained, as shown in Equation (34):

$$F_n = \begin{cases} \frac{2}{3} h_{max}'^2 tan\theta \sqrt{tan^2 + 2} H_v' N & \left(h'_{max} \le a'_{gc}\right) \\ \frac{2}{3} a_{gc}'^2 tan\theta \sqrt{tan^2 + 2} H_v' N_d + N_b (\frac{a_p^{\frac{3}{2}} bv_w R^{\frac{1}{2}} K_{ID}^{\frac{3}{4}} H_v'^{\frac{1}{2}}}{\sqrt{2}Kl_{TUVAG}v_s N})^{\frac{4}{5}} & \left(h'_{max} > a'_{gc}\right) \end{cases} \tag{34}$$

During the grinding process, normal grinding force and tangential grinding force have a linear relationship:

$$F_t = {}^{-}F_n \tag{35}$$

where, μ is friction coefficient.

4. Experiment Equipment and Condition

4.1. Experiment Equipment

The grinding machine is TC500R high-precision vertical machining center, the maximum spindle speed is 20,000 r/min, the two-dimensional ultrasonic vibration device is fixed below the fixture, the frequency chosen in the test is 24,600 Hz, and the vibration direction is tangential and radial, as shown in Figures 9 and 10. The grinding tool used in the test is single layer electroplated diamond grinding tool with a diameter of 10 mm, and an abrasive size of 200 #, as shown in Figure 11. The dynamometer is Kistler 9257 dynamometer, and the sampling frequency is 7000 Hz. The cutting force is collected by Kistler 9257 dynamometer and processed by Dynoware software to obtain the average grinding force in the grinding process, as shown in Figures 9 and 10.

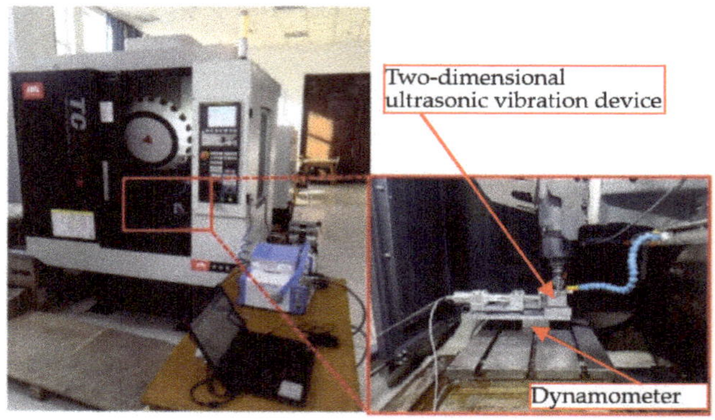

Figure 9. Experiment equipment.

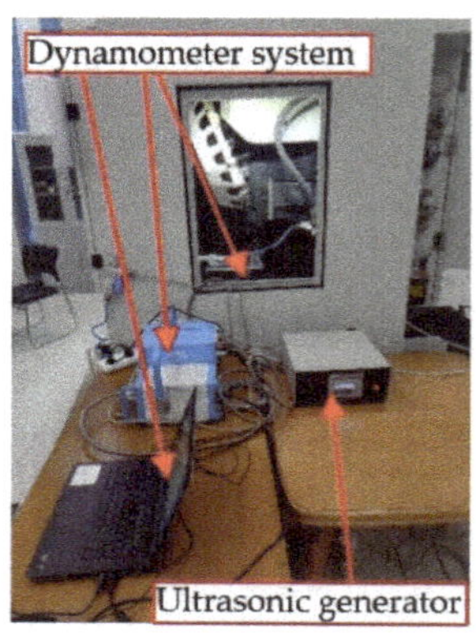

Figure 10. Kistler 9257 Dynamometer.

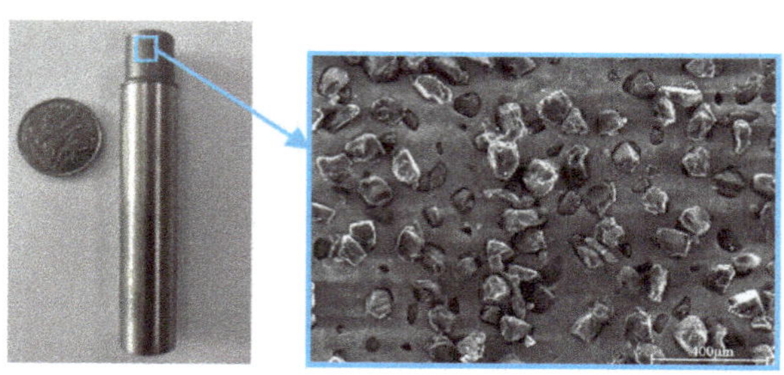

Figure 11. Micro morphology of grinding tool.

4.2. Experiment Material and Scheme

In this experiment, the experimental material used is 2.5D-C/SiC composite material, the size of 8 mm × 10 mm, the mechanical properties of 2.5D-C/SiC composite are shown in Table 1. The fiber structure composition is shown in Figure 12, including radial fibers, weft fibers, and a needle punched structure perpendicular to these two fibers.

Table 1. Mechanical properties of the 2.5D-C/SiC composite.

Parameter	2.5D-C/SiC Composite
Density (g/cm^3)	1.9
Fracture Toughness (Mpa.m$^{1/2}$)	10
Bending strength/MPa	240–300
Tensile strength/MPa	75–100
Compressive strength/MPa	420–500
Shear strength/MPa	15–25
Vickers hardness/GPa	25.5
Temperature resistance/vacuum °C	1600

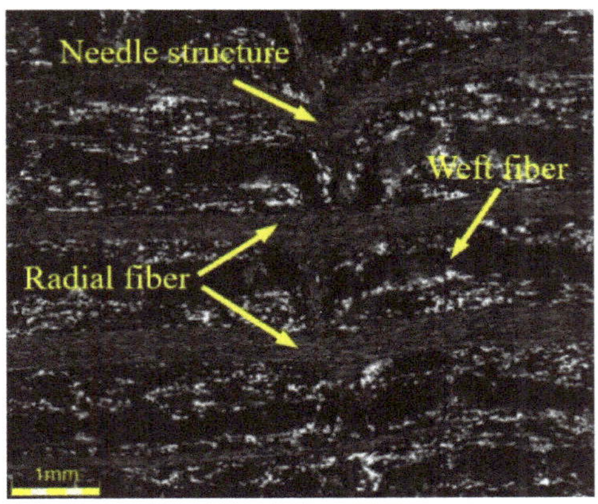

Figure 12. Micromorphology of 2.5D-C/SiC composite.

As shown in Table 2, to verify the correctness of the grinding force model of TUVAG and analyze the impact of grinding parameters (grinding velocity v_s, feeding rate v_w and grinding depth a_p) and ultrasonic parameters (ultrasonic amplitude A) on grinding force, the single factor experiment was carried out. To facilitate the analysis of experiment data, the ultrasonic amplitude was adjusted to make it consistent in the X and Y directions, therefore, the ultrasonic amplitude values in the two directions are no longer listed in Table 2. At the same time, the contrast experiment was carried out as the control to analyze the difference between CG and TUVAG. In the contrast experiment, the grinding parameters were consistent with in the single factor experiment, but the ultrasonic amplitude was set to 0. Therefore, the tables of the contrast experiment scheme were not listed in this paper.

Table 2. Single factor experiment scheme.

Serial Number	v_s (m/s)	v_w (mm/min)	a_p (mm)	A (μm)
1	6.28	50	0.02	3.7
2	6.28	250	0.02	3.7
3	6.28	450	0.02	3.7
4	6.28	650	0.02	3.7
5	1.00	250	0.02	3.7
6	3.66	250	0.02	3.7
7	8.90	250	0.02	3.7
8	6.28	250	0.01	3.7
9	6.28	250	0.03	3.7
10	6.28	250	0.04	3.7
11	6.28	250	0.02	2.4
12	6.28	250	0.02	5.3
13	6.28	250	0.02	6.6

5. Experimental Result and Discussion

5.1. Obtaining the Proportional Coefficient K

In the above section, $K = \xi_1\xi_2K_1K_2$. Here, the value of K was discussed.

Lawn et al. [27] obtained the calculation equations of microcrack size for hard and brittle material through research. Wang et al. [26] idealized the abrasives according to the Vickers indenter, and finally obtained the microcrack calculation equations of Equations (21) and (22). ξ_1, ξ_2 as the geometric factor of indenter, the specific value is only related to the shape of the abrasive, and when the shape of the abrasive is determined, ξ_1, ξ_2 are a fixed value. In the above hypothesis, K_1 and K_2 are proportional coefficients, indicating the impact of the presence of the fiber toughening phase on microcrack propagation. Theoretically, their values are only related to the structure of 2.5D-C/SiC composites.

Therefore, in this test, the above four proportional coefficients are all definite values, so the proportional coefficient K should also be a definite value. To obtain the value of K, the critical cutting depth of CG and TUVAG was obtained according to Equations (6) and (7), and the maximum undeformed chip thickness corresponding to each machining parameters combination was obtained according to Equations (8) and (15), from which the machining parameters combination under brittle removal was selected to obtain the grinding force at this time. Multiple groups of related grinding parameters are substituted into Equation (33) to obtain the value of K, which is about 1.80 in this test.

5.2. Obtaining the Friction Coefficient μ

Theoretically, tangential grinding force and normal grinding force should have a linear relationship. Therefore, the single factor test results fitting was carried out, as shown in Figure 13. Then, the friction coefficient μ is about 0.221, and the tangential force can be expressed as:

$$F_t = 0.221F_n \tag{36}$$

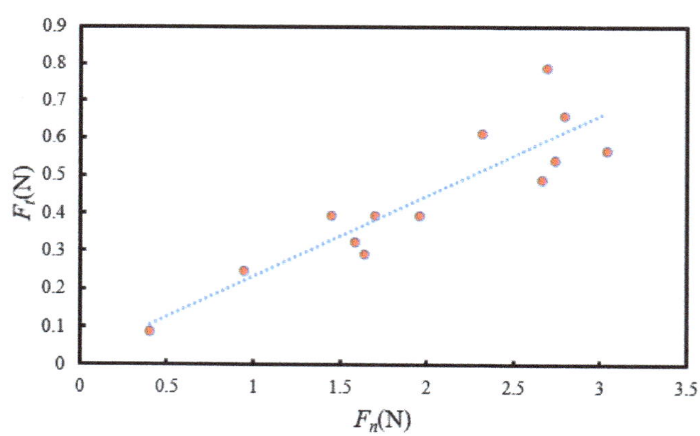

Figure 13. Experiment value and fitting value for tangential grinding force.

5.3. Impact of Machining Parameters on Grinding Force

5.3.1. Impact of Grinding Velocity on Grinding Force

The impact of grinding velocity on normal grinding force and tangential grinding force is shown in Figures 14 and 15. As can be concluded from these figures, when the grinding velocity increases from 1.00 m/s (n = 2000 r/min) to 8.90 m/s (n = 17,000 r/min), the normal and tangential grinding force in TUVAG and CG have an overall downward trend. The grinding force of TUVAG is smaller than that of CG. This situation is analyzed.

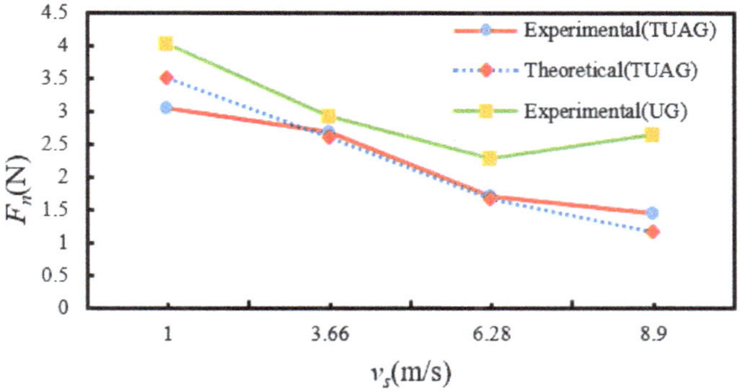

Figure 14. Impact of grinding velocity on the normal grinding force.

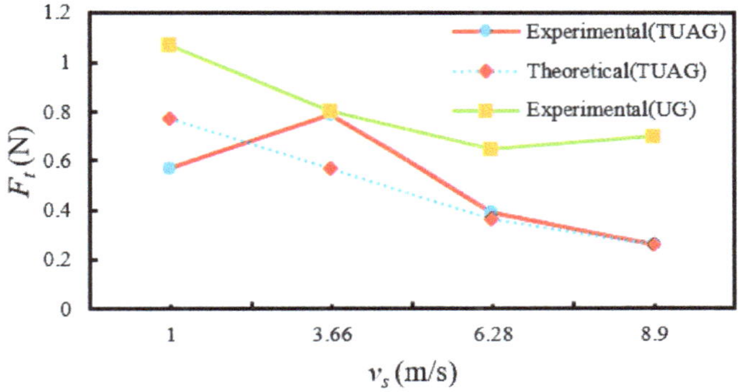

Figure 15. Impact of grinding velocity on the tangential grinding force.

When the feeding rate and grinding depth are constant, according to the above analysis, with the increase of grinding velocity, the maximum undeformed chip thickness of a single abrasive decrease. At the same time, during the cutting process, the cutting length of a single abrasive increases. According to Equation (20), the number of effective abrasives in the contact area between the workpiece and the grinding tool increases, which means that the material removal volume of abrasives decrease and the chip thickness of abrasives decrease. In addition, the increase in grinding velocity also means that the cutting time t of single abrasive decreases, which reduces the impact force of the abrasive on the workpiece, improves the chip removal conditions between the workpiece and the abrasive, slows down the wear of the abrasive, and better maintains the good cutting conditions. However, by comparing Figures 13 and 14, it can be seen that during CG, when the grinding velocity is 8.90 m/s, the tangential grinding force and normal grinding force will show an upward trend, which is speculated to be because the spindle speed is 17,000 r/min. It is close to the limit speed of the machine tool spindle (the limit speed is 20,000 r/min). In addition, the radial runout of the machine tool spindle, the vibration of the machine tool itself and other factors lead to the increase in grinding force.

5.3.2. Impact of Feeding Rate on Grinding Force

The impact of the feeding rate on normal grinding force and tangential grinding force is shown in Figures 16 and 17. It can be concluded from these figures, when the feeding rate increases from 50 mm/min to 650 mm/min, the normal and tangential grinding force in TUVAG and CG both show an upward trend, and the grinding force in TUVAG are significantly less than those in conventional grinding. This situation is analyzed.

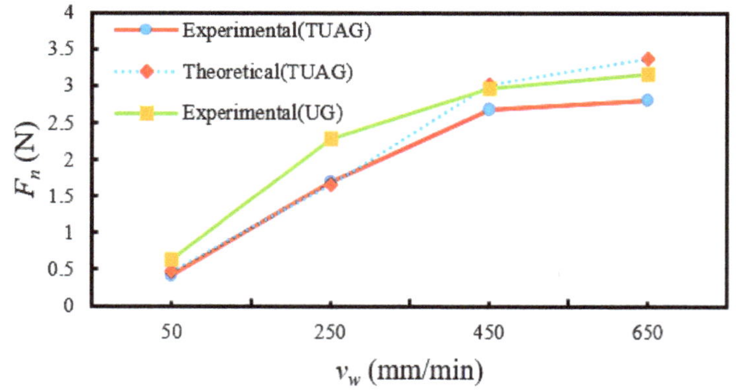

Figure 16. Impact of feeding rate on the normal grinding force.

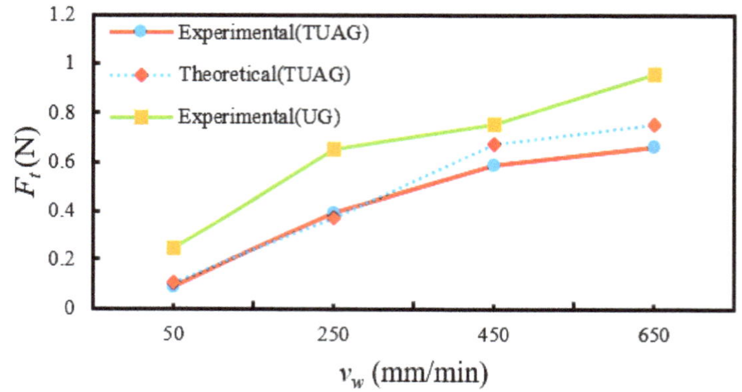

Figure 17. Impact of feeding rate on the tangential grinding force.

When the grinding velocity and grinding depth are constant, when the feeding rate is low, according to Equation (8), the maximum undeformed chip thickness of a single abrasive is low, which means that the ductile removal mode accounts for a large proportion in the material removal process at this time, reducing the overall grinding force. At the same time, the lower feeding rate means that the number of active abrasives that can participate in the grinding per unit time is higher, the chip thickness of a single abrasive is reduced, and the chips can be more smoothly discharged with the rotation of the grinding tool, maintain good heat dissipation condition, and reduce the wear degree of abrasives. Under the impact of the above conditions, the grinding force value is lower. When the feeding rate continues to increase, the above conditions continue to deteriorate, and because the grinding chips are difficult to discharge smoothly, the gap between the grinding abrasives on the grinding tool is blocked by the chips, making it difficult for the grinding tool to maintain the normal cutting process, the grinding heat accumulation gradually, resulting in the increase in grinding force.

5.3.3. Impact of Grinding Depth on Grinding Force

The impact of grinding depth on normal grinding force and tangential grinding force is shown in Figures 18 and 19. It can be concluded from these figures that when the grinding depth increases, the normal and tangential grinding force of TUVAG and CG both show an increasing trend. This situation is analyzed.

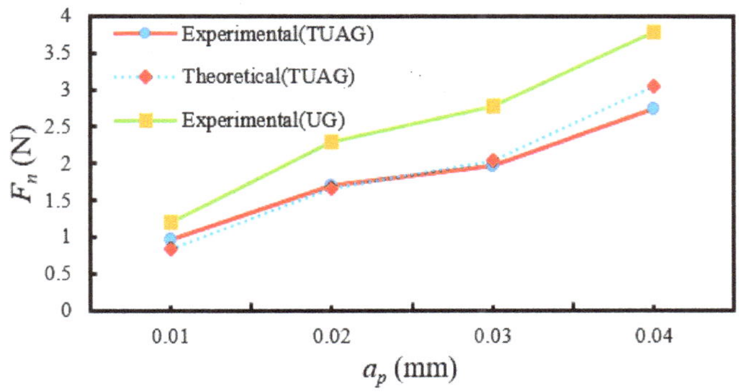

Figure 18. Impact of grinding depth on the normal grinding force.

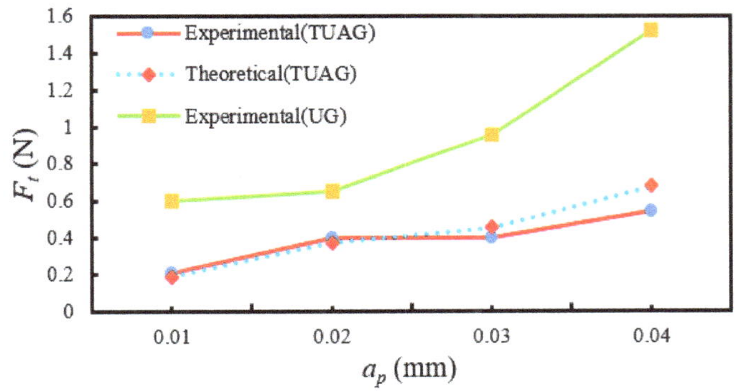

Figure 19. Impact of grinding depth on the tangential grinding force.

It can be concluded from the above analysis that the grinding depth mainly affects the cutting time t of the abrasives, the maximum undeformed chip thickness h_{max} and other important parameters. Therefore, when the grinding depth is low, the cutting time t of abrasives is shorter, and the cutting heat is not easy to accumulate, and the maximum undeformed chip thickness is reduced, and the material is mainly plastic removed. At the same time, when the grinding depth is low, the microscopic separation between a single abrasive and the workpiece caused by radial vibration is more obvious, which further reduces the impact between a single abrasive and the workpiece, making the overall grinding at this time lower. With the increase in grinding depth, the cutting time t of abrasives increases and the impact of abrasives on the workpiece increases. In addition, the maximum undeformed chip thickness of a single abrasive is larger than the critical cutting depth of the workpiece. The material is dominated by brittle removal. According to Equation (34), the grinding force is proportional to the grinding depth during brittleness removal, making the grinding force increase continuously.

5.3.4. Impact of Ultrasonic Vibration on Grinding Force

By comparing Figures 14–19, it can be concluded that the grinding force in CG is significantly higher than that in TUVAG, and it can be concluded from Figure 19 that with the increase of ultrasonic amplitude, the grinding force shows a downward trend. The impact of ultrasonic vibration and ultrasonic amplitude on the grinding force was analyzed.

It can be concluded from the above analysis that, from a microscopic point of view, the cutting length of a single abrasive in TUVAG is larger than that in CG. Under the same grinding parameter combination, the maximum undeformed chip thickness of a single abrasive in TUVAG is smaller than that in CG. Moreover, due to the weakened effect of ultrasonic vibration on the surface hardness of material, the surface hardness of material decreases and the critical cutting depth of the material increases. Therefore, under the same grinding parameter combination, it is easier to achieve the ductile removal of material during TUVAG, which not only obtains better surface quality but also reduces the grinding force during machining. In addition, due to the existence of ultrasonic vibration, there is always a small impact between the abrasive and the workpiece, the long and continuous chips into short but not continuous chips, reducing the friction coefficient between the grinding tool and the workpiece, promote the chip discharge, slow down the wear of abrasives, and further reduce the grinding force.

The impact of ultrasonic amplitude on grinding force is shown in Figures 20 and 21. It can be concluded from these figures that with the increase of ultrasonic amplitude, normal grinding force and tangential grinding force show a downward trend overall. This is because when the ultrasonic amplitude increases, the cutting length of a single abrasive further increases, which makes the maximum undeformed chip thickness of a single abrasive decrease; the ductile removal mode accounts for a large proportion of the material removal process, and the grinding force decreases. At the same time, due to the increase in ultrasonic amplitude, the micro-impact between the abrasives and the workpiece is increased, resulting in a large amount of micro-cracks on the surface. The micro-cracks are interleaved with each other, which promotes the removal of material and further reduces the grinding force.

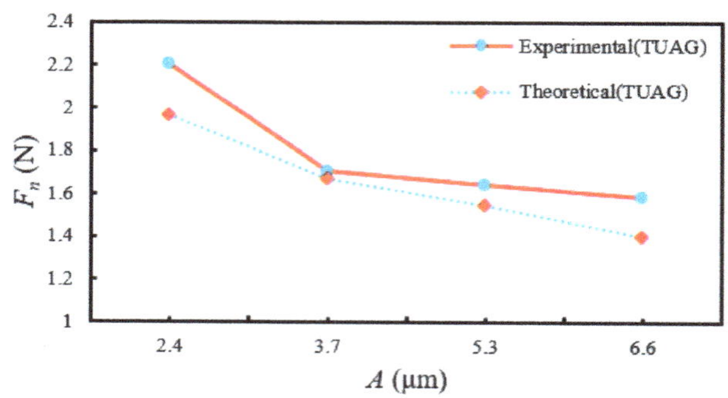

Figure 20. Impact of ultrasonic amplitude on the normal grinding force.

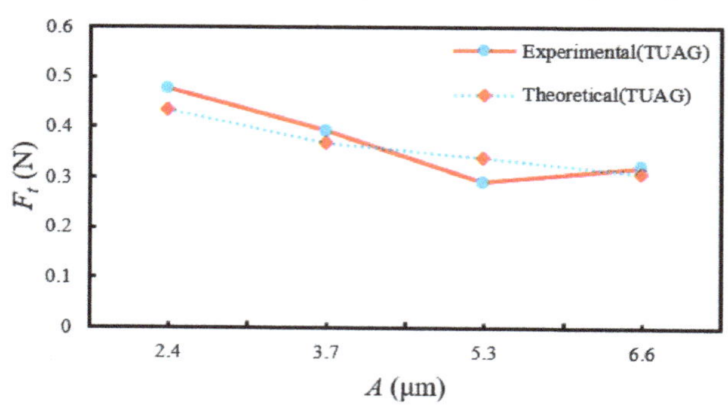

Figure 21. Impact of ultrasonic amplitude on the tangential grinding force.

5.4. Verification of Grinding Force Model

The errors between theoretical values and measured values corresponding to different ultrasonic parameters and grinding parameters are calculated, as shown in Table 3. It can be concluded from the table that the maximum relative error of normal grinding force is 11.99%, the minimum relative error is 4.45% and the average relative error is 8.49%. The maximum relative error of tangential grinding force is 18.17%, the minimum relative error is 8.91% and the average relative error is 13.59%. It can be seen that the relative errors are all within 20%, indicating that the model has certain feasibility. At the same time, it is concluded that the error value of tangential grinding force is large, which may be because the friction coefficient μ between the grinding tool and the workpiece in TUVAG is obtained by fitting the test data. Due to the small sample size of the test data, the fitting friction coefficient μ has a certain error compared with the real friction coefficient. The relative error of the tangential grinding force is larger.

Table 3. Relative errors between theoretical and measured values of grinding force.

Relative Errors (%)	v_s (m/s)	v_w (mm/min)	a_p (mm)	A (μm)	Average Error
F_n	9.98	11.99	4.45	7.55	8.49
F_t	18.17	13.40	13.88	8.91	13.59

6. Conclusions

Based on the maximum undeformed chip thickness of a single abrasive, the removal mechanism of 2.5D-C/SiC composites during grinding was analyzed. From the perspectives of ductile removal and brittle removal, the grinding force model of TUVAG for 2.5D-C/SiC composites was developed. Through the single factor test, the impact law of ultrasonic parameters and grinding parameters on grinding force and the feasibility of grinding force model were analyzed, and the following conclusions were drawn:

(1) The relationship between the critical cutting depth and the maximum undeformed chip thickness of a single abrasive determines the material removal mechanism. If the critical cutting depth is greater than the maximum undeformed chip thickness of a single abrasive, the material removal mode is ductile removal. On the contrary, the material removal mode is mainly brittle removal.

(2) During the grinding process, when the feed rate and grinding depth increase, the grinding force increases. When the grinding velocity and ultrasonic amplitude increase, grinding force decreases.

(3) Ultrasonic vibration can reduce the surface hardness of C/SiC composites, increase the critical cutting depth, and reduce the maximum undeformed chip thickness of a single abrasive. Therefore, compared with CG, the grinding force of TUVAG decreases obviously.

(4) The grinding force calculated by the theoretical grinding force model are consistent with the experimental value. The average relative error of the normal grinding force is 8.49% and the average relative error of the tangential grinding force is 13.59%. Therefore, the model can be used to predict the grinding force of TUVAG of C/SiC composites.

Author Contributions: Formal analysis, C.T. and S.J.; funding acquisition, Y.Z.; project administration, Y.Z.; validation, G.Y.; writing—original draft, C.T.; writing—review and editing, Y.Z., L.M. and Y.G. All authors have read and agreed to the published version of the manuscript.

Funding: This research is funded by the National Natural Science Foundation of China: 51905083, 51975113; the Natural Science Foundation of Hebei Province: E2022501004, E2021501027; the Fundamental Research Funds for the Central Universities: N2123025, N2203015, and the China Postdoctoral Science Foundation: 2021MD703912.

Data Availability Statement: The data and materials that support the findings of this study are available from the corresponding author upon reasonable request.

Conflicts of Interest: The authors declare no conflict of interest.

References

1. Qu, S.S.; Gong, Y.D.; Yang, Y.Y. An investigation of carbon nano fluid minimum quantity lubrication for grinding unidirectional carbon fibre-reinforced ceramic matrix composites. *J. Clean. Prod.* **2020**, *249*, 119353. [CrossRef]
2. Li, C.; Piao, Y.; Meng, B.; Hu, Y.; Li, L.; Zhang, F. Phase transition and plastic deformation mechanisms induced by self-rotating grinding of GaN single crystals. *Int. J. Mach. Tools Manuf.* **2022**, *172*, 103827. [CrossRef]

3. Li, C.; Piao, Y.C.; Zhang, F.H.; Zhang, Y.; Hu, Y.X.; Wang, Y.F. Understand anisotropy dependence of damage evolution and material removal during nanoscratch of MgF_2 single crystals. *Int. J. Extrem. Manuf.* **2023**, *5*, 015101. [CrossRef]
4. Ding, K.; Fu, Y.C.; Su, H.H. Study on surface/subsurface breakage in ultrasonic assisted grinding of C/SiC composites. *Int. J. Adv. Manuf. Technol.* **2017**, *91*, 3095–3105. [CrossRef]
5. Li, Z.; Jiao, Y.; Deines, T.W.; Pei, Z.J.; Treadwell, C. Rotary ultrasonic machining of ceramic matrix composites: Feasibility study and designed experiments. *Int. J. Mach. Tools Manuf.* **2005**, *45*, 1402–1411. [CrossRef]
6. Ramulu, M.; Jenkins, M.G.; Guo, Z. Abrasive water jet machining mechanisms incontinuous-fiber ceramic composites. *J. Compos. Technol. Res.* **2001**, *23*, 82–91.
7. Sun, Y.; Jin, L.Y.; Gong, Y.D. Experimental evaluation of surface generation and force time-varying characteristics of curvilinear grooved micro end mills fabricated by EDM. *Manuf. Process.* **2022**, *73*, 799–814. [CrossRef]
8. Li, C.; Hu, Y.X.; Zhang, F.H. Molecular dynamics simulation of laser assisted grinding of GaN crystals. *Int. J. Mech. Sci.* **2023**, *239*, 107856. [CrossRef]
9. Ding, K.; Fu, Y.C.; Su, H.H. Study on Matching Performance of Ultrasonic Vibration and Grinding Parameters Based on a Single Abrasive Grinding. *J. Mech. Eng.* **2017**, *19*, 59–65. [CrossRef]
10. Cao, J.G.; Nie, M.; Liu, Y.M.; Li, J.Y. Ductile-brittle transition behavior in the ultrasonic vibration-assisted internal grinding of silicon carbide ceramics. *Int. J. Adv. Manuf. Technol.* **2018**, *96*, 3251–3256. [CrossRef]
11. Chen, J.; An, Q.L.; Ming, W.W.; Chen, M. Investigation on machined surface quality in ultrasonic-assisted grinding of Cf/SiC composites based on fracture mechanism of carbon fibers. *Int. J. Adv. Manuf. Technol.* **2020**, *109*, 1583–1599. [CrossRef]
12. Guo, B.; Zhao, Q.L. Ultrasonic vibration assisted grinding of hard and brittle linear micro-structured surfaces. *Precis. Eng.* **2016**, *48*, 98–106. [CrossRef]
13. Cao, J.G.; Wu, Y.B.; Lu, D.; Fujimoto, M.; Nomura, M. Material removal behavior in ultrasonic-assisted scratching of SiC ceramics with a single diamond tool. *Int. J. Mach. Tools Manuf.* **2014**, *79*, 49–61. [CrossRef]
14. Liang, Z.Q.; Wang, X.B.; Wu, Y.B.; Xie, L.J.; Jiao, L.; Zhao, W.X. Experimental study on brittle–ductile transition in elliptical ultrasonic assisted grinding (EUAG) of monocrystal sapphire using single diamond abrasive grain. *Int. J. Mach. Tools Manuf.* **2013**, *7*, 41–51. [CrossRef]
15. Yan, Y.Y.; Zhang, Z.Q.; Zhao, B.; Liu, J.L. Study on prediction of three-dimensional surface roughness of nano-ZrO_2 ceramics under two-dimensional ultrasonic-assisted grinding. *Int. J. Adv. Manuf. Technol.* **2021**, *112*, 2623–2638. [CrossRef]
16. Sun, G.Y.; Zhao, L.L.; Ma, Z.; Zhao, Q.L. Force prediction model considering material removal mechanism for axial ultrasonic vibration-assisted peripheral grinding of Zerodur. *Int. J. Adv. Manuf. Technol.* **2018**, *98*, 2775–2789. [CrossRef]
17. Yang, Z.C.; Zhu, L.D.; Lin, B.; Zhang, G.X.; Ni, C.B.; Sui, T.Y. The grinding force modeling and experimental study of ZrO_2 ceramic materials in ultrasonic vibration assisted grinding. *Int. J. Mech. Sci.* **2019**, *45*, 8873–8889. [CrossRef]
18. Li, C.; Zhang, F.H.; Meng, B.B. Material removal mechanism and grinding force modelling of ultrasonic vibration assisted grinding for SiC ceramics. *Ceram. Int.* **2017**, *43*, 2981–2993. [CrossRef]
19. Xiao, X.Z.; Zheng, K.; Liao, W.H.; Meng, H. Study on cutting force model in ultrasonic vibration assisted side grinding of zirconia ceramics. *Int. J. Mach. Tools Manuf.* **2016**, *104*, 58–67. [CrossRef]
20. Cheng, J.; Gong, Y.D. Experimental study of surface generation and force modeling in micro-grinding of single crystal silicon considering crystallographic effects. *Int. J. Mach. Tools Manuf.* **2014**, *77*, 1–15. [CrossRef]
21. Chen, M.J.; Zhao, Q.L.; Dong, S.; Li, D. The critical conditions of brittle–ductile transition and the factors influencing the surface quality of brittle materials in ultra-precision grinding. *J. Mater. Process. Technol.* **2005**, *168*, 75–82. [CrossRef]
22. Wang, Y.; Lin, B.; Wang, S.L.; Cao, X.Y. Study on the system matching of ultrasonic vibration assisted grinding for hard and brittle materials processing. *Int. J. Mach. Tools Manuf.* **2014**, *77*, 66–73. [CrossRef]
23. Dai, C.W.; Ding, W.F.; Zhu, Y.J.; Xu, J.H.; Yu, H.W. Grinding temperature and power consumption in high speed grinding of Inconel 718 nickel-based superalloy with a vitrified CBN wheel. *Precis. Eng.* **2018**, *52*, 192–200. [CrossRef]
24. Mao, C.; Liang, C.; Zhang, Y.C.; Zhang, M.J.; Hu, Y.L.; Bi, Z.M. Grinding characteristics of cBN-WC-10Co composites. *Ceram. Int.* **2017**, *43*, 16539–16547. [CrossRef]
25. Zhang, X.H.; Kang, Z.X.; Li, S.; Shi, Z.Y.; Wen, D.D.; Jiang, J.; Zhang, Z.C. Grinding force modelling for ductile-brittle transition in laser macro-micro-structured grinding of zirconia ceramics. *Ceram. Int.* **2019**, *45*, 18487–18500. [CrossRef]
26. Wang, S.F. *Research on Grinding Mechanism of Elliptical Ultrasonic Assisted Diamond Grinding Wheel for K9 Glass*; Northeastern University: Shenyang, China, 2021.
27. Marshall, D.; Lawn, B.; Evans, A. Elastic/plastic indentation damage in ceramics:the lateral crack system. *J. Am. Ceram. Soc.* **1982**, *65*, 561–566. [CrossRef]

crystals

MDPI

Article

Ultraviolet Nanosecond Laser-Ablated Groove Analysis of 2.5D C_f/SiC Composites

Tangyong Zhang [1], Fei Liu [1], Yao Liu [2], Chongjun Wu [1,*] and Steven Y. Liang [3]

1 College of Mechanical Engineering, Donghua University, Shanghai 201620, China
2 Shanxi Key Laboratory of Advanced Manufacturing Technology, North University of China, Taiyuan 030051, China
3 Manufacturing Research Center, Georgia Institute of Technology, Atlanta, GA 30332, USA
* Correspondence: wcjunm@dhu.edu.cn; Tel.: +86-18817334066

Abstract: The 2.5D C_f/SiC composite is a typical heterogeneous material with the characteristics of anisotropy, which makes it difficult to predict the size and damage removed by the traditional contact removal process. This paper adopted the ultraviolet nanosecond laser to ablate the C_f/SiC composites by considering the heterogeneous structure's effect. The ablated groove topography and size prediction are effective in revealing the machined quality with predictable groove sizes. The effects of laser processing parameters on the groove morphology and surface thermally affected zone are investigated with the thermal removal mechanism. A regression model is established by considering the scanning times, scanning speed, laser power and pulse width as the main variables. In the regression models, the relative error values are all below 10%. It is revealed that the groove width diminishes with the scanning speed and increases as the laser power increases. However, the influence of the scanning times and pulse width is small, and the overall variation range is within ±10 μm. The results show that the arrangement direction of carbon fibers has an impact on laser processing, especially when the pulse width is 0.25 μs, upon which the opposite change occurs. Carbon fiber grooves are not obvious and are barely observed in the laser processing of the parallel carbon fiber direction, and the grooves are slightly uneven. This study could be helpful in analyzing the grooves of C_f/SiC affected by the laser processing process, which could support the hybrid machining of the C_f/SiC composites.

Keywords: 2.5-dimensional C_f/SiC composites; nanosecond laser; groove morphology; thermally affected zone

Citation: Zhang, T.; Liu, F.; Liu, Y.; Wu, C.; Liang, S.Y. Ultraviolet Nanosecond Laser-Ablated Groove Analysis of 2.5D C_f/SiC Composites. *Crystals* **2023**, *13*, 223. https://doi.org/10.3390/cryst13020223

Academic Editor: Jing Guo

Received: 13 November 2022
Revised: 13 January 2023
Accepted: 18 January 2023
Published: 25 January 2023

1. Introduction

Ceramic matrix composites (CMCs) are widely used in high-temperature, wear resistance applications because of their extremely high impact and fatigue resistance performance [1–3]. Generally, the Carbon Fiber-reinforced Silicon Carbide ceramics (C_f/SiC) composites are composed of carbon fiber reinforcement, silicon carbide ceramic matrix material and a pyrolytic carbon interface [4]. Compared with traditional structural ceramics or CFRP material, the C_f/SiC composites have demonstrated improved mechanical properties in engineering applications [5–8], such as a high strength, light mass, high flexural resistance, etc.

However, the heterogeneous structure of carbon fiber and SiC makes traditional machining [9–12] more difficult in dealing with the fracture cracks, delamination, burrs, etc. Yu et al. [13] showed that 2.5D C/SiC composites have a good fatigue resistance strength of 202.5 MPa, which is equivalent to 78.1% of the ultimate tensile strength. Due to the excellent properties of C_f/SiC composite materials, it is difficult to remove them efficiently and accurately [3,14]. Nowadays, Electric Discharge Machining (EDM), ultrasonic-assisted machining and laser machining are relatively effective methods. Researchers have reported on the EDM of the structure ceramic materials [15]; it has been believed that the microstructure is more suitable for this method [16]. However, functional ceramic materials are polycrystalline materials, which are generally ionic, covalent or both ionic and covalent. The materials are insulative and non-electrical conductive. Therefore, EDM technology has a relatively large limitation in the processing and application of functional ceramic materials. Ultrasonic-assisted machining can process both conductive and insulating materials, and it has a faster processing speed without the generation of heat-affected zones, so it can process engineering structure materials [17]. However, limited by the processing amplitude, the precision of ultrasonic-assisted machining is low. It is only suitable for the surface processing of complex contours, and it is not suitable for processing small holes with a high precision. The abrasive wear during ultrasonic-assisted grinding is twice as high as the material wear, and the cost is higher. Laser machining utilizes the interaction of laser beams with matter to process materials. Laser machining has shown the advantages of a high machining efficiency, a non-contact nature and a lack of tool wear, which make it an effective way to process C_f/SiC composites. The non-contact laser ablation process is very helpful in promoting its engineering applications and machining quality.

Understanding the ablation mechanism and quality evaluation for C_f/SiC composites is very important for its further applications. Zhai et al. [18] explored the effect of fiber orientation on surface morphology. It was found that, depending on different fiber orientations, the surface roughness decreases in different degrees. Lambiase et al. [19] conducted experiments on C_f/SiC composites by changing the laser parameters to explore the effect of the laser parameters on the composites. It is believed that the defects, such as delamination and microcracks, are caused by the heat-affected zone (HAZ). Wei et al. [20] explored a novel method for the underwater femtosecond laser ablation of SiC/SiC composites, which can obtain a high cleanliness and low-oxidation microporous surfaces. Yang and Zhang et al. [21,22] conducted the laser processing of a microstructure for surface creation, which has demonstrated the typical advantages of the contact removal method [23]. Zhai et al. [24] compared the morphologies under different laser processing parameters. It was found that the original surface roughness of C/SiC had a significant effect on the laser ablation morphology. Pan et al. [25] revealed that the surface morphology of the C/SiC composite under laser irradiation is usually divided into a central region, a transition region and a boundary region. Hu et al. [26] studied SiC/SiC through-holes and blind holes processed by a microsecond laser and analyzed their processing characteristics. Liu and Zhang et al. [27] first revealed the 'ablation evolution behavior' in the micro-hole machining of 2.5D C_f/SiC composites with a millisecond laser. Zhang et al. [28] studied 2.5D C/SiC composites at room temperature and 900 °C by applying cyclic loads and found that the fracture was closely related to the cracks generated by the crossing of fiber bundles. Deng et al. [29] established a thermal model, which investigates the thermal behavior of single-crystal SiC at different wavelengths and pulse durations and predicts the size and cross-sectional profile of the ablation hole. Liu and Wang et al. [30–32] found that when a high-power picosecond laser ablates the C/SiC composites, a large number of gas-phase substances and strong shock waves are generated, and fragments are ejected at a high speed. However, due to the strong thermal effect, the materials are prone to oxidation, stratification and grooving defects [33] in the laser processing process, and the microgrooves are prone to appear in fine processing.

At present, the laser ablation of unidirectional and 2D C_f/SiC composites is widely studied. The 2.5D C_f/SiC composite is different from the unidirectional and 2D C_f/SiC, which mainly uses the weft yarn through the warp to form a braided structure so as to form the interlock. Therefore, 2.5D C_f/SiC has a stronger interlayer bonding strength and better mechanical properties. In the research of laser processing 2.5-dimensional C_f/SiC composites, the surface morphology and its produced groove should be extensively focused on, which is helpful for laser-ablated size and quality prediction. Thus, this paper is devoted to investigating the experimental results of the variation in laser parameters with the groove depth and width prediction model. The research work reflects the changing pattern of the C_f/SiC groove morphology, and the surface thermally affected zone will be further analyzed through systematic experiments.

2. Experiment Setup

2.1. Material Preparation

The 2.5D C_f/SiC composites used in this experiment are provided by Shenyang Liming Engine Co., LTD. (Shenyang, China) and prepared by the chemical vapor infiltration (CVI) method. The preparation method is as follows. First, the T-300 carbon fiber was used as the reinforcing material. The T-300 carbon fiber was prepared into the fiber preform. Second, the carbon fiber precast was placed in a closed reaction chamber with a high temperature; the reaction gases such as H_2 and methyltrichlorosilane (MTS) are infiltrated into the interior or surface of the preform to form the ceramic matrix. Table 1 shows the key properties of the 2.5D C_f/SiC. The main components are carbon fiber, SiC ceramics and microplasma silicon. Table 2 shows the physical and mechanical properties of SiC ceramics and carbon fiber. Figure 1 shows the composite weaving structure. The texture morphology and preparation characteristics of carbon fiber can be observed through the surface. The workpiece has a dimensional size of 15 × 15 × 15 mm^3.

Table 1. Material characteristics of the 2.5D C_f/SiC [34].

Properties (at Room Temperature)		Parmeters	
Diameter of carbon fiber	7.6 μm	**Porosity**	17%
Density	1.7 g/cm^3	**Size**	15 × 15 × 15 mm^3
Fiber volume fraction	40~50%	**Thermal conductivity**	8–10 W/m·k
Fiber mass fraction	27.27~36%	**Thermal expansivity**	2–6 × 10^{-6} K^{-1}

Table 2. Characteristics of the carbon fiber and SiC matrix [35,36].

Properties	SiC Ceramics	Carbon Fiber
Thermal conductivity (W/m·k)	185	Radial: 5; axial: 50
Specific heat (J/kg·K)	800	710
Vaporization temperature (K)	2700	3550
Latent heat for vaporization (MJ/kg)	19.83	43
Density (kg/m^3)	3220	1780
Young's modulus (GPa)	450	Radial: 15; axial: 230
Shear modulus (GPa)	193	Radial: 7; axial: 27

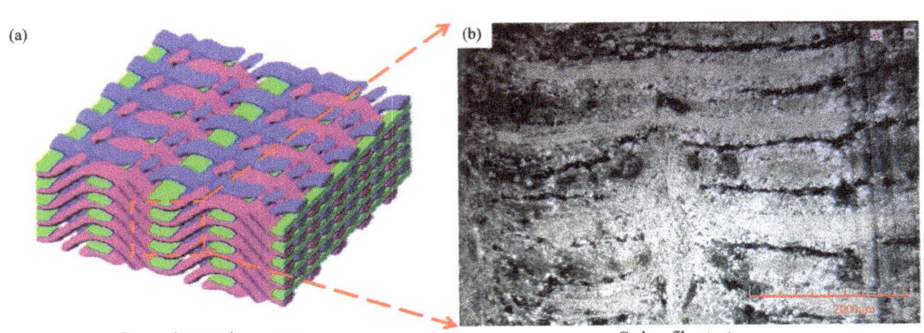

Figure 1. Structure of the 2.5-dimensional C_f/SiC composite material ((**a**). 2.5D structure diagram; (**b**). microscope figure).

2.2. Experimental Equipment

In the experiments, the C_f/SiC composites are processed by the Huari Poplar nanosecond laser Poplar-355. The temperature of the water cooler is set to be constant at 22 °C before the experiment. The laser z-axis is used to set the defocus distance of the laser processing, and the x-axis and y-axis are used to adjust the experimental platform to determine the laser processing position. The laser beam is emitted from the light outlet to process C_f/SiC composites.

Table 3 shows the parameters of the nanosecond laser. According to the parameter table, the output wavelength of the laser is 355 nm. The output power range is 5–20 W. The pulse width range is 0.15–0.3 μs, and the operating voltage is the standard 220 V with the water-cooling method.

Table 3. Nanosecond laser parameters.

Laser Equipment		Poplar-355	
wavelength	355 nm	**beam directivity**	<25 μrad
output power	>5 W@50 kHZ	**laser head size**	500 × 185 × 147 mm^3
single pulse energy	>125 μJ@40 kHZ	**divergence angle of light spots**	≤1 mrad (with a beam expansion)
repeat frequency rate	20–200 kHZ	**working temperature**	10–35 °C
pulse width	16 ± 2 ns@50 kHZ	**working humidity**	<65%
power stability	≤3% rms	**cooling-down method**	Water
pulse stability	≤3% rms	**beam diameter**	<8 mm

2.3. *Overall Experiments Design*

Figure 2 is the experimental flow chart of the laser irradiation of C_f/SiC ceramic matrix composites, which is composed of surface grinding and polishing, ultrasonic cleaning, dryer drying, laser irradiation processing, microscope observation and the interpretation of the results. Through changing the scanning times, scanning speed, laser power and pulse width, the observation and analysis of the groove topography and surface thermally affected width of the C_f/SiC composites are thoroughly conducted. The experimental parameters are shown in Table 4. The idea of control variables is adopted, and the other processing parameters remain unchanged in a laser processing experiment with variable parameters. The effects of different processing parameters on the surface groove and surface thermally affected width of the C_f/SiC composites are analyzed, and the causes of these effects are analyzed according to the relevant mechanism.

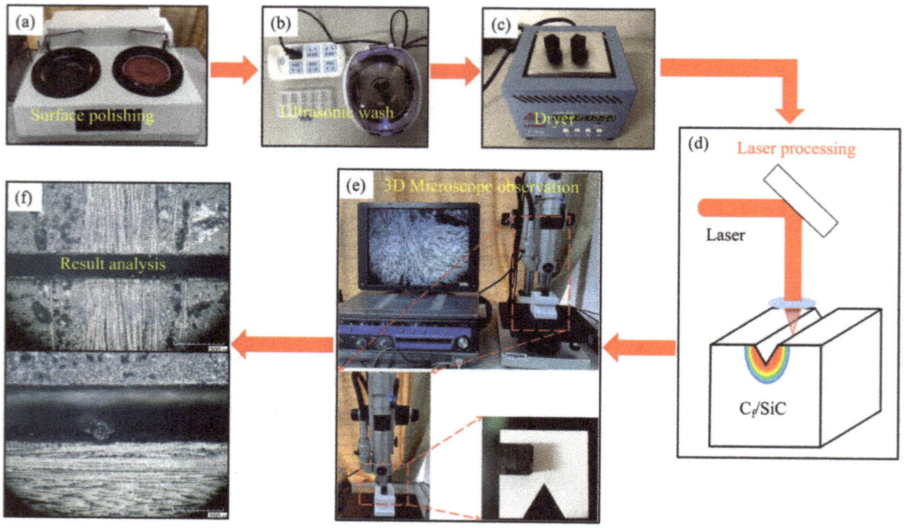

Figure 2. Experimental flow chart ((**a**). surface polishing; (**b**). ultrasonic wash; (**c**). dryer; (**d**). laser processing; (**e**). 3D microscope observation; (**f**). result analysis).

Table 4. Experimental processing parameters.

Serial Number	Scanning Times (N)	Scanning Speed ($V/mm.s^{-1}$)	Laser Power (P/W)	Pulse Width (Q/μs)	Wavelength (nm)	Cooling-Down Method
1	100	200	15	0.3	355	water
2	200	200	15	0.3	355	water
3	400	200	15	0.3	355	water
4	800	200	15	0.3	355	water
5	1200	200	15	0.3	355	water
6	400	100	15	0.3	355	water
7	400	200	15	0.3	355	water
8	400	400	15	0.3	355	water
9	400	800	15	0.3	355	water
10	400	200	5	0.3	355	water
11	400	200	10	0.3	355	water
12	400	200	15	0.3	355	water
13	400	200	20	0.3	355	water
14	400	200	15	0.15	355	water
15	400	200	15	0.2	355	water
16	400	200	15	0.25	355	water
17	400	200	15	0.3	355	water

The surface roughness of composites before processing has a great impact on the processing results. Therefore, the C_f/SiC ceramic matrix composites require pre-experimental treatment before laser irradiation processing. In Wei's experiments [20], in order to eliminate the effect of surface defects on the material, all samples were polished using standard micron-sized sandpaper and then ultrasonically cleaned in absolute alcohol before processing. In this experiment, the polishing paste of particles of a diameter of 28–1.5 μm is applied to the test mill rotary platform, and C_f/SiC is pressed to the mill rotary platform for surface treatment. Due to the poor removal consistency of C_f/SiC, uneven areas still appear in the surface treatment. With holes in the C_f/SiC composites, the residual grinding fluid may affect the experimental results of the laser processing. Therefore, after the surface treatment, the residual grinding particles and the grinding solution are cleaned with the ultrasonic cleaner and heating dryer. C_f/SiC is soaked in 75% ethanol solution and cleaned for 7 min. The dryer temperature is set at 100 °C for 5 min to ensure that the residual grinding fluid does not remain in the composites, as shown in Figure 2.

A metal plate is placed on the processing platform to avoid direct damage to the processing platform during laser processing. The thickness is 15 mm, and the setting value is 34.5 cm (the initial defocus distance of the laser is 33 cm). The software EzCad2.14.5 is used to set the laser processing parameters and draw the laser processing route. The composite materials are placed in the processing position for laser irradiation processing, as shown in Figure 3.

To ensure the reproducibility of the experiment, each group of experiments is repeated twice. Each observation target datum is the average value obtained from the measured data under that parameter. For example, when the variable parameter is 100 scanning times, three measurement points are taken from the processing lines. The maximum and minimum values are removed to obtain the average value. The surface morphology and data recording of the C_f/SiC composites are observed by a Japanese Haoshi KH-7700 three-dimensional microscope. The main objectives of the observation data after the laser irradiation processing of C_f/SiC composites are the surface groove width, depth and thermally affected zone width, as shown in Figure 4. During the measurement, the data of the main observation target are completely recorded, and the composite material morphology at the point of measurement is saved with a three-dimensional microscope. By observing and recording the groove morphology and the thermally affected zone width, the change law of the C_f/SiC grooves is analyzed.

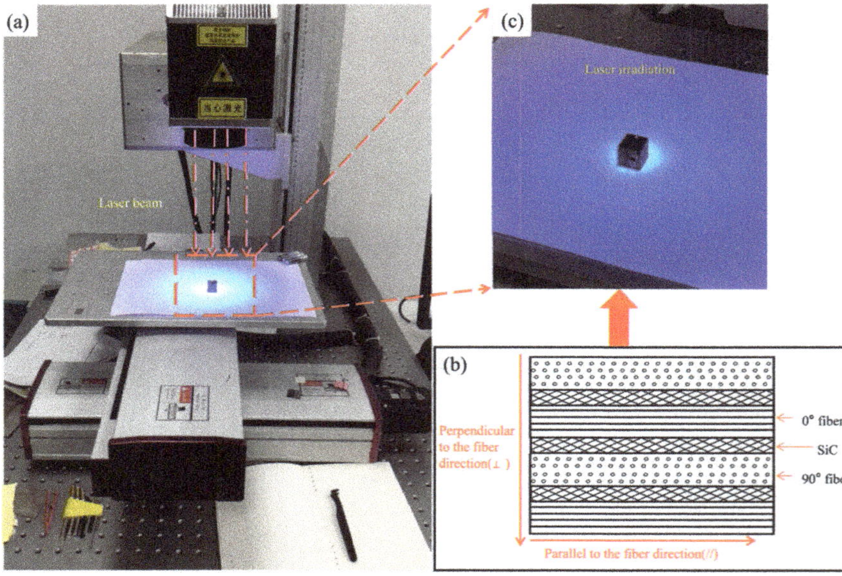

Figure 3. Laser processing of C_f/SiC composites ((**a**). processing macrograph; (**b**). processing direction diagram; (**c**). pictures during experimental processing).

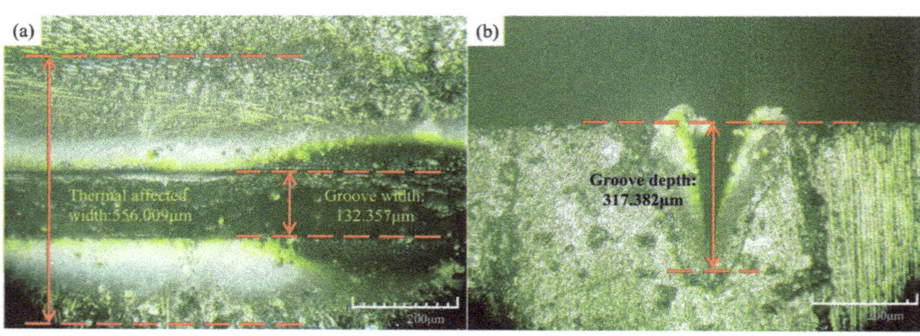

Figure 4. 3D-microscopic image with measured sizes ((**a**). thermal affected width and groove width; (**b**). groove depth).

3. Microgroove Analysis

The laser ablation of C_f/SiC composites is controlled by chemical and physical erosion, mainly the physical erosion of the center [37]. The laser energy presents a Gaussian distribution in space, with the largest spot in the center, and the laser energy on both sides of the center of the spot will be smaller and smaller. During the laser irradiation processing of the composites, most of the energy is absorbed by the material, and the surface of the material is ablated, resulting in grooves. The absorbed energy will be conducted and diffused in the C_f/SiC, causing different degrees of thermal damage to the processed material, forming the thermally affected zone.

3.1. Effect of Laser Scanning Times

Figure 5 shows a regular diagram of the effect of laser scanning times on the surface grooves of the C_f/SiC composites (V = 200 mm/s, P = 15 W, Q = 0.3 μs). The experimental results show that with the increment of laser passing numbers, the surface groove width reaches the lowest value when the number is 400, but the overall variation range is small, within ±10 μm. This indicates that scanning times have little effect on the surface groove width of the composites. With the increment of scanning times, the groove depth increases, and the width of the surface thermally affected zone first increases and then decreases. This is because, with the increment of laser scanning times, the energy radiated on the composites increases, increasing the groove depth. However, when the scanning times continue to increase, on the one hand, the laser irradiation energy will ablate and fracture the carbon fibers. On the other hand, the thermal conductivity of carbon fiber is reduced, preventing the spread of the laser heat. Therefore, the surface thermally affected zone tends to decrease.

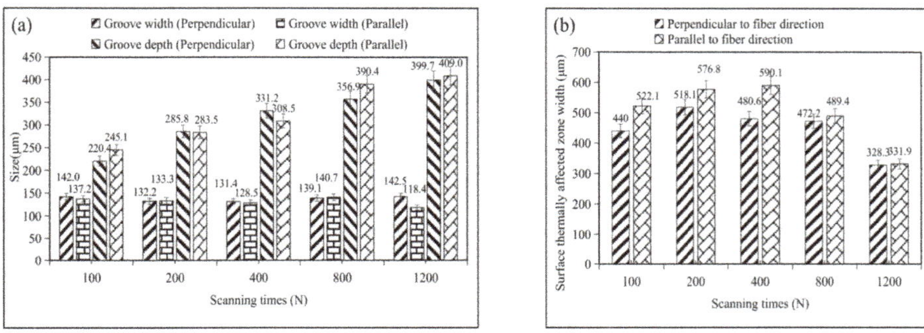

Figure 5. Effect of laser scanning times on the composite surface groove ((**a**). variation in groove width and depth; (**b**). variation in thermally affected zone width).

3.2. *Effect of the Laser Scanning Speed*

The influence of the laser scanning speed on the surface groove of C_f/SiC composites is shown in Figure 6 (N = 400, P = 15 W, Q = 0.3 μs). The experiments show that with the increment of scanning speed, the surface groove width gradually decreases, and the overall groove depth also decreases. The experimental results of the laser scanning times and laser scanning speed are consistent with the changing trend of the ablative experiments of Jiao and Zhang et al. [38,39], according to the following Equation (1):

$$ED = \frac{P}{V \times \varphi} \quad (1)$$

where ED is the energy density; P is the laser power; V is the laser scanning speed; φ is a spot diameter [40].

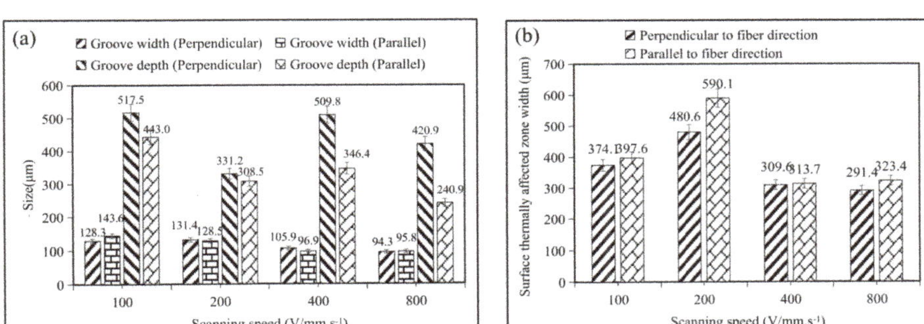

Figure 6. Effect of the scanning speed on the composite surface groove ((**a**). variation in groove width and depth; (**b**). variation in thermally affected zone width).

Therefore, with the increment of the laser scanning speed, the laser energy density decreases, and the composites under the same area are irradiated with less laser energy. In addition, the reduction in irradiation time results in the processing area not receiving enough energy to bring the material to the removal temperature. So, the groove width and groove depth decrease. The influence of the laser scanning speed on the width of the thermally affected zone also shows a similar variation law. With the increment of the laser scanning speed, the thermally affected zone width on the surface of the composites first increases and then decreases. When the laser scanning speed increases, the laser has a preheating effect on the composite, and the surface thermally affected zone width increases. With the continuous increment of the scanning speed, the laser energy density decreases, and the laser ablation heat decreases. Therefore, the surface thermally affected zone width becomes smaller.

3.3. Effect of Laser Power

Figure 7 shows the effect of laser power on the surface grooves of C_f/SiC composites (N = 400, V = 200 mm/s, Q = 0.3 μs). The experiments show that, with the increment of power, the surface groove width and groove depth of C_f/SiC gradually increase, according to the following Equation (2):

$$E = P \times \Delta\tau \quad (2)$$

where E is the laser pulse energy, P is the pulse power and $\Delta\tau$ is the pulse time width. It can be seen that, with the increment of laser pulse power, the laser pulse energy increases. The higher the energy, the higher the temperature during irradiation processing. The temperature of the laser center area exceeds the sublimation temperature of the SiC matrix and carbon fiber. As the heat is transferred, the material outside the central area also reaches the melting temperature. Therefore, the groove width and groove depth gradually increase. However, the surface thermally affected zone width gradually decreases. The greater the laser power, the higher the energy efficiency. The lesser the impact of the laser radiation, the less ablation heat is generated.

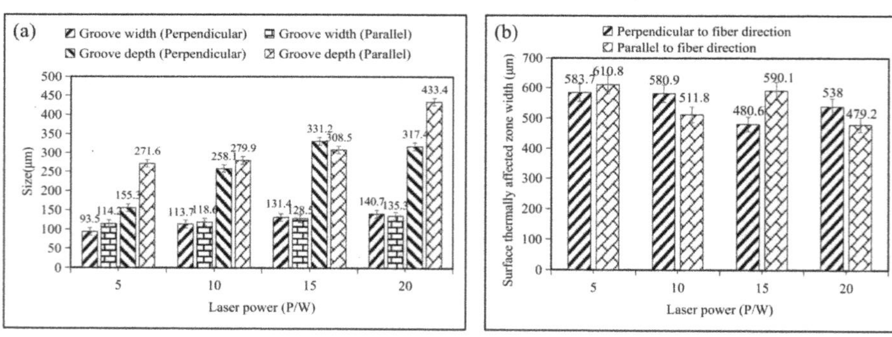

Figure 7. Effect of laser power on the groove of the composite surface ((**a**). variation in groove width and depth; (**b**). variation in thermally affected zone width).

3.4. Effect of Fiber Direction

By analyzing the experimental results, it is found that the two laser-processing directions of the perpendicular-to-fiber direction and parallel-to-fiber direction show similar laws. The width of the thermally affected zone, in the parallel-to-the-fiber-arrangement direction, is slightly larger than that in the perpendicular-to-the-fiber arrangement direction. However, the groove depth in the parallel direction is smaller than that in the perpendicular direction. This is because the C_f/SiC composites have anisotropy, and the heat generated by laser processing is easy to propagate along the carbon fiber direction. The composite materials at the groove openings are subject to significant machining stress due to the heat and the high-velocity jet of the molten material. When the processing path is parallel to the carbon fiber texture, all parts of the entire fiber will be subjected to processing stress. Due to the good adhesion between the fibers and the matrix, the carbon fibers are not easily broken. When the processing path is perpendicular to the carbon fiber texture, the stress effect on the composite material can only be borne by the local material. Therefore, the carbon fiber will be broken, and the groove width will become larger. According to the theorem of energy conservation, when the heat propagates along the fiber direction, the depth of the crack become smaller. When the heat propagates along the direction of the crack depth, the surface thermally affected zone width become smaller.

In the processing, the laser heat absorbed by the composites will be propagated, along with the carbon fiber. When the temperature rises to a certain level, the high bending resistance of the composite materials will be greatly reduced. As a result, the composite fibers at the openings of the trenches are largely fractured, as shown in Figure 8. As the temperature increases, the thermal diffusivity of C_f/SiC decreases, and the thermal conductivity of the composites also decreases. When the laser scanning times increase, the temperature of the composite processing zone increases, and the thermal conductivity of the C_f/SiC decreases. Therefore, whether it is perpendicular to the fiber direction or parallel to the fiber direction, with the increment of the scanning times, the width of the thermally affected zone decreases. During processing, heat transfer is mainly carried out by diffusion, and the heat transfer speed in the direction of the fiber is significantly faster than that in other directions. Therefore, when the variable parameter is the pulse width, the width of the thermally affected zone perpendicular direction is larger than that of the parallel direction. However, the thermal diffusivity of the 2.5-dimensional braided C_f/SiC composites is mainly affected by the radial direction of the carbon fiber and the SiC matrix [41]. The thermal conductivity of the material is mainly affected by thermal diffusivity. So, in other variable parameter experiments, the width of the thermally affected zone in the parallel direction is slightly larger than that in the perpendicular direction.

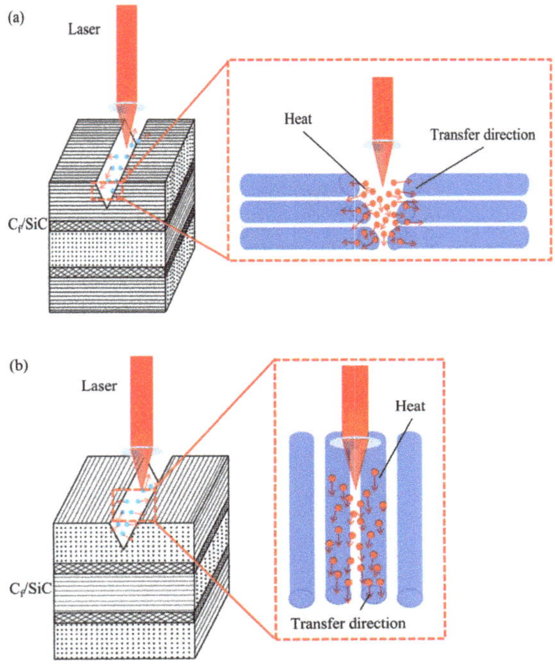

Figure 8. The heat transfer affected by the anisotropy structure ((**a**). perpendicular to fiber direction; (**b**). parallel to fiber direction).

As can be seen in Figure 9, the laser irradiation processing is quite different under the processing parameters of the same laser pulse width. The surface groove width shows a trend of decreasing first and then increasing in the perpendicular direction. The overall slight decreases in the parallel direction and the observed data in both processing directions do not fluctuate much, around ±10 μm. However, the changing trends of the surface thermally affected zone width of C_f/SiC composites are opposite in the two laser processing directions. The changing trend of the thermally affected zone width in the perpendicular direction increases first and then decreases. In the parallel direction, the changing trend decreases first and then increases. The changing nodes are all at the parameter of 0.25. This is because the 2.5-dimensional C_f/SiC composites have a strongly anisotropic structure. Therefore, it can be concluded that the fiber arrangement direction has an impact on laser processing, but only when individual parameters change will it have completely opposite effects.

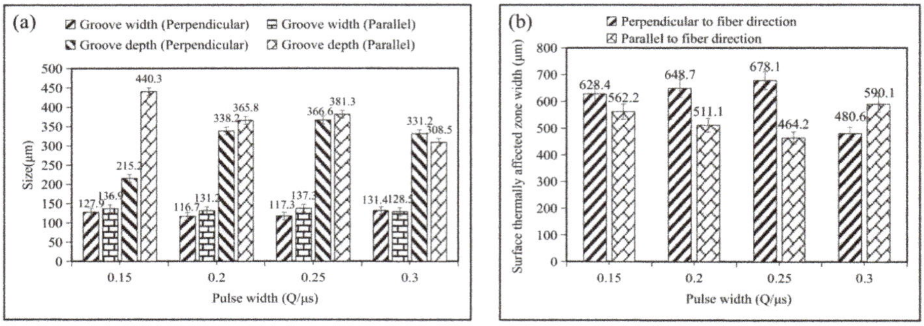

Figure 9. Effect of the pulse width on the composite surface groove ((**a**). variation in groove width and depth; (**b**). variation in thermally affected zone width).

3.5. *Effect of Pulse Width*

The influence of the pulse width on the surface groove of C_f/SiC composites is shown in Figure 9 (N = 400, V = 200 mm/s, P = 15 W). The experiments show that with the increment of the pulse width, the surface groove width reaches the minimum in the stage of 0.2–0.25 μs, which first decreases and then increases. However, the variation range is small, around ±10 μm. With the increment in the pulse width, the groove depth shows an upward trend in the stage of 0.15–0.25 μs, and the increment is obvious. The range of the thermally affected zone shows an increasing trend in the stage of 0.15–0.25 μs, but the changing trend is not obvious. However, when the laser irradiation direction is parallel to the carbon fiber arrangement direction, the width and depth of the surface grooves in the composites show a decreasing trend as a whole. It is obviously different from the change trend perpendicular to the direction of the fiber arrangement direction. The larger the pulse width of the laser is, the smaller the laser energy density will be. The composite is irradiated with less energy, and the groove width and depth are smaller. The reason for the increment in the groove width is that when the laser pulse width gradually increases, the thermal conductivity of the C_f/SiC composites is increased. With the increment of the pulse width, the width of the perpendicular and parallel fiber arrangement directions shows an opposite trend. This is due to the preheating of the laser processing and the insufficient cooling of the complex, thus increasing the width of the thermally affected zone. Different directions of fibers have different forces on the composite materials: a local fiber is stressed, and the entire carbon fiber is stressed. During processing, heat transfer is mainly carried out by diffusion, but the heat transfer is significantly faster along the fiber direction than in other directions. During laser processing, the laser path will cover the area with more heat transfer when parallel to the carbon fiber texture processing path, resulting in more of a heat overlap in this part. Compared with the perpendicular fiber arrangement direction and the parallel fiber arrangement direction, the groove width trend and the width of the thermally affected zone trend are obviously different, which shows that the fiber arrangement direction affects the grooves.

Pan et al. [25] provided six different power densities, as well as six levels of pulse numbers, when the ablation experiments are conducted for the C/SiC composite. However, they did not perform experimental pretreatment on composites. The laser ablation depth and the number of layers in the experimental results are basically consistent with these experiments, which also explains the reason for the change in the crack depth. Zhang et al. [39] discussed the effects of different processing parameters, including the helical line width and spacing, machining time and scanning speed. The influence trend of factors, such as the laser scanning speed, on the material laser ablation obtained by their experiments is consistent with this paper. However, the groove depth values obtained in the literature were as high as 700 μm and as low as 575 μm, which are much larger than the highest value of 517 μm obtained in this paper, which may be due to the influence of 2.5D composite materials.

4. Discussions

4.1. Regression Model of Groove Sizes

The exponential regression equation prediction model can not only describe the nonlinear characteristics but is also more accurate than the linear regression equation prediction in this analysis. Therefore, the exponential regression equation models are used in this paper to predict the groove width, groove depth and surface thermally affected zone width.

A quaternary regression equation is established, with the scanning times, scanning speed, laser power and pulse width as the main variable parameters in the nanosecond laser irradiation experiments:

$$B = C_T N^{a_1} V^{a_2} P^{a_3} Q^{a_4} \tag{3}$$

Taking the logarithm of both sides of the equation:

$$\ln B = \ln C_T + a_1 \ln N + a_2 \ln V + a_3 \ln P + a_4 \ln Q \tag{4}$$

Assuming $y = \ln B, \mathrm{a}_0 = \ln C_T, x_1 = \ln N, x_2 = \ln V, x_3 = \ln P, x_4 = \ln Q$, the following regression equation is obtained:

$$y = a_0 + a_1 x_1 + a_2 x_2 + a_3 x_3 + a_4 x_4 \tag{5}$$

Combining the 12 sets of experimental data, the regression models of the surface groove width, depth and surface thermally affected zone width perpendicular to the fiber direction and parallel to the fiber direction can be obtained, respectively. The models are as follows:

$$B_c = e^{5.14957} N^{-0.00746} V^{-0.18063} P^{0.30043} Q^{0.08039} \tag{6}$$

$$D_c = \mathrm{e}^{2.93539} N^{0.22138} V^{0.083208} P^{0.67275} Q^{0.55081} \tag{7}$$

$$T_c = e^{7.74621} N^{-0.07555} V^{-0.24357} P^{-0.25138} Q^{-0.63478} \tag{8}$$

$$B_p = e^{5.80636} N^{-0.03157} V^{-0.23114} P^{0.10858} Q^{-0.13260} \tag{9}$$

$$D_p = e^{4.79824} N^{0.15866} V^{-0.17743} P^{0.28071} Q^{-0.22467} \tag{10}$$

$$T_p = e^{7.01935} N^{-0.16713} V^{-0.11779} P^{0.24108} Q^{-0.10253} \tag{11}$$

where B_c is the regression model of the surface groove width perpendicular to the fiber direction, D_c is the regression model of the surface groove depth perpendicular to the fiber direction, T_c is the regression model of the surface thermally affected zone width perpendicular to the fiber direction, B_p is the regression model of the surface groove width in the parallel fiber direction, D_p is the regression model of the surface groove depth in the parallel fiber direction and T_p is the regression model of the surface thermally affected zone width parallel to the fiber direction.

Five sets of experimental data are brought in, and the accuracy is verified by calculating the relative error of these regression models. The values of the four parameters of the laser treatment variables are substituted into Equations (6) to (11), and the relative errors are obtained from the following two equations:

$$|e(x^*)| = |x^* - x| \tag{12}$$

where $|e(x^*)|$ is the absolute error, x^* is the back-substitution value and x is the true value.

$$\delta = \frac{|e(x^*)|}{x} \times 100\% \tag{13}$$

where δ is the relative error. The results are obtained, as shown in Figures 10–15.

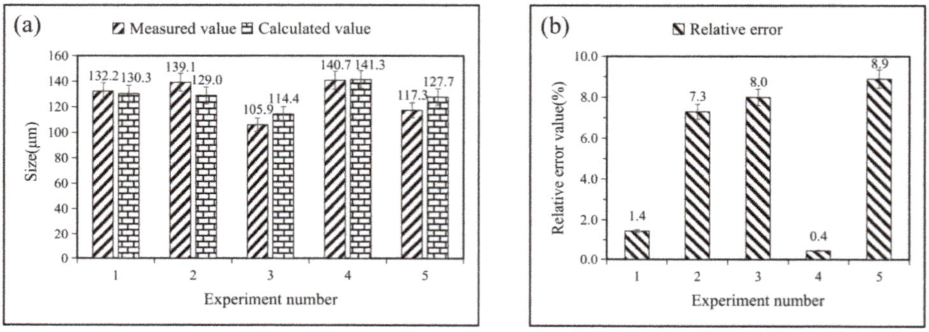

Figure 10. Error verification of Equation (6) ((**a**). comparison of measureed value and calculated value; (**b**). corresponding relative error).

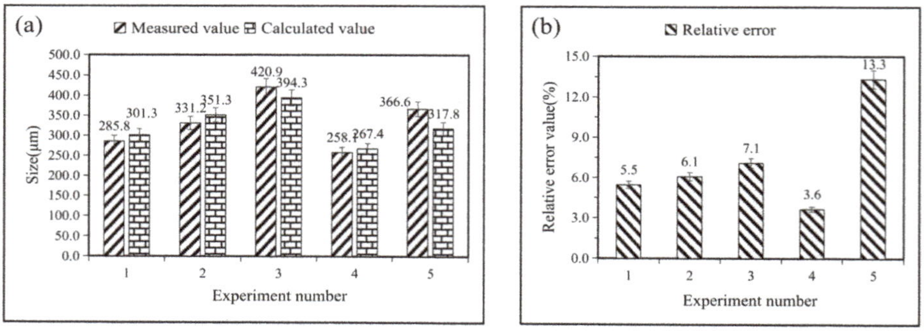

Figure 11. Error verification of Equation (7) ((**a**). comparison of measureed value and calculated value; (**b**). corresponding relative error).

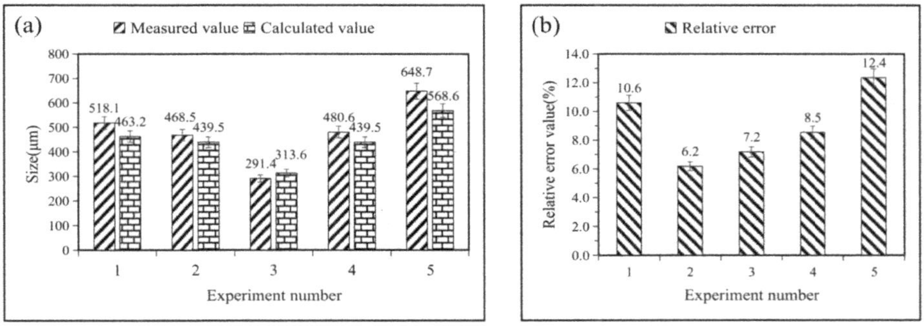

Figure 12. Error verification of Equation (8) ((**a**). comparison of measureed value and calculated value; (**b**). corresponding relative error).

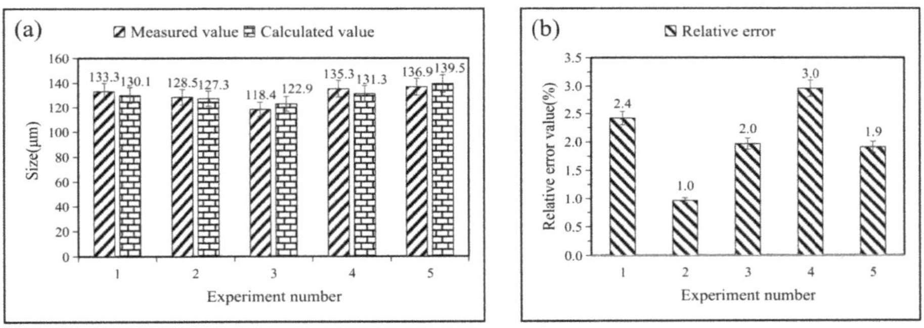

Figure 13. Error verification of Equation (9) ((**a**). comparison of measureed value and calculated value; (**b**). corresponding relative error).

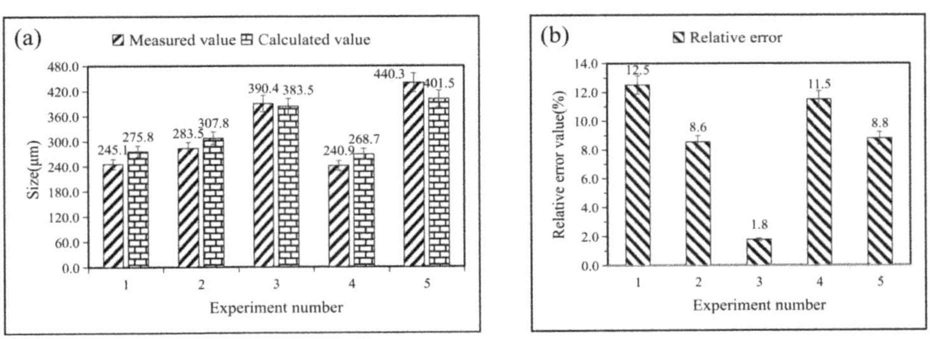

Figure 14. Error verification of Equation (10) ((**a**). comparison of measureed value and calculated value; (**b**). corresponding relative error).

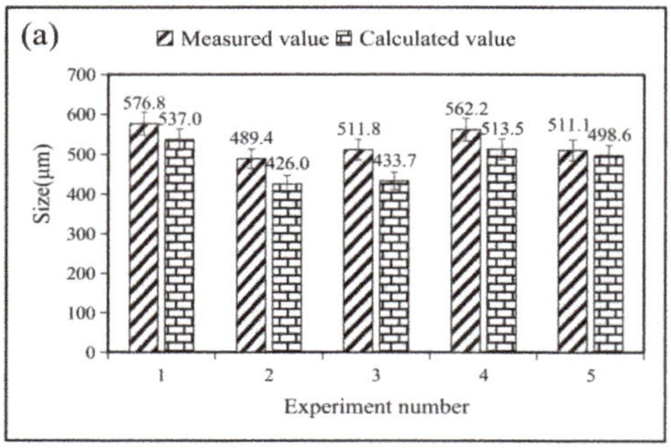

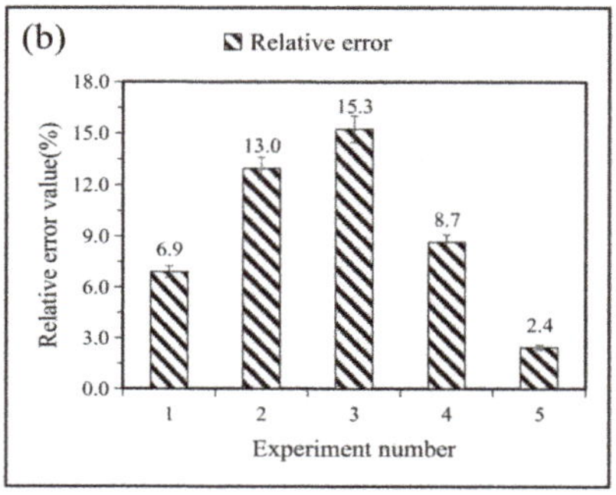

Figure 15. Error verification of Equation (11) ((**a**). comparison of measureed value and calculated value; (**b**). corresponding relative error).

After calculation, the average relative errors between the calculated value of the regression model equations and the actual measured value are shown in Table 5. In these regression models, the relative error values are all below 10%, which can truly reflect the influence of laser-processing parameters on the surface groove morphology. By analyzing the exponential value of these models, it can be concluded that the laser scanning speed and laser power have a greater impact on the surface groove width, followed by the laser pulse width, and the scanning times have the least impact.

Table 5. Relative error.

Regression Model	B_c	D_c	T_c	B_p	D_p	T_p
Relative error	5.2%	7.1%	9.0%	2.0%	8.6%	9.2%

4.2. Mechanism of Laser Ablation

The principle of the laser processing of C_f/SiC composite materials is mainly to use a laser beam to irradiate the surface of the composite material. The SiC and carbon fibers are melted and evaporated or sublimated by heat. Therefore, on the surface of the processed sample, we can observe the two important measurement objects: the processing grooves and the thermally affected zone. However, during the laser processing, the flowing of the heat source between the tool and the workpiece will bring forces into the materials, such as surface tension and gravity. The material will partially splash under the combined action of these forces. This part of the molten material splashed into the non-processing zone cools to form a residue.

Under laser irradiation, the material obtains laser energy through nonlinear absorption. The composite materials melt to generate plasma of Si, C and O. In the air, they are easy to combine with the plasma. With the increment of the O content, more carbon fibers and the SiC matrix are oxidized [27]. The main products of the carbon fiber and SiC matrix oxidation are COx and SiOx. COx is a gas and is easily discharged. SiOx remains on the surface. After cooling, a heavy condensate is formed at the opening of the groove. Equations (14)–(19) are the main chemical reaction equations that occur in the processing zone. Many scholars, such as Pan [25] and Tong [42], also mentioned similar chemical reactions during laser ablation. Substances, such as heavy condensates and residues, are easily removed after cleaning. Due to the brittle and hard nature of the composite material, it will be subjected to internal stress and processing stress during processing. So, some fiber fractures are also observed.

$$C(s) + \frac{1}{2}O_2(g) \rightarrow CO(g) \tag{14}$$

$$C(s) + O_2(g) \rightarrow CO_2(g) \tag{15}$$

$$Si(s) + 2O_2(g) \rightarrow SiO_2(l) + CO_2(g) \tag{16}$$

$$SiC(s) + O_2(g) \rightarrow SiO(g) + CO(g) \tag{17}$$

$$2SiC(s) + 3O_2(g) \rightarrow 2SiO_2(l) + 2CO(g) \tag{18}$$

$$SiO_2(l) \rightarrow SiO_2(g) \tag{19}$$

Because of the different laser processing paths, the surface topography of the processed samples is different. Heat is easily transferred along the direction of carbon fibers, and the ablation behavior of composite materials is more complicated.

4.3. Fiber Topography for Grooves

There are many influencing factors regarding the processed surface of C_f/SiC composites, which make it difficult to observe the fiber pattern of the material within the observation groove scale. After cleaning the heavy condensates and debris on the surface of the composites, the groove morphology and fiber pattern at the fiber fracture can be observed. When the laser processing direction is perpendicular to the carbon fiber, as shown in Figure 16, neat and obvious fiber grooves can be seen. At the fiber grooves, the composite surface grooves do not show obvious processing differences from the other areas, and the overall processing grooves are relatively smooth. Figure 17 shows the end surface of the parallel direction. The laser processing route is not completely covered by the fiber direction, and obvious fiber grooves cannot be observed at the processed surface groove. When the processing route involves the fiber direction, the grooves will exhibit small changes, with a slight uneven phenomenon.

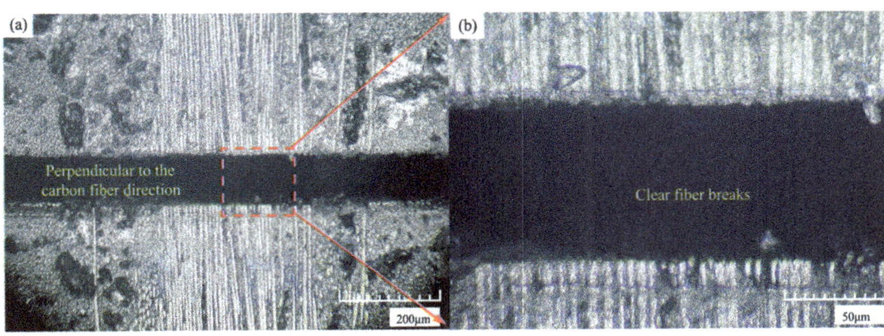

Figure 16. Vertical fiber direction processing end surface ((**a**). the groove morphology at the fiber fracture; (**b**). partial enlargement of the picture).

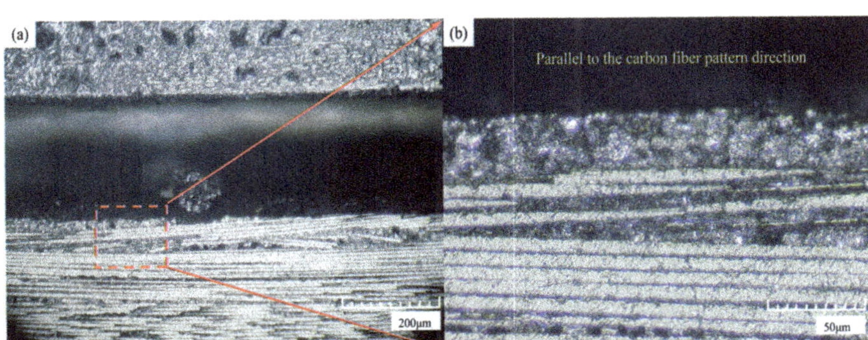

Figure 17. Parallel fiber direction processing end surface ((**a**). the groove morphology at the fiber fracture; (**b**). partial enlargement of the picture).

5. Conclusions

In this paper, the effects of nanosecond laser irradiation on C_f/SiC ceramic matrix composites have been investigated, and the effects of different laser processing parameters on the groove morphology and surface thermally affected zone have been analyzed in detail. Some of the conclusions drawn from the experimental results are as follows:

1. When the variation in laser parameters involves the laser scanning times and pulse width, the resulting groove width is small, at about ± 10 μm. When the variation in parameters involves the laser scanning speed and laser power, the groove width decreases with the increment of the laser scanning speed and increases with the increment of the laser power.
2. The groove depth increases with the increment of the parameters in the processing experiments where the parameters are the laser scanning times and laser power. However, when the variable parameter is the laser scanning speed, the faster the speed, the shallower the groove depth. In the machining experiments where the variable parameter is the pulse width, the groove depth shows different trends in different machining path directions.
3. The variation trend of the surface thermally affected zone width is similar to that of the laser irradiation processing experiment where the variable parameters of the laser are the scanning times and laser scanning speed. With the increment of the parameters, the width of the surface thermal influence area first increases and then decreases. When the variable parameter is the laser power, the surface thermally affected zone width decreases with the increment of power. When the variable parameter is the laser pulse width, different processing path directions show different trends.
4. The fiber direction has an impact on laser processing, but there is an opposite trend only when individual parameters change. In addition, the groove morphology at the fiber fracture is affected by the fiber texture. The appearance of carbon fibers in the laser processing of parallel fiber textures will make the processing groove slightly uneven. The fiber grooves are not obvious and can hardly be observed. During processing, the fiber grooves are obviously clear.
5. Based on the multivariate nonlinear regression equation, a regression model with the laser scanning times, scanning speed, laser power and pulse width as the main variables is established. The model can truly reflect the influence of laser processing parameters on the surface groove morphology of C_f/SiC. It is concluded that the laser scanning speed and laser power have a greater impact on the surface groove width, followed by the laser pulse width. The laser scanning times cause a minimal impact.
6. In this paper, the laser machining process is mainly understood from the perspective of the ablation mechanism. However, the effect of the mechanical properties of composite materials and force variation on laser-assisted processing should also be considered in practical laser processing. The ablated process can be applied as a pre-process for the laser-assisted grinding of this type of material, which can substantially improve the machining efficiency.

Author Contributions: C.W. designed the whole conception and theoretical analysis. T.Z. and F.L. conducted the experiments and wrote the paper. Y.L. and S.Y.L. helped in polishing and revising the manuscript. All authors have read and agreed to the published version of the manuscript.

Funding: This work is supported by the National Natural Science Foundation of China (No. 52005098 & 51905498) and the Shanghai Natural Science Foundation (22ZR1402400). The authors wish to express their gratitude for the generous support.

Data Availability Statement: Not applicable.

Acknowledgments: Not applicable.

Conflicts of Interest: The authors declare that they have no known competing financial interest or personal relationships that could have appeared to influence the work reported in this paper.

References

1. Liu, Y.; Quan, Y.; Wu, C.J.; Ye, L.Z.; Zhu, X.J. Single diamond scribing of SiCf/SiC composite: Force and material removal mechanism study. *Ceram. Int.* **2021**, *47*, 27702–27709. [CrossRef]
2. Hu, J.H.; Tang, J.L.; Li, T.; Xu, L.L.; Lin, B.; Yang, M.J.; Wang, Y.Y.; Li, H.Y.; Zhang, Z.H. Research Progress on Etching Modification of Carbon Fiber and Aramid Fiber and the Interface Bonding Performance of Their Composite Materials. *Surf. Technol.* **2021**, *50*, 94–116.
3. Guo, M.X.; Tao, J.B.; Wu, C.J.; Luo, C.; Lin, Z.J. High-speed grinding fracture mechanism of C_f/SiC composite considering interfacial strength and anisotropy. *Ceram. Int.* **2023**, *49*, 2600–2612. [CrossRef]
4. Zhou, W.W.; Wang, J.Q.; Zhao, J.; Liu, Y. Experimental research on single abrasive grain scratch SiC_f/SiC ceramic matrix composite. *Diam. Abras. Eng.* **2021**, *41*, 51–57.
5. Diaz, G.G.; Luna, Z.; Liao, Z.; Axinte, D. The new challenges of machining Ceramic Matrix Composites (CMCs): Review of surface integrity. *Int. J. Mach. Tools Manuf.* **2019**, *139*, 24–36. [CrossRef]
6. Wang, J.Q.; Yan, Y.D.; Li, C.; Geng, Y.Q. Material removal mechanism and subsurface characteristics of silicon 3D nanomilling. *Int. J. Mech. Sci.* **2023**, *242*, 108020. [CrossRef]
7. Wu, C.J.; Pang, J.Z.; Li, B.Z. High Speed Grinding of HIP-SiC Ceramics on Transformation of Microscopic Features. *Int. J. Adv. Manuf. Technol.* **2019**, *102*, 1913–1921. [CrossRef]
8. Liu, G.S.; Liu, Y.X.; Bian, D.; Zhao, Y.W. Tribological Properties of Graphene Oxide Grafted Carbon Fiber and Its Resin Coating under Different Loads. *Surf. Technol.* **2021**, *50*, 62–69.
9. Liu, S.; Xiao, G.J.; Lin, O.C.; He, Y.; Song, S.Y. A new one-step approach for the fabrication of microgrooves on Inconel 718 surface with microporous structure and nanoparticles having ultrahigh adhesion and anisotropic wettability: Laser belt processing. *Appl. Surf. Sci.* **2023**, *607*, 15510801–15510818. [CrossRef]
10. Qu, S.S.; Yao, P.; Gong, Y.D. Modelling and grinding characteristics of unidirectional C-SiCs. *Ceram. Int.* **2022**, *48*, 8314–8324. [CrossRef]
11. Li, C.; Hu, Y.X.; Zhang, F.H.; Geng, Y.Q.; Meng, B.B. Molecular dynamics simulation of laser assisted grinding of GaN crystals. *Int. J. Mech. Sci.* **2023**, *239*, 107856. [CrossRef]
12. Shu, C.S.; Su, Q.T.; Li, M.H.; Wang, Z.B.; Yin, S.H.; Huang, S. Fabrication of extreme wettability surface for controllable droplet manipulation over a wide temperature range. *Int. J. Extrem. Manuf.* **2022**, *4*, 045103. [CrossRef]
13. Yu, J.W.; Fei, Q.G.; Zhang, P.W.; Li, Y.B.; Chen, Q. Fatigue Life of a 2.5D C/SiC Composite Under Tension–Tension Cyclic Loading: Experimental Investigation and Sensitivity Analysis. *Acta Mech. Solida Sin.* **2021**, *34*, 277–285. [CrossRef]
14. Li, C.; Piao, Y.C.; Zhang, F.H.; Zhang, Y.; Hu, Y.X.; Wang, Y.F. Understand anisotropy dependence of damage evolution and material removal during nanoscratch of MgF2 single crystals. *Int. J. Extrem. Manuf.* **2023**, *5*, 015101. [CrossRef]
15. Puertas, I.; Luis, C.J. A study on the electrical discharge machining of conductive ceramics. *J. Mater. Process. Technol.* **2004**, *153–154*, 1033–1038. [CrossRef]
16. Sun, Y.; Jin, L.Y.; Gong, Y.D.; Wen, X.L.; Yin, G.Q.; Wen, Q.; Tang, B.J. Experimental evaluation of surface generation and force time-varying characteristics of curvilinear grooved micro end mills fabricated by EDM. *J. Manuf. Process.* **2022**, *73*, 799–814. [CrossRef]
17. Wang, Z.; Wang, J.T.; Song, H.W.; Yuan, W.; Liu, Y.W.; Ma, T.; Huang, C.G. Laser ablation behavior of C/SiC composites subjected to transverse hypersonic airflow. *Corros. Sci.* **2021**, *183*, 109345. [CrossRef]
18. Zhai, C.T.; Xu, J.K.; Hou, Y.G.; Sun, G.B.; Zhao, B.B.; Yu, H. Effect of fiber orientation on surface characteristics of C/SiC composites by laser-assisted machining. *Ceram. Int.* **2022**, *48*, 6402–6413. [CrossRef]
19. Lambiase, F.; Genna, S.; Leone, C.; Paoletti, A. Laser-assisted direct-joining of carbon fibre reinforced plastic with thermosetting matrix to polycarbonate sheets. *Opt. Laser Technol.* **2017**, *94*, 45–58. [CrossRef]
20. Wei, J.Y.; Yuan, S.M.; Zhang, J.Q.; Zhou, N.; Zhang, W.; Li, J.B.; An, W.Z.; Gao, M.X.; Fu, Y.Z. Removal mechanism of SiC/SiC composites by underwater femtosecond laser ablation. *J. Eur. Ceram. Soc.* **2022**, *42*, 5380–5390. [CrossRef]
21. Yang, G.F.; Liu, L.; Xia, H.Y.; Cui, J. Experimental Study of Ice Suppression Characteristics of TC4 Microstructure Surface Induced by Nanosecond Laser. *Surf. Technol.* **2021**, *50*, 93–102.
22. Zhang, X.; Huang, T.; Xiao, R.S. Effect of Crystal Orientation on High Power Green Femtosecond Laser Processing of Single Crystal Silicon. *Surf. Technol.* **2021**, *50*, 362–371.
23. Dong, Z.; Yan, Y.; Peng, G.; Li, C.; Geng, Y.Q. Effects of sandwiched film thickness and cutting tool water contact angle on the processing outcomes in nanoskiving of nanowires. *Mater. Des.* **2023**, *225*, 111438. [CrossRef]
24. Zhai, Z.Y.; Zhang, Y.C.; Cui, Y.H.; Zhang, Y.F.; Zeng, Q.R. Investigations on the ablation behavior of C/SiC under femtosecond laser. *Optik* **2020**, *224*, 165719. [CrossRef]
25. Pan, S.N.; Li, Q.Y.; Xian, Z.K.; Su, N.G.; Zeng, F.Z. The Effects of Laser Parameters and the Ablation Mechanism in Laser Ablation of C/SiC Composite. *Materials* **2019**, *12*, 3076. [CrossRef]
26. Hu, W.Q.; Shin, Y.C.; King, G.B. Micromachining of Metals, Alloys, and Ceramics by Picosecond Laser Ablation. *J. Manuf. Sci. Eng.* **2010**, *132*, 011009. [CrossRef]
27. Liu, C.; Zhang, X.Z.; Wang, G.F.; Wang, Z.F.; Gao, L. New ablation evolution behaviors in micro-hole drilling of 2.5D C_f/SiC composites with millisecond laser. *Ceram. Int.* **2021**, *47*, 29670–29680. [CrossRef]

28. Zhang, C.Y.; Wang, X.W.; Liu, Y.S. Tensile fatigue of a 2.5D-C/SiC composite at room temperature and 900 degrees C. *Mater Design.* **2013**, *49*, 814–819. [CrossRef]
29. Deng, Y.; Zhou, Y.F.; Zhang, Y.M.; Chen, D.K.K.; Zhou, X.L. Numerical and experimental analysis of nanosecond laser ablation of SiC. *Mater. Sci. Semicond. Process.* **2022**, *151*, 107020. [CrossRef]
30. Liu, Y.S.; Wang, J.; Li, W.N.; Wang, C.H.; Zhang, Q.; Yang, X.J.; Cheng, L.F. Effect of energy density and feeding speed on micro-holes drilling in SiC/SiC composites by picosecond laser. *Int. J. Adv. Manuf. Technol.* **2016**, *84*, 1917–1925. [CrossRef]
31. Wang, J.T.; Ma, Y.Z.; Liu, Y.W.; Yuan, W.; Song, H.W.; Huang, C.G.; Yin, X.W. Experimental investigation on laser ablation of C/SiC composites subjected to supersonic airflow. *Opt. Laser Technol.* **2019**, *113*, 399–406. [CrossRef]
32. Liu, Y.S.; Wang, C.H.; Li, W.N.; Yang, X.J.; Zhang, Q.; Cheng, L.F.; Zhang, L.T. Effect of energy density on the machining character of C/SiC composites by picosecond laser. *Appl. Phys. A* **2014**, *116*, 1221–1228. [CrossRef]
33. Vadim, A.; Fabrice, L.; Alfred, E.; Lionel, M.; Gildas, L.; Bernard, D. An experimental method to assess the thermo-mechanical damage of CFRP subjected to a highly energetic 1.07 μm-wavelength laser irradiation. *Compos. Part B* **2016**, *92*, 326–331.
34. Zhai, Z.Y.; Wang, W.J.; Zhao, J.; Mei, X.S.; Wang, K.D.; Wang, F.C.; Yang, H.Z. Influence of surface morphology on processing of C/SiC composites via femtosecond laser. *Compos. Part A Appl. Sci. Manuf.* **2017**, *102*, 117–125. [CrossRef]
35. Wu, C.J.; Li, B.Z.; Liu, Y.; Liang, S. Surface roughness modeling for grinding of silicon carbide ceramics considering co-existence of brittleness and ductility. *Int. J. Mech. Sci.* **2017**, *133*, 167–177. [CrossRef]
36. Liu, C.; Zhang, X.Z.; Gao, L.; Jiang, X.Z.; Wang, X.D.; Yang, T. Study on damage characteristics and ablation mechanism in fiber laser trepan drilling of 2.5D Cf/SiC composites. *Int. J. Adv. Manuf. Technol.* **2021**, *117*, 3647–3660. [CrossRef]
37. Dang, X.L.; Yin, X.W.; Fan, X.M.; Ma, Y.Z.; Wang, J.T.; Ju, P.F.; Song, H.W. Microstructural evolution of carbon fiber reinforced SiC-based matrix composites during laser ablation process. *J. Mater. Sci. Technol.* **2019**, *35*, 2919–2925. [CrossRef]
38. Jiao, H.W.; Chen, B.; Deng, Z.H. Influence of laser parameters on processing microgrooves of 2.5-dimensional C/SiC composites via nanosecond laser. *Int. J. Adv. Manuf. Technol.* **2022**, *118*, 85–101. [CrossRef]
39. Zhang, R.H.; Li, W.N.; Liu, Y.S.; Wang, J.; Yang, X.J.; Cheng, L.F. Machining parameter optimization of C/SiC composites using high power picosecond laser. *Appl. Surf. Sci.* **2015**, *330*, 321–331. [CrossRef]
40. Liu, Y.S.; Wang, C.H.; Li, W.N.; Wang, C.H.; Zhang, Q.; Yang, X.J.; Cheng, L.F. Effect of energy density and feeding speed on micro-hole drilling in C/SiC composites by picosecond laser. *J. Mater. Process. Technol.* **2014**, *214*, 3131–3140. [CrossRef]
41. Cheng, L.F.; Xu, Y.D.; Zhang, Q.; Zhang, L.T. Thermal diffusivity of 3D C/SiC composites from room temperature to 1400 °C. *Carbon* **2003**, *41*, 701–711.
42. Tong, Y.G.; Bai, S.X.; Zhang, H.; Ye, Y.C. Laser ablation behavior and mechanism of C/SiC composite. *Ceram. Int.* **2013**, *39*, 6813–6820. [CrossRef]

crystals MDPI

Review

Review on Research and Development of Abrasive Scratching of Hard Brittle Materials and Its Underlying Mechanisms

Huina Qian [1], Mengkai Chen [2], Zijian Qi [3], Qi Teng [4,5], Huan Qi [4,5], Li Zhang [4,5] and Xiaohang Shan [4,5,*]

1 Fair Friend Institute of Intelligent Manufacturing, Hangzhou Vocational & Technical College, Hangzhou 310018, China
2 College of Information and Intelligent Engineering, Zhejiang Wanli University, Ningbo 315100, China
3 School of Energy Science and Engineering, Central South University, Changsha 410083, China
4 College of Mechanical Engineering, Zhejiang University of Technology, Hangzhou 310023, China
5 Key Laboratory of Special Purpose Equipment and Advanced Processing Technology, Zhejiang University of Technology, Ministry of Education, Hangzhou 310023, China
* Correspondence: sxh@zjut.edu.cn

Abstract: Hard brittle materials such as ceramics and crystals are commonly utilized in various industries, including information technology, mechanical engineering, and semiconductors. These materials, known for their high brittleness and hardness but low fracture toughness, pose challenges in efficient and high-quality machining. Current abrasive machining techniques involve rough grinding, fine grinding, and polishing processes, with the latter being the most time-consuming and accounting for over half of the total machining time. Improving processing parameters in rough and fine grinding can increase machining efficiency, reduce surface and subsurface damage, and improve workpiece quality, ultimately reducing the polishing time. This paper explores the abrasive scratching of hard brittle materials, examining the nucleation and propagation of cracks causing surface and subsurface damage, and the underlying mechanisms. The research provides suggestions for enhancing abrasive machining efficiency and ensuring the surface quality of hard brittle materials.

Keywords: hard brittle materials; ceramics and crystals; abrasive scratching; brittle–ductile transition; surface integrity

Citation: Qian, H.; Chen, M.; Qi, Z.; Teng, Q.; Qi, H.; Zhang, L.; Shan, X. Review on Research and Development of Abrasive Scratching of Hard Brittle Materials and Its Underlying Mechanisms. *Crystals* **2023**, *13*, 428. https://doi.org/10.3390/cryst13030428

Academic Editors: Chen Li, Chongjun Wu, Binbin Meng and Shanshan Chen

Received: 16 January 2023
Revised: 5 February 2023
Accepted: 6 February 2023
Published: 2 March 2023

1. Introduction

Hard brittle materials, such as ceramics and crystals, have the characteristics of lo
density, high hardness, and good chemical stability [1], which make them widely used i
mechanical engineering, civil engineering, energy, information technology, and other field
as shown in Figure 1. These materials, known for their high brittleness and hardness b
low fracture toughness, pose challenges in efficient and high-quality machining. Abras
machining technology is one of the commonly used methods to machine these materia

Currently, abrasive machining of hard brittle materials typically involves rough gr
ing, fine grinding, and polishing processes. These materials present a complex rem
mechanism during grinding due to their disordered network structure at the micro
and homogeneity, continuity, and isotropy at the macro level. Fine grinding and poli
play a crucial role in determining the accuracy and surface quality of the workpiece,
in turn affects its performance. Polishing, which takes up over 60% of the total proc
time, is essential for removing the surface and subsurface damage caused by rough a
grinding [2]. However, hard brittle materials are prone to introducing further dam
ing grinding, leading to reduced performance and stability, as well as lower coatin
Hence, it is crucial to obtain workpieces with high surface integrity and low su
damage during rough and fine grinding. To achieve this, repeated experiment
parameter adjustments are required, which can be time-consuming and costly. A
tive approach is abrasive scratching technology, which can efficiently explore th

removal mechanism involved in abrasive grinding and provide guidance for realizing good surface integrity. Molecular dynamics simulation and experimental studies have been used to examine subsurface damage and surface damage, respectively, to validate numerical results and optimize processing parameters [3,4].

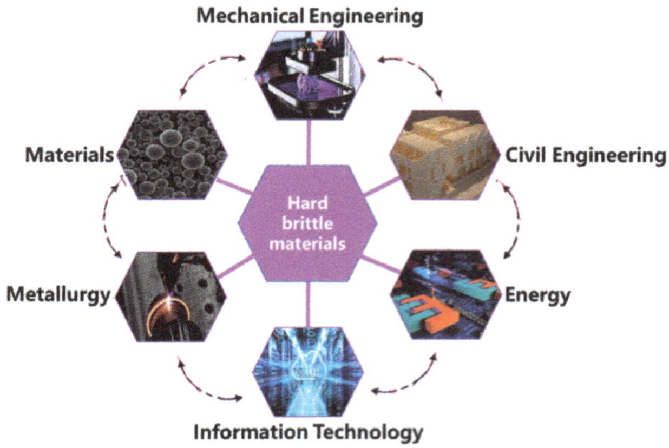

Figure 1. Application of hard brittle materials.

Therefore, this paper aims to enhance the machining accuracy and efficiency of hard brittle materials during rough and fine grinding. To achieve this goal, it is essential to comprehend the mechanism behind the formation of surface and subsurface damage in abrasive scratching of these materials, analyze the nucleation and propagation of cracks, investigate the effect of the strain rate and temperature on grinding, and determine the feasible processing parameters for high-speed machining.

2. Abrasive Grinding and Scratching Mechanism of Hard Brittle Materials

The abrasive grinding process, assisted by the grinding wheel, can be seen as a multi-edge cutting process with multiple randomly distributed abrasives on the wheel. However, the different sizes, shapes, uneven distribution, and protrusion heights of these abrasive grains make it challenging to observe and analyze the grinding process through experiments. To understand the mechanism and complexity of the grinding process, a model that discretely views the contact, interaction, and cutting between a single abrasive grain and the workpiece can be utilized to analyze the various physical phenomena involved in the grinding process.

In the grinding of hard brittle materials, there are three stages of contact between the abrasive grain and the workpiece: scratching, ploughing, and cutting [5], as depicted in Figure 2. During the scratching stage, the material experiences elastic deformation without material removal. In the ploughing stage, the material is removed through the plastic removal mode, and in the cutting stage, it is removed through the brittle removal mode. The two forms of material removal are the plastic removal mode and brittle removal mode. The brittle removal mode involves the formation of microcrack chips, resulting in interwoven cracks, but it leaves micro cracks on the workpiece surface and subsurface. Despite this, the brittle removal mode is efficient, as it provides a high material removal rate, particularly for glass materials. Conversely, the plastic removal mode occurs in the plastic domain of the material where the grinding depth is less than the material's brittle-plastic transition depth. This mode of removal resembles metal processing, producing continuous chips that discharge from the rake face. The plastic removal mode does not cause surface or subsurface damage, making it an ideal method for processing hard brittle materials. In actual grinding processes, both modes are often utilized to reap the benefits of each.

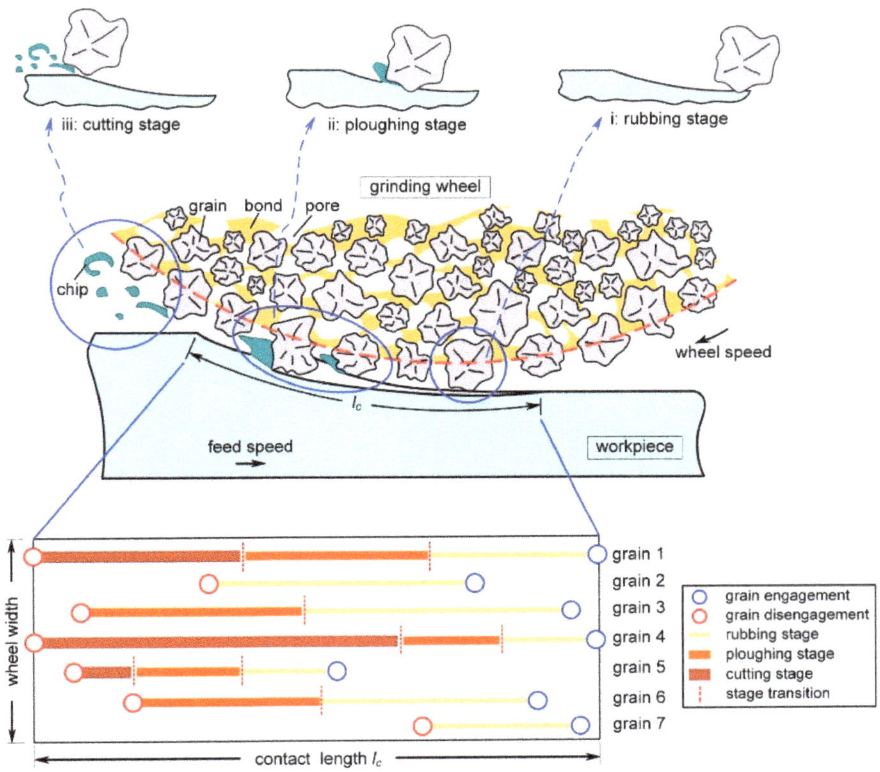

Figure 2. Schematics of contact stages between the abrasive grit and workpiece during grinding: rubbing, ploughing, and cutting [5].

Currently, the investigation of the grinding mechanism for hard brittle materials such as ceramics and crystals is divided into two approaches, one being the mechanical approximation of the indentation process and the other being the approximation of the scratching process [6].

The mechanical approximation of the indentation process views the grinding process between abrasive particles and the workpiece as a pressure by the indenter on the surface of the hard brittle material. This approach, based on indentation fracture mechanics, states that if the normal load applied by the indenter surpasses the workpiece's critical load capacity, the material will deform and break, resulting in the appearance of brittle cracks. Studies on the generation of cracks during indenter pressing and critical normal loads leading to brittle fractures have been conducted. B. Lawn and D. Marshall established the relationship between the normal load of indenter, material properties, and crack system using contact mechanics and indentation fracture mechanics, and the lateral crack and median/radial crack system were found to induce the elastic/plastic indentation damage in ceramics [7,8]. X. Shi and colleagues obtained the critical normal load value of brittle–ductile transition during indentation experiments on hard brittle materials with a diamond indenter [9]. W. Zhang and team established an Elastic–Plastic–Cracking (EPC) constitutive model to simulate the fracture characteristics of hard brittle materials during indenter pressing, using the Vickers indenter [10]. A. Muchtar and team studied the fracture process of alumina during indentation by the finite element method and analyzed the crack nucleation and propagation under different normal loads, finding that the elastic mismatch between the crack and uncracked zones during unloading is critical for hard brittle materials with minimal plastic deformation [11]. B. Zhang et al. carried out indentation experiments on soda–lime glass using a Vickers indenter and an in-house developed device, discovering that the spacing of the experiments and stress field affect crack nucleation and propagation [12].

Figure 3a shows that during the indenter pressing process, a plastic deformation zone will form under the indenter, and a median crack (M) and a lateral crack (L) will appear at the edge of the plastic zone [7,8]. Propagation of the median crack will reduce material strength, while migration of the lateral crack to the free surface will result in brittle removal of the material. However, this research approach only takes normal load into account and does not reflect the actual grinding process.

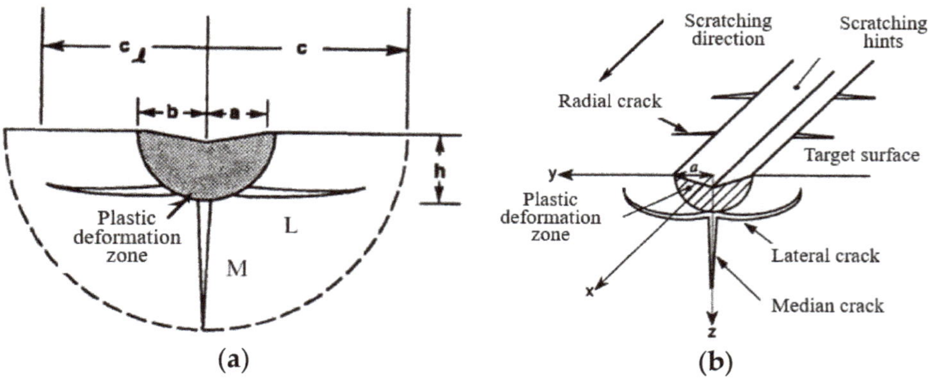

Figure 3. Grinding mechanism of hard brittle materials: (**a**) mechanical approximation of indentation process [7,8], (**b**) mechanical approximation of scratch process [13].

The mechanical approximation of the grinding process views the interaction between the abrasive particles and the workpiece as a process of scratching the hard indenter along the surface of the workpiece with a certain normal force and tangential force. The method is used to study the elastic deformation, plastic flow, and brittle fracture behavior of materials, including hard brittle materials, polymers, and metal materials. Figure 4b shows that the scratches formed by the scratching of the indenter or a single abrasive particle are surrounded by a semi-cylindrical plastic zone, and this zone creates median cracks and lateral cracks at the bottom and radial cracks on the free surface [13].

Research methods for single abrasive grain scratching can be broadly categorized into two types. The first involves simulating abrasive grains on a grinding wheel using regular indenters, such as the Vickers indenter [14–16] or Berkovich indenter [17,18], using specialized scratching experimental instruments. The second involves using actual processing tools, such as turning tools, on lathes or five-axis machining centers, to simulate the abrasive grains.

In 1997, V.H. Bulsara et al. [19] conducted scratch experiments on brittle materials, such as soda–lime glass and sapphire, using a Vickers indenter and observed the contact area using a high-definition optical microscope. The experiments revealed the formation of plastic scratches, slip lines, and banded wear particles, and that cracks form in the loading stage and evolve during the loading and unloading stages. K. Li et al. [20] performed similar experiments on soda–lime glass using spherical and conical indenters, finding that the scratch hardness increased and crack density decreased with the increasing scratching speed for the conical indenter. The results for the spherical indenter indicated that the friction was mainly caused by adhesion, with the damage from transitioning from plastic deformation to brittle cracking with the increasing scratch depth. M. Nakamura et al. [21] conducted scratch experiments with varying scratch speeds and depths using a Vickers indenter, observing that subsurface cracks formed before surface cracks and the crack depth increased with the scratch depth. M. Yoshino et al. [22] performed scratch experiments on hard brittle materials, such as soda–lime glass and quartz glass, under hydrostatic pressure using a special experimental device and found that a larger hydrostatic pressure reduced brittle fracture and enhanced plasticity machinability. W. Gu et al. [23] studied the scratch morphology, material removal volume, and depth during scratching by single and double abrasive grains and found that the material removal mechanism varied depending on the removal mode, plastic or brittle. Z. Qiu et al. [24] conducted single-grain and double-grain scratching experiments on glass ceramics and found that the lateral crack extended towards the free surface and the interaction between lateral and radial cracks were the main reasons for material removal. J. Feng et al. [25] performed a single abrasive scratch experiment on BK7 optical glass and found that the half vertex angle of the abrasive particle influenced the number and depth of lateral and radial cracks formed during the scratching process.

3. Elastic Stress Field of Hard Brittle Materials by Single Abrasive Scratching

The single abrasive scratching process is the fundamental of the abrasive machining technology [26]. The relationship between processing parameters in the grinding of hard brittle materials with a grinding wheel and the material removal mechanism is complex. However, the stress state in the grinding wheel–workpiece interaction region is the fundamental factor limiting the process. Researchers frequently employ single-grain scratching tests to examine this process; thus, gaining insight into the stress field during single-grain scratching can deepen our understanding of material removal, crack initiation, and progression in the grinding process.

The elastic stress field generated during scratching can be viewed as an extension of the elastic stress field produced during indentation, incorporating the effect of tangential load. In indentation tests, materials are typically assumed to be incompressible. However, evidence suggests that for hard brittle materials, indenter indentation results in a reduction of material volume through shear deformation and compaction, leading to material densification. The plastic deformation of these materials must consider plastic flow and densification deformation. Plastic flow results in an elastic–plastic mismatch and residual stress, while densification deformation causes volume reduction but no residual stress [27].

Several studies have explored the elastic stress field during the loading and unloading process of normal loads in contact mechanics. B. Lawn and D. Marshall et al. [7,8,28] considered the contact between a point load and the material surface as an isotropic half-space problem, also known as the Boussinesq problem. E. Yoffe [29] later introduced a model for the elastic stress field in indentation experiments, which takes into account residual stress from densification and shear flow. The model proposes the formation of a rigid plastic core in the material, which generates a strain source, referred to as the Blister stress field. R. Cook and G. Pharr [30] used Yoffe's model to examine the different types of cracks in brittle materials. Y. Ahn et al. [13] expanded Yoffe's model to propose the Sliding Blister Field Model (SBFM) for the scratching process of hard brittle materials, considering the superposition of the Boussinesq stress field, the Cerruti stress field, and the moving Blister stress field. B. Lin [31] proposed a similar elastic stress field model for the grinding process of engineering ceramics. X. Jing et al. [32] combined the SBFM with the Expanding Cylindrical Cavity Model (ECCM) to predict the median and lateral cracks in single-abrasive scratching. W. Wang [27] studied the stress state in the contact area during the single abrasive particle scratching experiment. The crack nucleation position and sequence of quartz glass and BK7 glass and its influence on the material removal mechanism were comprehensively studied and the research results of the crack distribution were shown in Figure 4. X. Yang et al. [33] conducted multiple scratching experiments on glass ceramics to examine the effect of the scratching sequence on the material removal mechanism and established a multiple scratching stress field model. They also studied material flow behavior and chip formation during the grinding process of glass-ceramic materials [34].

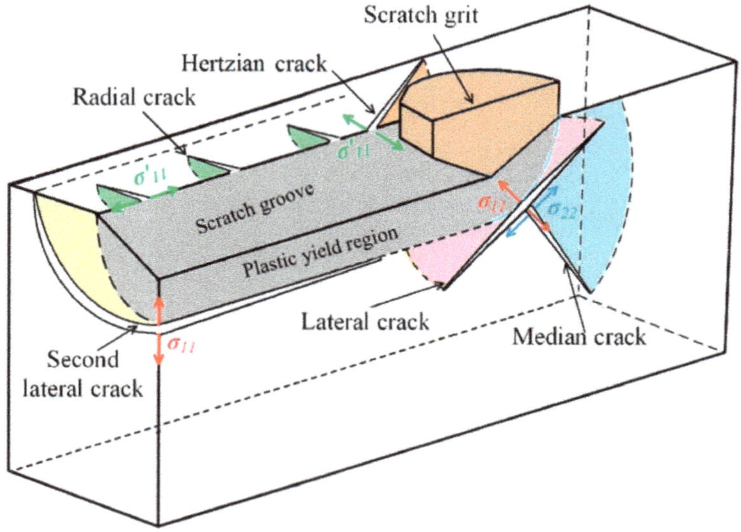

Figure 4. Distribution of the stresses and cracks in the single abrasive scratching of fused silica [27].

4. Effect of Strain Rate and Temperature on the Single Abrasive Scratching of Hard Brittle Materials

4.1. Strain Rate

To improve the processing efficiency of hard brittle materials in the grinding stage, it is necessary to improve the grinding speed. Although the existing single-abrasive scratch theoretical models have made great efforts in studying deformation characteristics and the basic mechanism of crack evolution, these analytical models are based on static loading conditions and ignore the influence of the strain rate on plastic deformation, crack nucleation, and propagation. However, the strain rate caused by the scratching speed has a great influence on the material properties, especially the material strength [35,36], and the strength largely determines the deformation behavior and crack evolution of the material [37–39].

The strain rate caused by the scratching speed has a significant effect on the material properties. In 2000, C. Gauthier et al. [40] carried out scratching experiments on polymethyl methacrylate (PMMA) at different scratching speeds. The experimental results show that the faster the scratching speed, the higher the scratching hardness of the material. K. Wasmer [41] and P. Haasen [42] consider that the strain rate is related to the plastic deformation of the material and affects the hardness of the material during scratching. In 2020, P. Huang et al. [43] carried out scratch experiments on silicon carbide ceramics at different speeds to study the effect of the strain rate on material behavior and brittle–ductile transition depth. The results show that the microhardness, elastic modulus, and brittle–ductile transition depth of the material increase at high scratching speeds, and the brittle chip and subsurface damage depend on the lateral crack and the median crack, respectively.

Many scholars have also studied the brittle–ductile transition of materials under different strain rates. In 2004, G. Cai et al. [44] carried out ultra-high speed grinding experiments on brittle materials. The results show that under ultra-high speed impacting conditions, brittle materials will undergo plastic flow and lead to material failure. With the increase of grinding speed, the grinding force decreases gradually. They also established a chip formation model under ultra-high speed grinding conditions based on the experimental results. In 2013, P. Feng [45] carried out nano-scratching experiments on Al_2O_3 and found that the scratch speed has a significant effect on the plastic and brittle deformation characteristics of the material. In 2014, B. Li et al. [46] showed that high-speed grinding can change the contact behavior between brittle materials and abrasive grains, so that the penetration depth increases without brittle cracks. It is proved by simulations and experiments that the brittle–ductile transition depth at a high strain rate is larger than that under quasi-static conditions. In 2017, K. Mukaiyama et al. [47] studied the brittle–ductile transition of monocrystalline silicon at different scratching speeds and found that increasing the scratching speed resulted in a decrease in the transverse force during the brittle–ductile transition.

The strain rate has a significant impact on the surface and subsurface damage in hard, brittle materials. In 2017, Yang et al. [48] proposed a new stress field model that considers the relationship between the strain rate and material properties, revealing the effect of scratching speed and strain rate on plastic deformation and crack formation. Scratching experiments on glass ceramics showed that with increasing scratching speed, the radius of plastic deformation and median crack length decrease, while the depth of the brittle–ductile transition increases. In 2019, Li et al. [49] conducted nano-scratching experiments at different speeds on GGG single crystal to study the effect of the strain rate on surface and subsurface deformation, as shown in Figure 5. The results showed that higher scratching speed leads to a shallower penetration depth and larger continuous chips, improving the plasticity of GGG single crystal and reducing the depth of the brittle–ductile transition. The authors also established a penetration depth prediction model incorporating strain rate and material elastic rebound, with results that agree well with experiments.

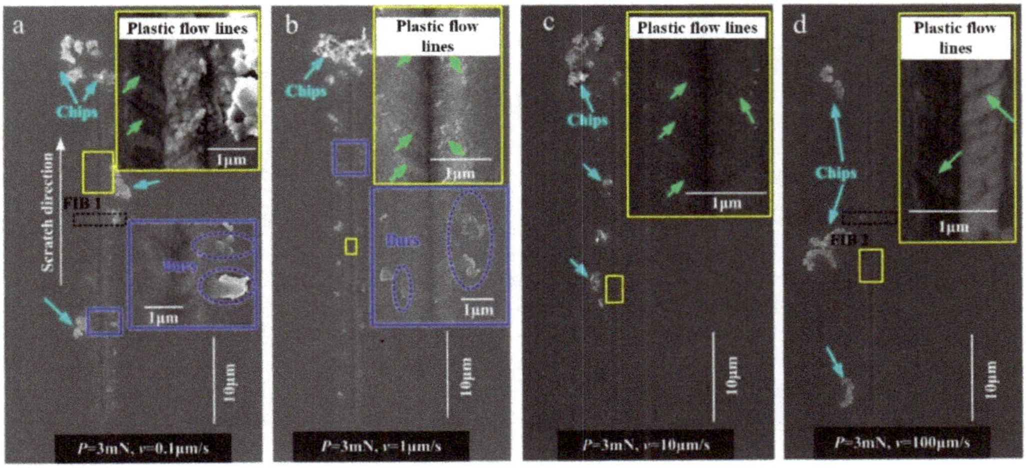

Figure 5. Effect of strain rates caused by different scratching speeds on surface morphology [49].

B. Zhang [50] comprehensively studied the strain rate effect in ultra-high speed machining. They proposed that the depth of the material processing damage decreased with the increase of the processing strain rate, showing a 'skin effect'. Ultra-high speed machining can improve the strain rate of the material in the machining area, reduce the depth of machining damage, improve the machining surface integrity, and greatly improve the material processing efficiency. The increase of strain rate during loading will lead to the increase of yield strength and hardness of the material, and the decrease of toughness of the material, which leads to the embrittlement of the material. The embrittlement of brittle materials at high strain rates is related to dislocation pile-ups, and the damage introduced in the grinding of ceramic materials at high strain rates is small [51].

4.2. Temperature

The grinding process is essentially a random comprehensive process of material removal caused by scratching, ploughing, and cutting of the material on the target surface by countless randomly distributed abrasive grains on the grinding wheel. Grinding heat will be generated during the contact between these abrasive grains and the workpiece, and the grinding heat will cause the workpiece surface temperature to rise. At this time, all the abrasive grains involved in the grinding of the grinding wheel–workpiece contact area can be regarded as a single point heat source that continuously emits heat, and the result of the combined action of these point heat sources will cause the increase of temperature in the grinding wheel–workpiece contact area. The physical quantities on the grinding temperature include the total temperature rise of the workpiece, temperature of the grinding wheel–workpiece contact surface, the grinding surface temperature of the workpiece, and the grinding point temperature of the abrasive grain. The overall temperature rise of the workpiece is determined by the part of the grinding heat introduced into the workpiece. This part of the grinding heat will cause the thermal expansion and distortion of the workpiece, resulting in a decrease in size and shape accuracy, and may lead to changes in the mechanical properties of the material. The grinding wheel–workpiece contact surface temperature will affect the quality of the workpiece surface, and the surface temperature of the workpiece grinding affects the surface metamorphic layer of parts, such as the surface residual stress, crack distribution, etc. The grinding point temperature of the abrasive grain is the temperature of the small area directly acting on the rake face of the abrasive grain and the workpiece. It is often the area with the highest grinding temperature, which will affect the life of the abrasive grain and the chemical reaction between the abrasive grain and the target material [52].

Many scholars have proposed theoretical models for the heat transfer in the grinding zone to explore the heat generation and transfer process during the grinding process and predict the workpiece temperature to avoid thermal damage. Many theoretical models of grinding heat are based on the moving heat source model proposed by J. C. Jaeger [53]. In 1962, R. Hahn [54] found that most of the grinding heat was generated in the scratching stage during the contact between the abrasive particles and the workpiece. He believed that the contact area between the grinding wheel and the workpiece was a heat source, and the heat flux evenly distributed in the grinding area moved on the adiabatic surface at the workpiece in a given feed rate. In 1995, C. Guo et al. [55] proposed a heat source model with a linear increase in heat flux, and this model assumes that the heat flux is approximately proportional to the undeformed chip thickness, where the heat flux at the rear of the contact area is greater than the heat flux at the front. For high-efficiency deep grinding, in 2001, W.B. Rowe and T. Jin [56] considered that the wheel–workpiece contact surface is a moving heat source with a linear increase in heat flux or a horizontal circular arc distribution, because in high-efficiency deep grinding, the grinding depth tends to reach about 10 mm, and the wheel–workpiece contact area is no longer a horizontal plane. The model proposed by T. Jin and G. Cai [57] regards the heat flux distribution of the moving heat source as an inclined uniform distribution. Compared with other models, this model is more suitable for creep feed grinding, and Rowe's inclined triangular heat flux density heat source model is more accurate for the temperature prediction of the workpiece surface in high-efficiency deep grinding.

Scholars have also studied the distribution of grinding heat, which includes the percentage of heat flowing into cutting fluid, chips, grinding wheels, and workpieces. In 1970, N. DesRuisseaux and R. Zerkle [58] used the convective heat flux uniformly distributed on the entire workpiece surface to describe the convective cooling effect of the cutting fluid. In 1971, S. Malkin and N. Cook [59] found that the chip will not melt before leaving the workpiece during the grinding process, so the heat flow into the chip will be limited by the melting energy of the chip. In 1994, C. Guo and S. Malkin [60] performed a transient thermal analysis based on the critical temperature of cutting fluid boiling to predict burn heat flux. In 2004, Z. Hou et al. [61] also proposed a micro-scale thermodynamic model, which takes into account the distribution of abrasive grains on the surface of the grinding wheel. The accuracy of the model was verified by experiments as well. In 2013, W. Rowe [62] conducted a shallow grinding experiment to measure these values and verify the related models.

The processing of hard brittle materials at high temperatures is often considered as an effective way to improve the machinability of hard brittle materials. In 2004, M. Michel et al. [63] conducted indentation experiments on soda–lime glass at different temperatures and different loads with a Vickers indenter. The results show that the hardness of the material decreases with the increase of temperature, and the radial crack length increases with the increase of temperature, but high temperatures can inhibit the propagation of the median crack. In 2006, the scholar carried out high temperature indentation experiments and analysis on fused silica glass. The research showed that with the increase of temperature, the load required for radial crack nucleation gradually increased [64]. In 2016, W. Wang et al. [17] carried out high temperature indentation experiments and high temperature grinding wheel experiments on fused silica. The results show that the molecular structure of the fused silica is dense under suitable high temperature conditions. Under the same indentation load, the critical brittle–ductile transition load of the fused silica at a high temperature is higher than that at room temperature, which is beneficial to the plastic domain processing of materials. High temperatures can improve the shear flow of fused silica and prevent crack nucleation and propagation. Under high temperature conditions, a larger grinding thickness is beneficial to directly remove the subsurface cracks generated in the previous machining process at one time, so as to obtain lower surface roughness and better surface integrity. However, excessive grinding thickness will cause grinding wheel burns as well. In 2019, X. Rao et al. [65] carried out the indentation test of reaction sintered silicon carbide RB-SiC ceramics from room temperature to 1200 °C by laser heating, and studied the influence of temperature on the hardness, elastic modulus, and fracture toughness of the material. The results show that with the increase of temperature, the permanent deformation of the material leads to the decrease of the hardness of the material, and the elastic modulus decreases and the fracture toughness increases with the increase of contact depth. After that, they also carried out scratching experiments on a RB-SiC ceramic Vickers indenter at a high temperature, and observed the material removal behavior, scratch hardness, critical depth of the brittle–ductile transition, scratch force, and friction force during scratching. The results show that under high temperature processing conditions, the material deformation and adhesion behavior promote the removal of the plastic zone and increase the friction coefficient [15]. Z. Li et al. [66] used laser-assisted grinding to study the material removal mechanism, grinding force ratio, and surface integrity of RB-SiC ceramics during grinding. The research shows that a higher processing temperature changes the material structure, decreases the hardness, promotes the plastic removal of the material, and obtains better surface integrity. In 2020, P. Li et al. [14] studied the effect of temperature on the subsurface damage of optical glass grinding. The results show that with the increase of abrasive temperature and strain rate in the grinding zone, the fracture toughness of the material increases, the microhardness decreases, and the subsurface damage decreases, as shown in Figure 6. In 2021, Y. Niu et al. [67] simulated the scratch behavior of monocrystalline silicon at different temperatures by molecular dynamics, and studied the subsurface damage mechanism, scratch force, and phase transition. The results show that the scratch force at a high temperature is smaller and the material removal rate is higher. In summary, the above studies on the effect of the strain rate and temperature on the single abrasive scratching of hard brittle materials are generally based on theoretical and experimental methods, limited studies have contributed to the numerical modelling and analysis of this effect, and the numerical method has been extensively used in exploring the abrasive machining process and its underlying mechanisms [68–72]. It is also noted that with the development of the high-tech machining and measuring technologies understanding the abrasive scratching of hard brittle materials and associated mechanisms is beneficial for extending its applications in other aspects [73–76].

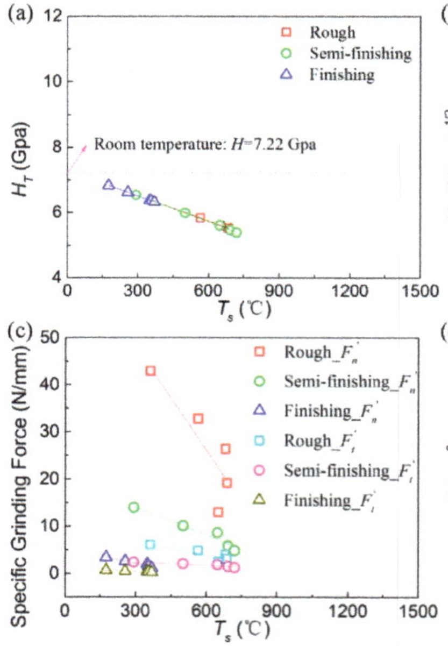

Figure 6. Effect of temperature on the single abrasive scratching of optical glass: (**a**) microhardness, (**b**) fracture toughness, (**c**) specific grinding force, (**d**) specific grinding energy [14].

5. Conclusions

Hard brittle materials such as ceramics and crystals are known for their high hardness, brittleness, and low fracture toughness, making them challenging to machines. Common processing methods include rough grinding, fine grinding, and polishing, and it is important to reduce surface and subsurface damage during the grinding stage to improve the processing efficiency and reduce costs. This study examines the stress field distribution in the single abrasive scratching of hard brittle materials and investigates the effect of the scratching speed and grinding temperature on surface and subsurface damage. The feasibility of high-speed grinding of hard brittle materials is also evaluated.

This paper presents an investigation into the stress field distribution in the single abrasive grain scratching of hard brittle materials. It seeks to reduce surface and subsurface damage in the grinding stage to improve the processing efficiency and reduce costs. The study explores the impact of scratching speed and grinding temperature on damage and verifies the feasibility of high-speed grinding. The results suggest that the efficient reduction of damage in the grinding stage can minimize the need for polishing and reduce the time and cost of this process. The paper proposes an elastic stress field model and a principal stress and hardness ratio model to simulate the distribution of stress and the formation of surface and subsurface cracks. The study also indicates that high-speed or ultra-high-speed grinding can improve processing efficiency, reduce surface roughness, and limit crack formation. Finally, the paper proposes a subsurface damage prediction model and highlights the benefits of increasing the grinding speed in reducing normal and tangential grinding forces and extending the processing life of abrasive grains or grinding wheels.

In this review, a stress field model, a ratio model of principal stress to hardness in rectangular coordinates, and a subsurface damage prediction model taking into account the strain rate and grinding temperature for hard brittle materials during single abrasive scratching has been presented. However, there are still areas for further investigation, which include:

(1) In the calculation of the stress field model and principal stress–hardness ratio model for the single abrasive scratching of hard brittle materials, the abrasive was idealized as a sharp indenter and the force on the workpiece was treated as a point load. However, in actual single abrasive scratching, the contact between the abrasive and the workpiece is usually surface contact between the abrasive rake face and the workpiece surface. Further study is necessary to better understand the nature of this contact and improve the model accordingly.

(2) This study considered the impact of grinding temperature on the abrasive grain–workpiece contact area. However, the grinding area temperature was not directly measured during the experiments, and the temperature calculation model was not verified, leading to potential inaccuracies in the subsurface damage model calculation. Further research should focus on verifying the temperature calculation model through experimentation.

Author Contributions: Conceptualization, H.Q. (Huina Qian) and Q.T.; data curation, M.C. and H.Q. (Huina Qian); formal analysis, M.C. and Q.T.; funding acquisition, H.Q. (Huan Qi) and X.S.; investigation, M.C. and Z.Q.; project administration, H.Q. (Huan Qi) and L.Z.; supervision, X.S. and L.Z.; writing—original draft, Z.Q. and H.Q. (Huina Qian); writing—review and editing, X.S. All authors have read and agreed to the published version of the manuscript.

Funding: This research was funded by the Key Research and Development Program of Zhejiang Province (2021C04011), and National Natural Science Foundation of China (51575493).

Data Availability Statement: Data sharing not applicable.

Conflicts of Interest: The authors declare that the research was conducted in the absence of any commercial or financial relationships that could be construed as a potential conflict of interest.

References

1. Li, C.; Piao, Y.; Meng, B.; Hu, Y.; Li, L.; Zhang, F. Phase transition and plastic deformation mechanisms induced by self-rotating grinding of GaN single crystals. *Int. J. Mach. Tools Manuf.* **2022**, *172*, 103827. [CrossRef]
2. Evans, C.; Paul, E.; Dornfeld, D.; Lucca, D.; Byrne, G.; Tricard, M.; Klocke, F.; Dambon, O.; Mullany, B. Material removal mechanisms in lapping and polishing. *CIRP Ann.* **2003**, *52*, 611–633. [CrossRef]
3. Li, C.; Hu, Y.; Zhang, F.; Geng, Y.; Meng, B. Molecular dynamics simulation of laser assisted grinding of GaN crystals. *Int. J. Mech. Sci.* **2023**, *239*, 107856. [CrossRef]
4. Li, C.; Piao, Y.; Zhang, F.; Zhang, Y.; Hu, Y.; Wang, Y. Understand anisotropy dependence of damage evolution and material removal during nanoscratch of MgF_2 single crystals. *Int. J. Extrem. Manuf.* **2023**, *5*, 015101. [CrossRef]
5. Li, H.N.; Axinte, D. On a stochastically grain-discretised model for 2D/3D temperature mapping prediction in grinding. *Int. J. Mach. Tools Manuf.* **2017**, *116*, 60–76. [CrossRef]
6. Qi, H.; Wang, Y.; Qi, Z.; Shi, L.; Fang, Z.; Zhang, L.; Riemer, O.; Karpuschewski, B. A Novel Grain-Based DEM Model for Evaluating Surface Integrity in Scratching of RB-SiC Ceramics. *Materials* **2022**, *15*, 8486. [CrossRef]
7. Marshall, D.; Lawn, B.; Evans, A. Elastic/plastic indentation damage in ceramics: The lateral crack system. *J. Am. Ceram. Soc.* **1982**, *65*, 561–566. [CrossRef]
8. Lawn, B.R.; Evans, A.G.; Marschall, D. Elastic/plastic indentation damage in ceramics: The median/radial crack system. *J. Am. Ceram. Soc.* **1980**, *63*, 574–581. [CrossRef]
9. Shi, X.K.; Teng, L.; Li, Y.Q. The critical condition of ductile grinding of hard brittle materials. *Aviat. Precis. Manuf. Technol.* **1996**, *4*, 10–13.
10. Zhang, W.; Subhash, G. An elastic–plastic-cracking model for finite element analysis of indentation cracking in brittle materials. *Int. J. Solids Struct.* **2001**, *38*, 5893–5913. [CrossRef]
11. Muchtar, A.; Lim, L.C.; Lee, K.H. Finite element analysis of vickers indentation cracking processes in brittle solids using elements exhibiting cohesive post-failure behaviour. *J. Mater. Sci.* **2003**, *38*, 235–243. [CrossRef]
12. Zhang, B.; Yoshioka, M.; Hira, S. Experimental observation and numerical analysis of cracking phenomena in adjoining Vickers indentations on glass surface—ScienceDirect. *J. Mater. Process. Technol.* **2008**, *208*, 171–178. [CrossRef]
13. Ahn, Y.; Farris, T.; Chandrasekar, S. Sliding microindentation fracture of brittle materials: Role of elastic stress fields. *Mech. Mater.* **1998**, *29*, 143–152. [CrossRef]
14. Li, P.; Chen, S.; Xiao, H.; Chen, Z.; Qu, M.; Dai, H.; Jin, T. Effects of local strain rate and temperature on the workpiece subsurface damage in grinding of optical glass. *Int. J. Mech. Sci.* **2020**, *182*, 105737. [CrossRef]
15. Rao, X.; Zhang, F.; Luo, X.; Ding, F.; Cai, Y.; Sun, J.; Liu, H. Material removal mode and friction behaviour of RB-SiC ceramics during scratching at elevated temperatures. *J. Eur. Ceram. Soc.* **2019**, *39*, 3534–3545. [CrossRef]

16. Sellappan, P.; Rouxel, T.; Celarie, F.; Becker, E.; Houizot, P.; Conradt, R.J. Composition dependence of indentation deformation and indentation cracking in glass. *Acta Mater.* **2013**, *61*, 5949–5965. [CrossRef]
17. Wang, W.; Yao, P.; Wang, J.; Huang, C.; Zhu, H.; Zou, B.; Liu, H.; Yan, J. Crack-free ductile mode grinding of fused silica under controllable dry grinding conditions. *Int. J. Mach. Tools Manuf.* **2016**, *109*, 126–136. [CrossRef]
18. Meng, B.; Zhang, F.; Li, Z. Deformation and removal characteristics in nanoscratching of 6H-SiC with Berkovich indenter. *Mater. Sci. Semicond. Process.* **2015**, *31*, 160–165. [CrossRef]
19. Bulsara, V.H.; Chandrasekar, S. Direct observation of contact damage around scratches in brittle solids. In Proceedings of the SPIE—The International Society for Optical Engineering, Orlando, FL, USA, 20–25 April 1997; Volume 3060.
20. Li, K.; Shapiro, Y.; Li, J.C.M. Scratch tests of soda-lime glass. *Acta Mater.* **1998**, *46*, 5569–5578. [CrossRef]
21. Nakamura, M.; Sumomogi, T.; Endo, T. Evaluation of surface and subsurface cracks on nano-scale machined brittle materials by scanning force microscope and scanning laser microscope. *Surf. Coat. Technol.* **2003**, *169*, 743–747. [CrossRef]
22. Yoshino, M.; Ogawa, Y.; Aravindan, S. Machining of hard-brittle materials by a single point tool under external hydrostatic pressure. *J. Manuf. Sci. Eng.* **2005**, *127*, 837–845. [CrossRef]
23. Gu, W.; Yao, Z.; Liang, X. Material removal of optical glass BK7 during single and double scratch tests. *Wear* **2011**, *270*, 241–246. [CrossRef]
24. Qiu, Z.; Liu, C.; Wang, H.; Yang, X.; Fang, F.; Tang, J. Crack propagation and the material removal mechanism of glass–ceramics by the scratch test. *J. Mech. Behav. Biomed. Mater.* **2016**, *64*, 75–85. [CrossRef] [PubMed]
25. Feng, J.; Wan, Z.; Wang, W.; Ding, X.; Tang, Y. Crack behaviors of optical glass BK7 during scratch tests under different tool apex angles. *Wear* **2019**, *430–431*, 299–308. [CrossRef]
26. Ji, R.Q.; Qi, Z.J.; Chen, J.C.; Zhang, L.; Lin, K.F.; Lu, S.S.; Li, Y.B. Numerical and experimental investigation on the abrasive flow machining of artificial knee joint surface. *Crystals* **2023**, *13*, 430.
27. Wang, W. High Efficiency Controllable Precision Grinding Mechanisms of Fused Silica. Ph.D. Thesis, Shandong University, Jinan, China, 2017.
28. Lawn, B.; Swain, M. Microfracture beneath point indentations in brittle solids. *J. Mater. Sci.* **1975**, *10*, 113–122. [CrossRef]
29. Yoffe, E. Elastic stress fields caused by indenting brittle materials. *Philos. Mag. A* **1982**, *46*, 617–628. [CrossRef]
30. Cook, R.F.; Pharr, G.M. Direct observation and analysis of indentation cracking in glasses and ceramics. *J. Am. Ceram. Soc.* **1990**, *73*, 787–817. [CrossRef]
31. Lin, B. Theoretical and Experimental Studies on the Ultra-precision Grinding of Engineering Ceramics. Doctoral Thesis, Tianjin University, Tianjin, China, 1999.
32. Jing, X.; Maiti, S.; Subhash, G. A new analytical model for estimation of scratch-induced damage in brittle solids. *J. Am. Ceram. Soc.* **2007**, *90*, 885–892. [CrossRef]
33. Yang, X.; Qiu, Z.; Li, X. Investigation of scratching sequence influence on material removal mechanism of glass-ceramics by the multiple scratch tests. *Ceram. Int.* **2019**, *45*, 861–873. [CrossRef]
34. Yang, X.; Qiu, Z.; Wang, Y. Investigation of material flow behaviour and chip formation mechanism during grinding of glass-ceramics by nanoscratch. *Ceram. Int.* **2019**, *45*, 15954–15963. [CrossRef]
35. Yuan, H.Y.; Yang, W.B.; Zhang, L.; Hong, T. Model development of stress intensity factor on 7057T6 aluminum alloy using extended finite element method. *Coatings* **2023**, *13*, 581.
36. Xie, X.D.; Zhang, L.; Zhu, L.L.; Li, Y.B.; Hong, T.; Yang, W.B.; Shan, X.H. State of the art and perspectives on surface strengthening process and associated mechanisms by shot peening. *Coatings* **2023**, *13*.
37. Qi, H.; Shi, L.; Teng, Q.; Hong, T.; Tangwarodomnukun, V.; Liu, G.; Li, H.N. Subsurface damage evaluation in the single abrasive scratching of BK7 glass by considering coupling effect of strain rate and temperature. *Ceram. Int.* **2021**, *48*, 8661–8670. [CrossRef]
38. Shi, H.; Liu, G.; Yang, G.; Bi, Q.; Zhao, Y.; Wang, B.; Sun, X.; Liu, X.; Qi, H.; Xu, W.; et al. Analytical modelling of edge chipping in scratch of soda-lime glass considering strain-rate hardening effect. *Ceram. Int.* **2021**, *47*, 26552–26566. [CrossRef]
39. Hu, W.; Teng, Q.; Hong, T.; Saetang, V.; Qi, H. Stress field modeling of single-abrasive scratching of BK7 glass for surface integrity evaluation. *Ceram. Int.* **2021**, *48*, 12819–12828. [CrossRef]
40. Gauthier, C.; Schirrer, R. Time and temperature dependence of the scratch properties of poly (methylmethacrylate) surfaces. *J. Mater. Sci.* **2000**, *35*, 2121–2130. [CrossRef]
41. Wasmer, K.; Parlinska-Wojtan, M.; Gassilloud, R.; Pouvreau, C.; Tharian, J.; Micher, J. Plastic deformation modes of gallium arsenide in nanoindentation and nanoscratching. *Appl. Phys. Lett.* **2007**, *90*, 031902. [CrossRef]
42. Haasen, P. Zur plastischen Verformung von Germanium und InSb. *Z. Phys.* **1962**, *167*, 461–467. [CrossRef]
43. Huang, P.; Zhang, J. Strain rate effect on the ductile brittle transition in grinding hot pressed SiC ceramics. *Micromachines* **2020**, *11*, 545. [CrossRef]
44. Zhao, H.H.; Cai, G.Q. Model establishment and mechanism research of chip-formation due to shock under ultra-high speed grinding. *China Mech. Eng.* **2004**, *12*, 1038–1041.
45. Feng, P.F.; Zhang, C.L.; Wu, Z.J.; Zhang, J.F. Effect of scratch velocity on deformation features of C-plane Sapphire during nanoscratching. *J. Mech. Eng.* **2013**, *59*, 367–374. [CrossRef]
46. Li, B.; Ni, J.; Yang, J.; Liang, S.Y. Study on high-speed grinding mechanisms for quality and process efficiency. *Int. J. Adv. Manuf. Technol.* **2013**, *70*, 813–819. [CrossRef]

47. Mukaiyama, K.; Ozaki, M.; Wada, T. Study on ductile-brittle transition of single crystal silicon by a scratching test using a single diamond tool. In Proceedings of the 2017 8th International Conference on Mechanical and Aerospace Engineering (ICMAE), Prague, Czech Republic, 22–25 July 2017.
48. Yang, X.; Qiu, Z.; Lu, C.; Li, X.; Tang, J. Modelling the strain rate sensitivity on the subsurface damages of scratched glass ceramics. *Ceram. Int.* **2017**, *43*, 12930–12938. [CrossRef]
49. Li, C.; Zhang, F.; Piao, Y. Strain-rate dependence of surface/subsurface deformation mechanisms during nanoscratching tests of GGG single crystal. *Ceram. Int.* **2019**, *45*, 15015–15024. [CrossRef]
50. Yang, X.; Zhang, B. Material embrittlement in high strain-rate loading. *Int. J. Extrem. Manuf.* **2019**, *1*, 022003. [CrossRef]
51. Qi, H.; Xing, W.; Tan, W.; Lin, H.; Guo, H.J.; Chen, M.Z.; Tang, H.P. Effect of the sintering process on mechanical behaviors of Zirconia ceramics by NanoParticle Jetting. *Ceram. Int.* **2023**. *Submitted for Publication.*
52. Ji, P.J.; Zhang, J.J.; Xie, X.D.; Ying, R.M.; Zhang, L. Review on Wear Characteristics of Artificial Hip Joint and Associated Physical Training. *Front. Mater.* **2023**, *10*.
53. Jaeger, J.C. Moving sources of heat and the temperature of sliding contacts. *J. Proc. R. Soc.* **1942**, *76*, 203.
54. Hahn, R.S. On the nature of the grinding process. In Proceedings of the 3rd International Machine Tool Design & Research Conference, Manchester, UK, 18–20 September 1962; pp. 129–154.
55. Guo, C.; Malkin, S. Analysis of energy partition in grinding. *J. Eng. Ind.* **1995**, *117*, 55–61. [CrossRef]
56. Rowe, W.B.; Jin, T. Temperatures in high efficiency deep grinding. *CIRP Ann.* **2001**, *50*, 205–208. [CrossRef]
57. Jin, T.; Cai, G. Analytical thermal models of oblique moving heat source for deep grinding and cutting. *J. Manuf. Sci. Eng.* **2001**, *123*, 185–190. [CrossRef]
58. Desruisseaux, N.R.; Zerkle, R.D. Temperature in semi-infinite and cylindrical bodies subjected to moving heat sources and surface cooling. *J. Heat Transf.* **1970**, *92*, 456–464. [CrossRef]
59. Malkin, S.; Cook, N.H. The wear of grinding wheels: Part 2—Fracture wear. *J. Eng. Ind.* **1971**, *93*, 1129–1133. [CrossRef]
60. Guo, C.; Malkin, S. Analytical and experimental investigation of burnout in creep-feed grinding. *CIRP Ann.* **1994**, *43*, 283–286. [CrossRef]
61. Hou, Z.B.; Komanduri, R. On the mechanics of the grinding process, Part III-thermal analysis of the abrasive cut-off operation. *Int. J. Mach. Tools Manuf.* **2004**, *44*, 271–289. [CrossRef]
62. Rowe, W.B. *Principles of Modern Grinding Technology*, 2nd ed.; Elsevier Inc.: Amsterdam, The Netherlands, 2013.
63. Michel, M.; Mikowski, A.; Lepienski, C.; Foerster, C.; Serbena, F. High temperature microhardness of soda-lime glass. *J. Non-Cryst. Solids* **2004**, *348*, 131–138. [CrossRef]
64. Michel, M.D.; Serbena, F.C.; Lepienski, C.M. Effect of temperature on hardness and indentation cracking of fused silica. *J. Non-Cryst. Solids* **2006**, *352*, 3550–3555. [CrossRef]
65. Rao, X.; Zhang, F.; Luo, X.; Ding, F. Characterization of hardness, elastic modulus and fracture toughness of RB-SiC ceramics at elevated temperature by Vickers test. *Mater. Sci. Eng. A* **2019**, *744*, 426–435. [CrossRef]
66. Li, Z.; Zhang, F.; Luo, X.; Chang, W.; Cai, Y.; Zhong, W.; Ding, F. Material removal mechanism of laser-assisted grinding of RB-SiC ceramics and process optimization. *J. Eur. Ceram. Soc.* **2019**, *39*, 705–717. [CrossRef]
67. Niu, Y.; Zhao, D.; Wang, S.; Li, S.; Wang, Z.; Zhao, H. Investigations on thermal effects on scratch behavior of monocrystalline silicon via molecular dynamics simulation. *Mater. Today Commun.* **2021**, *26*, 102042. [CrossRef]
68. Zhang, L.; Yuan, Z.; Qi, Z.; Cai, D.; Cheng, Z.; Qi, H. CFD-based study of the abrasive flow characteristics within constrained flow passage in polishing of complex titanium alloy surfaces. *Powder Technol.* **2018**, *333*, 209–218. [CrossRef]
69. Qi, H.; Fan, J.; Wang, J.; Li, H. Impact erosion by high velocity micro-particles on a quartz crystal. *Tribol. Int.* **2015**, *82*, 200–210. [CrossRef]
70. Qi, H.; Fan, J.; Wang, J.; Li, H. On the erosion process on quartz crystals by the impact of multiple high-velocity micro-particles. *Tribol. Int.* **2016**, *95*, 462–474. [CrossRef]
71. Qi, H.; Qin, S.; Cheng, Z.; Teng, Q.; Hong, T.; Xie, Y. Towards understanding performance enhancing mechanism of micro-holes on K9 glasses using ultrasonic vibration-assisted abrasive slurry jet. *J. Manuf. Process.* **2021**, *64*, 585–593. [CrossRef]
72. Qi, H.; Cheng, Z.; Cai, D.; Yin, L.; Wang, Z.; Wen, D. Experimental study on the improvement of surface integrity of tungsten steel using acoustic levitation polishing. *J. Mater. Process. Technol.* **2018**, *259*, 361–367. [CrossRef]
73. Wang, Y.Y.; Wang, Z.; Ni, P.C.; Wang, D.J.; Guo, S.H.; Chen, Z.Z. Experimental and numerical study on regulation of cutting temperature during the circular sawing of 45 steel. *Coatings* **2023**, *13*.
74. Chen, Z.; Wen, D.; Lu, J.; Yang, J.; Qi, H. Surface quality improvement by using a novel driving system design in single-side planetary abrasive lapping. *Materials* **2022**, *14*, 1691. [CrossRef]
75. Wu, B.; Li, D.R.; Zhou, Y.; Zhu, D.; Zhao, Y.P.; Qiao, Z.K.; Chen, B.; Wang, X.L.; Lin, Q. Construction of a calibration field of absolute gravity in a cave using the Cold Atom Gravimeter. *Sensors* **2023**, *23*.
76. Qi, H.; Qin, S.; Cheng, Z.; Zou, Y.; Cai, D.; Wen, D. DEM and experimental study on the ultrasonic vibration-assisted abrasive finishing of WC-8Co cemented carbide cutting edge. *Powder Technol.* **2021**, *378*, 716–723. [CrossRef]

crystals

MDPI

Article

Numerical and Experimental Investigation on the Abrasive Flow Machining of Artificial Knee Joint Surface

Renquan Ji [1,2], Zijian Qi [3,*], Junchao Chen [1,2], Li Zhang [1,2], Kaifeng Lin [1,2], Shasha Lu [1,2] and Yanbiao Li [1,2]

1 College of Mechanical Engineering, Zhejiang University of Technology, Hangzhou 310014, China
2 Key Laboratory of Special Purpose Equipment and Advanced Processing Technology, Zhejiang University of Technology, Ministry of Education, Hangzhou 310014, China
3 School of Energy Science and Engineering, Central South University, Changsha 410083, China
* Correspondence: 203912043@csu.edu.cn

Abstract: The titanium alloy artificial knee joint is used extensively in the current medical industry due to its distinct characteristics and properties that are like the real human knee joint, but it does need to be polished to improve its performance and service life before it can be used. Due to the complicated surface profile, the traditional abrasive flow machining technique cannot achieve a good surface finish offering uniformity and quality. Thus, in this paper, a proper constrained flow channel is designed to conduct the abrasive flow machining of the titanium alloy artificial knee joint surface to overcome these issues. A numerical study is first conducted to explore the distribution of abrasive flow velocity and pressure near the target surface in the constrained flow channel by using the COMSOL Multiphysics software, and it is found from the distribution of the dimensionless material removal rate on the target surface that the exchange of the abrasive flow inlet and outlet during the machining process is recommended to improve the surface finish uniformity. Then, the corresponding experiments are conducted to analyze the surface morphology before and after the abrasive flow machining process. It is found that the surface roughness of the target surface decreases from approximately 394 nm to 171 nm with good uniformity as well. Therefore, the proposed abrasive flow machining method with a properly designed constrained flow channel is useful for the rough polishing and fine finishing of the titanium alloy artificial joint.

Keywords: abrasive flow machining; constrained flow channel; artificial knee joint; surface finish; titanium alloy

Citation: Ji, R.; Qi, Z.; Chen, J.; Zhang, L.; Lin, K.; Lu, S.; Li, Y. Numerical and Experimental Investigation on the Abrasive Flow Machining of Artificial Knee Joint Surface. *Crystals* **2023**, *13*, 430. https://doi.org/10.3390/cryst13030430

Academic Editor: Umberto Prisco

Received: 29 January 2023
Revised: 4 February 2023
Accepted: 6 February 2023
Published: 2 March 2023

1. Introduction

An artificial knee joint is an artificial organ that is implanted in the human body instead of human bone to keep the original function of the joint, with the most widely used material in artificial knee joints being the titanium alloy [1]. The titanium alloy possesses the distinct advantages of good biocompatibility, good corrosion resistance, and high mechanical strength, making it an ideal material for artificial knee joints [2,3]. In general, the shape of artificial knee joints can be obtained with casting, forging, additive manufacturing, and hot isostatic pressing [4,5]; however, the titanium alloy artificial knee joint cannot be implanted directly into the human body immediately after these molding processes, and needs to undergo a series of subsequent machining processes because the surface quality of the artificial knee joint determines the characteristics of the surface friction, thereby impacting adhesive wear [6,7]. Wear issues with the artificial knee joint not only threaten human health, but also stimulate macrophages and fibroblasts in the body, hence, causing these cells to release factors that cause osteolysis and leading to the aseptic loosening of the artificial knee joint, which then has to be replaced with a new joint [8–11]. Therefore, it is important and necessary to treat the molded artificial knee joint before it can be used as an implant in the human body.

Abrasive flow machining is an untraditional processing technology that uses fluid as a carrier, and in which abrasives are suspended to form an abrasive flow. Material is then removed from the target via the microcutting action of the abrasive relative to the target surface [12,13]. The abrasive flow usually employs a weak viscous fluid as a carrier, while the irregular turbulent motion of the abrasive particles within the flow impact the target surface in a disorderly manner to achieve a good surface finish [14–18]. Davies et al. [19] investigated the rheological and thermal properties of liquid carrier materials in abrasive flow machining and found that the variation in viscosity caused by the rheological and thermal properties of liquid affected the polishing efficiency and quality. Xu et al. [20] investigated the rheological properties and polishing performance of viscoelastic materials for the dilatancy pad, and found that the material underwent a shear thickening effect when subjected to a shearing force, exhibiting a significant shear-stiffening performance. Williams et al. [21] explored the properties associated with the abrasive flow machining process, modeled the abrasive flow machining process, and investigated the effect of processing parameters on the polishing performance. Qi et al. [22,23] proposed a novel hydrodynamic suspension polishing and acoustic levitation polishing to control particle impact erosion on the target surface and achieve an ultrasmooth surface.

The above research works mainly focused on the abrasive flow machining of flat surfaces by considering the effects of the fluid characteristics and properties, but in order to conduct the abrasive flow machining of structures with complex surfaces, a predesigned abrasive flow channel needs to be formed. Uhlmann et al. [24] found that there was an obvious problem in the abrasive flow machining of the exhaust edge of the additive manufacturing blade due to the complicated surfaces involved. Fu et al. [25] alleviated the overthrowing problem by regulating the flow field near the inlet and exhaust sides of the blade. Hence, it is important to properly design the desired channel between the abrasive flow and target complex surface for the abrasive flow machining of freeform surfaces. The numerical method seems to be a powerful tool for optimizing constrained structures before they are created for experiments.

Moreover, numerical studies seem to be an effective and efficient way to explore the complicated abrasive flow machining of freeform surfaces and guide experimental studies. Zhang et al. and Qi et al. [26,27] conducted a numerical study on abrasive flow characteristics, such as the distribution of the abrasive particle velocity, flow pressure, and particle impact erosion near the target surface, by using the computational fluid dynamics (CFD) method in ANSYS FLUENT software (ANSYS Inc., Canonsburg, PA, USA), where the standard k-ε turbulence model was employed to model the abrasive flow and could be used to optimize the design of the polishing structures effectively and efficiently. Peng et al. [28] carried out a CFD investigation of flow behavior and sand erosion patterns in a horizontal pipe bend under annular flow, where the coupling calculation of the multifluid VOF model and DPM model in ANSYS FLUENT software was employed to simulate the process and explore the underlying mechanisms. Currently, with the development of CFD-based numerical technology, the COMSOL Multiphysics software can also be employed to simulate abrasive flow. Kumar et al. [29] investigated the magnetorheological abrasive flow finishing of gear with complex surfaces and optimized the processing parameters for good performance. Thus, in this study, a titanium alloy artificial joint with a freeform surface is numerically explored using the CFD-based method and the developed COMSOL Multiphysics software.

In this study, a titanium alloy artificial knee joint model with complex surfaces is first designed by imitating a scanned model of a real human knee joint, and the corresponding abrasive flow machining in constrained flow channels is designed around the artificial knee joint model. Then, the distribution of the abrasive particles, velocity, and pressure of the abrasive flow field are numerically analyzed by using the CFD-based method in the COMSOL Multiphysics software, and the effects of the relative processing parameters on the abrasive flow field of the target surface are analyzed to optimize the polishing performance. Finally, corresponding experiments are conducted to validate the numerical model and analyze the surface morphology before and after the proposed abrasive flow machining process.

2. Modeling of Knee Joint Surface and Associated Abrasive Flow Machining Channel

Figure 1a shows the scanned structure of a real human knee joint produced with an industrial CT (ZEISS METROTOM 1), where its surface shape is truly disordered and complex; thus, it was difficult for us to conduct the abrasive flow machining of this model. Thus, this kind of scanned model was simplified to the artificial knee joint model in Figure 1b. Surface A and surface B can be represented by six curvature parameters, as shown in Figure 2, and Table 1 shows the values of the curvature parameters of surface A and surface B, respectively. Since surface C is a transition plane, formed with a straight line transitioning along the edges of surface A and surface B, in this study, it mainly focused on the analysis of the abrasive flow machining of the two surfaces A and B.

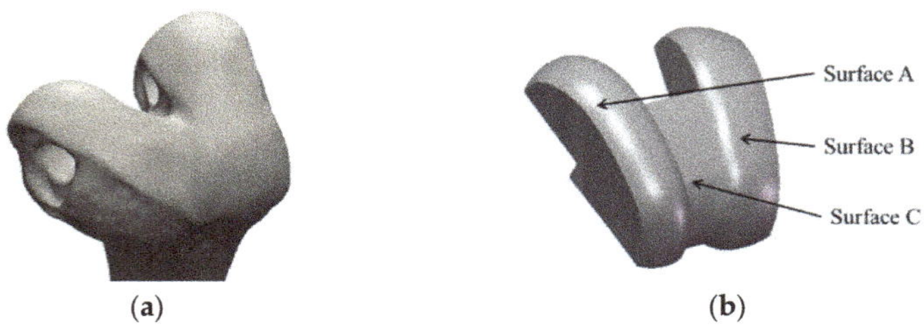

Figure 1. Knee joint model: (**a**) scanned knee joint model; (**b**) artificial knee joint model.

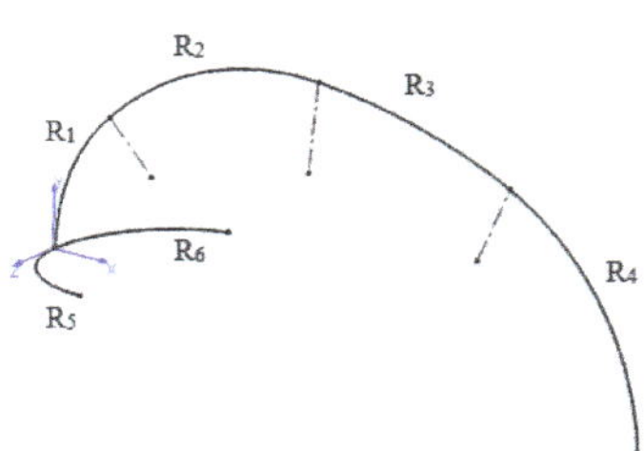

Figure 2. Schematic representation of curvature parameters in surface A and surface B.

Table 1. Values of curvature parameters in surface A and surface B.

Curvature Parameters	Surface A (mm)	Surface B (mm)
R_1	30	30
R_2	45	45
R_3	100	100
R_4	40	40
R_5	22.5	15
R_6	22.5	40

The abrasive flow machining channel for the artificial knee joint model (see Figure 1b) could be properly designed by configuring a constraining module near the target surface [30]. The abrasive flow machining channel, which can also be called the constrained flow channel, had good contact with the complex surface of the artificial knee joint model, hence, improving the overall machining performance. A schematic representation of the constrained flow channel is shown in Figure 3.

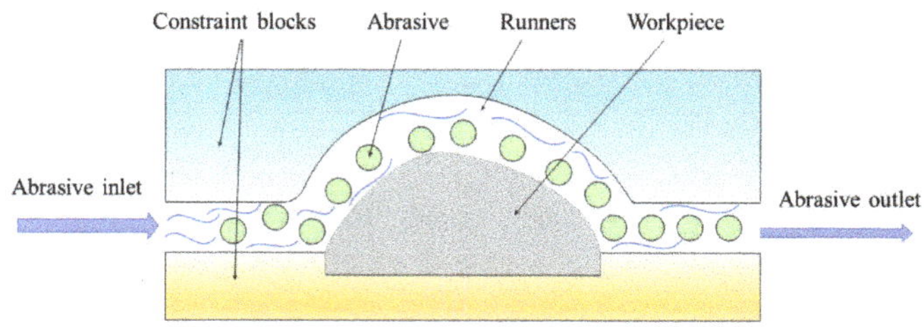

Figure 3. Schematic representation of constrained flow channel.

3. Numerical Work

3.1. Model Development

The CFD-based numerical model related to the structure and dimension illustrated in Table 1 and Figure 3 was developed in COMSOL Multiphysics software, as shown in Figure 4, where the abrasive flow entered through inlet one of the constrained flow channel and flowed out through outlet three after passing through the constrained flow channel. At the beginning, it was assumed that the abrasive flow filled the portion between one and two; then, two became the initial boundary interface of the abrasive flow. In addition, four is the upper surface of the artificial knee joint to be machined, and five is the constrained flow channel surface.

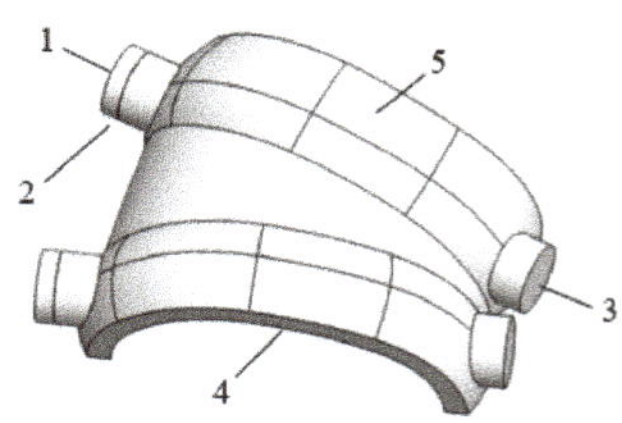

Figure 4. Model and boundary conditions: 1—runner inlet; 2—initial interface; 3—runner outlet; 4—upper surface of artificial knee joint; 5—constrained flow channel surface.

The abrasive flow considered in this study was a solid–liquid two-phase flow, where the solid phase used a silicon carbide (SiC) abrasive and the liquid phase used water. In the analysis of the overall characteristics of the abrasive flow field in the constrained flow channel, it was assumed that the diameter of the SiC abrasive was sufficiently small and that the two solid–liquid phases were mixed uniformly. In this case, the two-phase flow could be regarded as a special liquid flow, its density based on the solid-phase quality, and the major simulation parameters are given in Table 2. Further, the standard k-ε turbulence model was selected in COMSOL Multiphysics software to model the abrasive flow, similar to those used in previous studies [1,23,27]. The relative governing equations involved in this software can be found in the user manual, which are common and have been used extensively in similar works [31]. A grid independence test was also carried out before determining the simulation parameters in order to ensure simulation accuracy; all the following works are based on the optimized parameters within the simulation. Figure 5 shows the abrasive flow process with respect to the simulation time, t, from 0 s to 0.1 s, where the value in the colored figure legend represents the proportional volume of abrasive flow in the whole constrained flow channel. The initial speed of the abrasive flow from the left inlet was 5 m/s, which then gradually filled the entire constrained flow channel over time. As can be seen in Figure 5, at t = 0.1 s, the abrasive flow completely filled the entire constrained flow channel, indicating that the abrasive flow fully contacted the target surface to be machined at the time.

Table 2. Simulation parameters.

Parameters	Values	Unit
Temperature	22	°C
Environmental pressure	1.01325×10^5	Pa
Inlet speed	5	m/s
Abrasive flow density	1.333×10^3	kg/m^3
Abrasive flow viscosity	1.011×10^{-3}	Pa·s

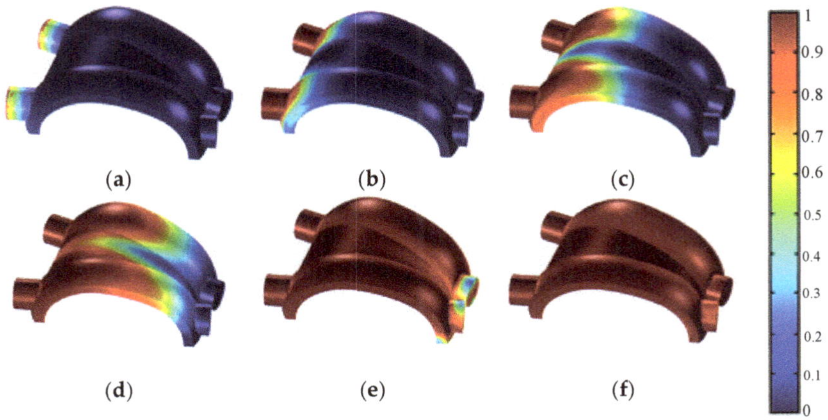

Figure 5. Abrasive flow process with respect to simulation time: (**a**) t = 0 s; (**b**) t = 0.005 s; (**c**) t = 0.01 s; (**d**) t = 0.015 s; (**e**) t = 0.025 s; (**f**) t = 0.1 s.

3.2. Numerical Simulation Results and Discussion

Since the characteristics of the abrasive flow on the upper surface of the artificial knee joint had a significant effect on the machining performance, the abrasive flow velocity and pressure in this area were numerically analyzed in this study. Figure 6 shows the distribution of the abrasive flow velocity on the upper surface of the artificial knee joint to be machined. It can be seen from this figure that after the abrasive flow entered the inlet of the channel, the abrasive flow velocity decreased as the cross-sectional area of the channel decreased. During the simulation time considered in this study, the maximum abrasive flow velocity reached approximately 11.9 m/s, and after passing through the constrained flow channel, the velocity decreased until the flow exited through the outlet of the channel. It was also noticed that the distribution of the abrasive flow velocity at the simulation time of 0.025 s was similar to that at the simulation time of 0.05 s, because the abrasive flow almost filled the entire channel at t = 0.025 s, as can be seen in Figure 5e, so it could be deduced that the distribution of the abrasive flow between t = 0.05 s and t = 0.1 s was stable; hence, we could predict that after t = 0.1 s, the distribution of abrasive particle velocity would not change. Therefore, the distribution of the abrasive flow velocity at the simulation time of 0.1 s could be considered as the steady-state during the entire abrasive flow machining process.

Similarly, Figure 7 shows the distribution of the abrasive flow pressure on the upper surface of the artificial knee joint. It can be seen from this figure that the pressure at the inlet of the constrained flow channel was the highest, reaching approximately 2.06×10^{10} Pa, whereas the pressure at the outlet of the channel was the lowest, which was closer to the range of environmental pressure. It was also interesting to note that, compared with the pressure distribution at t = 0.025 s and the pressure distribution at t = 0.05 s, only the pressure distribution at the outlet of the channel demonstrated a slight difference; meanwhile, compared with the pressure distribution at t = 0.05 s and t = 0.1 s, there was no obvious difference. Thus, similar to the findings in the analysis of the abrasive flow velocity distribution, it could be seen that the pressure distribution between t = 0.05 s and t = 0.1 s did not change, and we predicted that after the simulation time of 0.1 s, the distribution of the abrasive flow pressure would not change either. Therefore, the distribution of the abrasive flow pressure at the simulation time of 0.1 s could be considered the steady-state during the entire abrasive flow machining process.

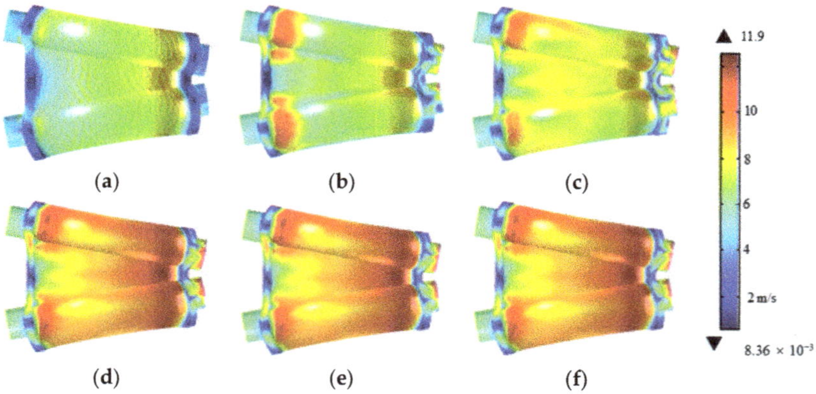

Figure 6. Distribution of abrasive flow velocity across upper surface of artificial knee joint: (**a**) t = 0 s; (**b**) t = 0.005 s; (**c**) t = 0.01 s; (**d**) t = 0.025 s; (**e**) t = 0.05 s; (**f**) t = 0.1 s.

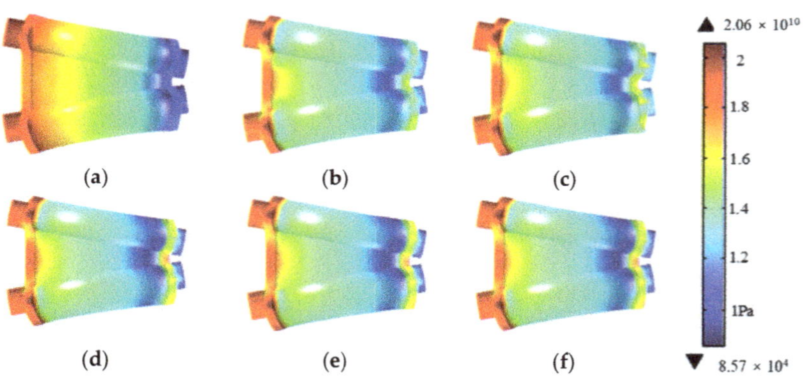

Figure 7. Distribution of abrasive flow pressure on upper surface of artificial knee joint: (**a**) t = 0 s; (**b**) t = 0.005 s; (**c**) t = 0.01 s; (**d**) t = 0.025 s; (**e**) t = 0.05 s; (**f**) t = 0.1 s.

In addition, the Preston equation states that the material removal rate on the workpiece surface is directly proportional to the cutting speed (V) and loading pressure (P). The material removal mechanism of the abrasive flow machining could be considered to be the microcutting action of abrasive particles on the target surface, where the cutting speed and loading pressure of abrasive particles relative to the target surface actually correspond to the velocity and pressure of the abrasive flow. Therefore, a dimensionless material removal rate model, $M = PV/\text{min}$, was developed to explore the relative machining performance according to the above distribution of abrasive flow velocity and pressure obtained from simulation results. Figure 8 shows the dimensionless material removal rates for the two main machined surfaces, i.e., surface A and surface B in Figure 1b. In order to accurately demonstrate the simulation results, two cases are illustrated in Figure 8, where, in case one, the lower part was set as the inlet and the upper part was set as the outlet, and in case two, the inlet and outlet were set in the opposite positions. It can be seen from Figure 8 that the relative material removal rate near the inlet of the constrained flow channel was relatively large, and the relative material removal rate at the outlet of the constrained flow channel was relatively small. It was also found that in the middle part of the constrained flow channel, the relative material removal rate seemed to be uniform, but in order to realize a good abrasive flow machining performance with a relatively uniform material removal rate across the whole surface, it was recommended that the exchange of the abrasive flow inlet and outlet was necessary.

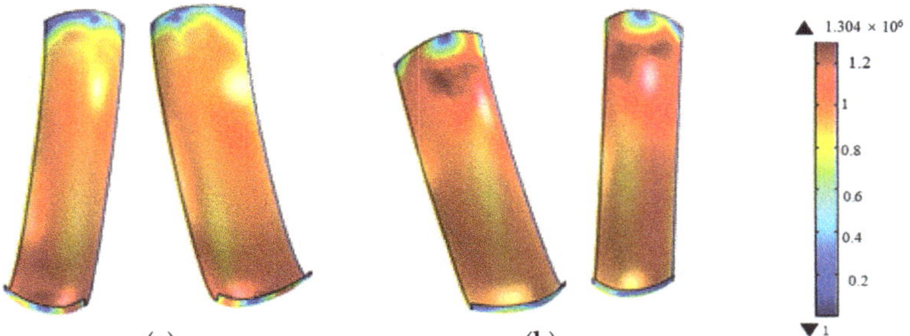

Figure 8. Dimensionless material removal rate: (**a**) case 1: lower inlet and upper outlet; (**b**) case 2: upper inlet and lower outlet.

4. Experimental Work

4.1. Experimental Setup

Figure 9 shows the experimental setup, which included the titanium alloy artificial knee joint model A and three restraining parts B, C, and D. The artificial knee joint model A was installed on restraining part B with screws and then both parts were embedded together on base plate C, which was then covered with cover plate D. The small square groove on part C was used for applying the sealant, cooperating with the square protrusion in D to realize the sealing function.

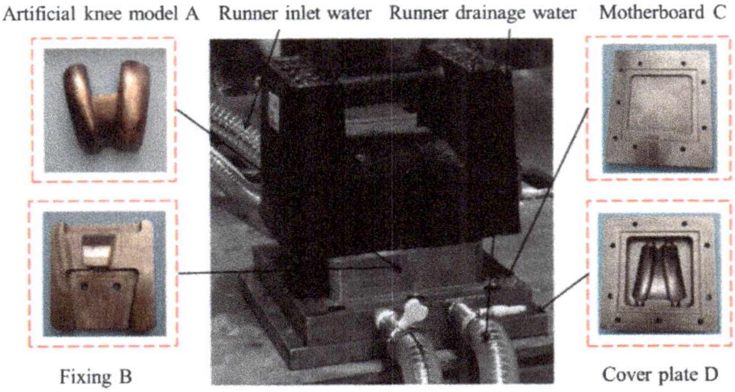

Figure 9. Experimental setup.

In this experiment, the SiC particles were selected as the abrasive and two different particle sizes of 120 mesh (with an average diameter of 120 μm) and 240 mesh (with an average diameter of 60 μm) were employed for the rough polishing and fine finishing, respectively. The material of the artificial knee joint was composed of a titanium alloy (TC4, Ti-6Al-4V), whose major properties are given in Table 3. The experimental procedure was mainly divided into two abrasive flow machining processes, namely, rough polishing and fine finishing, with the processing parameters given in Table 4. Considering the recommendation of the simulation results, that the exchange of the abrasive flow inlet and outlet was necessary to realize the uniform abrasive flow machining performance, the rough polishing and fine finishing processes were performed at an interval of 1 h with the exchange of the inlet and outlet. During this process, the SiC particles in the abrasive flow were replaced with new ones to guarantee machining quality. Further, three similar titanium alloy artificial knee joint models were used in the experiment, and the average data on surface morphology were taken for a further analysis.

Table 3. Major material properties of TC4.

Density, g/cm^3	Young's Modulus, GPa	Poisson's Ratio	Yield Stress, MPa
4.43	104	0.31	880

Table 4. Experimental processing parameters.

Processing Parameters	Values
Temperature	22 °C
Abrasive	SiC
Fluid	Water
Abrasive particle concentration	10%
Abrasive particle inlet velocity	5 m/s
Average diameter of SiC for rough polishing	120 μm
Rough polishing time	10 h
Average diameter of SiC for fine finishing	60 μm
Fine finishing time	8 h

4.2. Results and Discussion

After 16 h of cumulative abrasive flow machining, the experimental observations were qualitatively analyzed, and the results are shown in Figure 10. It can be seen from this figure that the changes in surface quality before and after the machining process were obvious, where surface A and surface B (see Figure 1b) of the titanium alloy artificial knee joint before the machining process essentially could not reflect the word "Mechanical" written on paper, whereas after the abrasive machining process, these two surfaces could reflect each letter more clearly. However, due to the shape of the workpiece, only the lower part of the surface shown in the figure could reflect these letters, and the upper part was blurred because it could not receive the light reflected from the paper. Thus, we demonstrated that abrasive flow machining with a constrained flow channel could significantly improve the surface quality of the titanium alloy artificial knee joint; a further quantitative analysis was conducted as follows in detail.

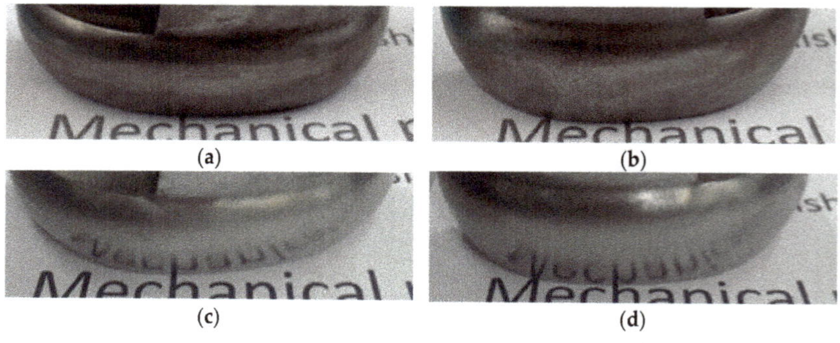

Figure 10. Comparison of surface quality before and after abrasive flow machining process: (**a**) original surface A; (**b**) original surface B; (**c**) surface A after machining process; (**d**) surface B after machining process.

With the development of the high-tech measuring technique [32], the surface roughness of the workpiece was measured with the assistance of the Veeco NT9800 optical profiler (Veeco Instruments Inc., Plainview, NY, USA). It can be seen from Figure 11 that the original surface roughness of the workpiece was approximately 394 nm, and the surface roughness reduced to approximately 300 nm after 3 h of rough polishing. Finally, after the fine finishing process, the surface roughness was approximately 171 nm. It was also noted that Figure 10a–c clearly show changes in terms of surface roughness during the abrasive flow machining process, in which the surface roughness of the titanium alloy artificial knee joint was continuously reduced, and the surface microstructure gradually changed from being obviously uneven to being relatively flat. This indicated that the abrasive flow machining process significantly improved the surface quality of the artificial knee joint surface.

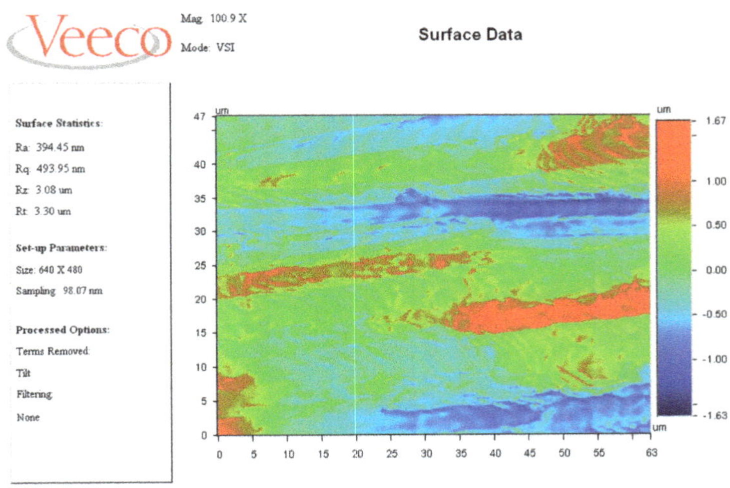

(a)

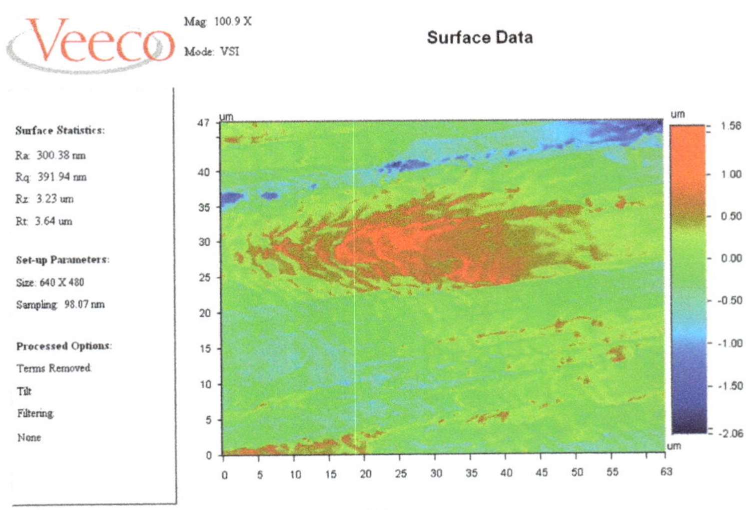

(b)

Figure 11. *Cont.*

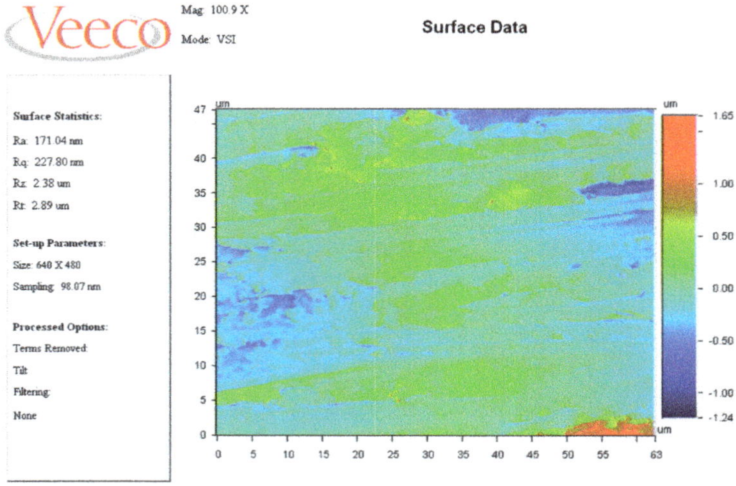

(c)

Figure 11. Comparison of surface roughness: (**a**) original surface of workpiece; (**b**) after 3 h rough polishing; (**c**) after fine finishing.

5. Conclusions

In this paper, a proper constrained flow channel was designed to conduct the abrasive flow machining of s titanium alloy artificial knee joint surface in order to improve the performance and service life of artificial knee joints in the human body. This abrasive flow machining process was first numerically investigated with the assistance of the CFD-based method in COMSOL Multiphysics software, and it was found from the distribution of the abrasive flow velocity that the maximum abrasive flow velocity in the constrained flow channel could reach approximately 11.9 m/s, with even the inlet velocity having the capacity to reach 5 m/s. From the distribution of the dimensionless material removal rate on the target surface, it was also found that the exchange of the abrasive flow inlet and outlet during the machining process could improve the surface finish uniformity. Then, corresponding experiments were conducted to analyze the surface morphology before and after the abrasive flow machining process using the proposed constrained flow channel. It was found from the qualitative study that abrasive flow machining with a constrained flow channel could significantly improve the surface quality of the titanium alloy artificial knee joint by clearly reflecting the word "Mechanical" after the machining process. Further, it was found from the quantitative study that after the fine finishing process, the surface roughness decreased from approximately 394 nm to 171 nm. Therefore, the constrained flow channel proposed in this paper for the abrasive flow machining of a knee joint model is a useful potential method for the fine finishing of titanium alloy artificial joints.

Author Contributions: Conceptualization, R.J. and L.Z.; Data curation, J.C., K.L. and S.L.; Formal analysis, K.L. and S.L.; Funding acquisition, Y.L.; Investigation, R.J.; Project administration, Y.L.; Software, Z.Q.; Supervision, L.Z.; Validation, J.C.; Writing—original draft, Z.Q.; Writing—review and editing, Y.L. All authors have read and agreed to the published version of the manuscript.

Funding: This research was funded by the National Natural Science Foundation of China (U21A20122 and 51575493) and the Zhejiang Provincial Natural Science Foundation of China (LGG19E050025).

Institutional Review Board Statement: Not applicable.

Informed Consent Statement: Not applicable.

Data Availability Statement: The data presented in this study are available on request from the corresponding author. The data are not publicly available due to privacy.

Acknowledgments: The authors would like to thank Zijian Qi for his contribution on the CFD-based simulation and analysis in this study under the cooperation project with Central South University.

Conflicts of Interest: The authors declare that the research was conducted in the absence of any commercial or financial relationships that could be construed as a potential conflict of interest.

References

1. Zhang, L.; Yuan, Z.M.; Qi, Z.J.; Cai, D.H.; Cheng, Z.C.; Qi, H. CFD-based study of the abrasive flow characteristics within constrained flow passage in polishing of complex titanium alloy surfaces. *Powder Technol.* **2018**, *333*, 209–218. [CrossRef]
2. Hu, Y.; Zhou, G.; Yuan, X.; Li, D.; Cao, L.; Zhang, W.; Wu, P. An artificial neural network-based model for roping prediction in aluminum alloy sheet. *Acta Mater.* **2023**, *245*, 118605. [CrossRef]
3. Zhang, L.; Zheng, B.J.; Xie, Y.; Ji, R.Q.; Li, Y.B.; Mao, W.B. Control mechanism of particle flow in the weak liquid metal flow field on non-uniform curvature surface based on lippmann model. *Front. Mater.* **2022**, *9*, 895263. [CrossRef]
4. Yuan, H.Y.; Yang, W.B.; Zhang, L.; Hong, T. Model development of stress intensity factor on 7057T6 aluminum alloy using extended finite element method. *Coatings*, **2023**, *13*, 581. [CrossRef]
5. Ji, R.Q.; Zhang, L.Y.; Zhang, L.; Li, Y.B.; Lu, S.S.; Fu, Y.F. Processing method for metallic substrate using the liquid metal lapping-polishing plate. *Front. Mater.* **2022**, *9*, 896346. [CrossRef]
6. Li, C.; Piao, Y.C.; Zhang, F.H.; Zhang, Y.; Hu, Y.X.; Wang, Y.F. Understand anisotropy dependence of damage evolution and material removal during nanoscratch of MgF_2 single crystals. *Int. J. Extrem. Manuf.* **2023**, *5*, 015101. [CrossRef]
7. Li, C.; Piao, Y.; Meng, B.; Hu, Y.; Li, L.; Zhang, F. Phase transition and plastic deformation mechanisms induced by self-rotating grinding of GaN single crystals. *Int. J. Mach. Tool. Manu.* **2022**, *172*, 103827. [CrossRef]
8. Qi, H.; Shi, L.; Teng, Q.; Hong, T.; Tangwarodomnukun, V.; Liu, G.; Li, H.N. Subsurface damage evaluation in the single abrasive scratching of BK7 glass by considering coupling effect of strain rate and temperature. *Ceram. Int.* **2021**, *48*, 8661–8670. [CrossRef]
9. Xie, X.D.; Zhang, L.; Zhu, L.L.; Li, Y.B.; Hong, T.; Yang, W.B.; Shan, X.H. State of the art and perspectives on surface strengthening process and associated mechanisms by shot peening. *Coatings*, 2023, *accepted*.
10. Li, C.; Hu, Y.X.; Zhang, F.H.; Geng, Y.Q.; Meng, B.B. Molecular dynamics simulation of laser assisted grinding of GaN crystals. *Int. J. Mech. Sci.* **2023**, *239*, 107856. [CrossRef]
11. Qi, H.; Wang, Y.; Qi, Z.; Shi, L.; Fang, Z.; Zhang, L.; Riemer, O.; Karpuschewski, B. A Novel Grain-Based DEM Model for Evaluating Surface Integrity in Scratching of RB-SiC Ceramics. *Materials* **2022**, *15*, 8486. [CrossRef]
12. Qi, H.; Xing, W.; Tan, W.; Lin, H.; Guo, H.J.; Chen, M.Z.; Tang, H.P. Effect of the sintering process on mechanical behaviors of Zirconia ceramics by NanoParticle Jetting. *Ceram. Int.* 2023, *submitted*.
13. Qi, H.; Fan, J.M.; Wang, J.; Li, H.Z. On the erosion process on quartz crystals by the impact of multiple high-velocity micro-particles. *Tribol. Int.* **2016**, *95*, 462–474. [CrossRef]
14. Wang, Y.Y.; Wang, Z.; Ni, P.C.; Wang, D.J.; Guo, S.H.; Chen, Z.Z. Experimental and numerical study on regulation of cutting temperature during the circular sawing of 45 steel. *Coatings*, 2023; *accepted*.
15. Ji, P.J.; Zhang, J.J.; Xie, X.D.; Ying, R.M.; Zhang, L. Review on wear characteristics of artificial hip joint and associated physical training. *Front. Mater.* 2023, *accepted*.
16. Xie, Y.; Gui, F.X.; Wang, W.J.; Chien, C.F. A two-stage multi-population genetic algorithm with heuristics for workflow scheduling in heterogeneous distributed computing environments. *IEEE Trans. Cloud Comp.* 2022, *in press*. [CrossRef]
17. Li, L.; Qi, H.; Yin, Z.; Li, D.; Zhu, Z.; Tangwarodomnukun, V.; Tan, D. Investigation on the multiphase sink vortex Ekman pumping effects by CFD-DEM coupling method. *Powder Technol.* **2020**, *360*, 462–480. [CrossRef]
18. Qian, H.N.; Chen, M.K.; Qi, Z.J.; Teng, Q.; Qi, H.; Zhang, L.; Shan, X.H. Review on research and development of abrasive scratching of hard brittle materials and its underlying mechanisms. *Crystals*, **2023**, *13*, 428. [CrossRef]
19. Davies, P.J.; Fletcher, A.J. The assessment of the rheological characteristics of various polyborosiloxane grit mixtures as utilized in the abrasive flow machining process. *Proc. Inst. Mech. Eng. Part C J. Mech. Eng. Sci.* **1995**, *209*, 409–418. [CrossRef]
20. Xu, L.; Chen, H.Y.; Lyu, B.H.; Hang, W.; Yuan, J.L. Study on rheological properties and polishing performance of viscoelastic material for dilatancy pad. *Precis. Eng.* **2022**, *77*, 328–339. [CrossRef]
21. Williams, R.E.; Rajurkar, K.P. Metal removal and surface finish characteristics in abrasive flow machining. *Mater. Sci.* **1989**, *38*, 93–106.
22. Qi, H.; Cheng, Z.C.; Cai, D.H.; Yin, L.Z.; Wang, Z.W.; Wen, D.H. Experimental study on the improvement of surface integrity of tungsten steel using acoustic levitation polishing. *J. Mater. Process. Technol.* **2018**, *259*, 361–367. [CrossRef]
23. Qi, H.; Xie, Z.; Hong, T.; Wang, Y.Y.; Kong, F.Z.; Wen, D.H. CFD modelling of a novel hydrodynamic suspension polishing process for ultra-smooth surface with low residual stress. *Powder Technol.* **2017**, *317*, 320–328. [CrossRef]
24. Uhlmann, E.; Schmiedel, C.; Wendler, J. CFD simulation of the abrasive flow machining process. *Procedia CIRP* **2015**, *31*, 209–214. [CrossRef]
25. Fu, Y.Z.; Wang, X.P.; Gao, H.; Wei, H.B.; Li, S.C. Blade surface uniformity of blisk finished by abrasive flow machining. *Int. J. Adv. Manuf. Technol.* **2016**, *84*, 1725–1735. [CrossRef]
26. Zhang, L.; Ji, R.Q.; Fu, Y.F.; Qi, H.; Kong, F.Z.; Li, H.N.; Tangwarodomnukun, V. Investigation on particle motions and resultant impact erosion on quartz crystals by the micro-particle laden waterjet and airjet. *Powder Technol.* **2020**, *360*, 452–461. [CrossRef]

27. Qi, H.; Qin, S.K.; Cheng, Z.C.; Teng, Q.; Hong, T.; Xie, Y. Towards understanding performance enhancing mechanism of micro-holes on K9 glasses using ultrasonic vibration-assisted abrasive slurry jet. *J. Manuf. Process.* **2021**, *64*, 585–593. [CrossRef]
28. Peng, W.S.; Ma, L.; Wang, P.; Cao, X.W.; Xu, K.; Miao, Y.C. Experimental and CFD investigation of flow behavior and sand erosion pattern in a horizontal pipe bend under annular flow. *Particuology* **2023**, *75*, 11–25. [CrossRef]
29. Kumar, M.; Kumar, V.; Kumar, A.; Yadav, H.N.S.; Das, M. CFD analysis of MR fluid applied for finishing of gear in MRAFF process. *Mater. Today Proc.* **2021**, *45*, 4677–4683. [CrossRef]
30. Ji, S.; Tang, B.; Tan, D.; Gong, B.; Yuan, Q.; Pan, Y. Structured surface softness abrasive flow precision finish machining and its abrasive flow dynamic numerical analysis. *Chin. J. Mech. Eng.* **2010**, *46*, 178–184. [CrossRef]
31. Luo, B.; Yan, Q.S.; Chai, J.F.; Song, W.Q.; Pan, J.S. An ultra-smooth planarization method for controlling fluid behavior in cluster magnetorheological finishing based on computational fluid dynamics. *Precis. Eng.* **2022**, *74*, 358–368. [CrossRef]
32. Wu, B.; Li, D.R.; Zhou, Y.; Zhu, D.; Zhao, Y.P.; Qiao, Z.K.; Chen, B.; Wang, X.L.; Lin, Q. Construction of a calibration field of absolute gravity in a cave using the Cold Atom Gravimeter. *Sensors*, 2023, *submitted*.

crystals

MDPI

Article

Study on the Surface Generation Mechanism during Ultra-Precision Parallel Grinding of SiC Ceramics

Shanshan Chen [1,2], Shuming Yang [1,*], Chi Fai Cheung [3,*], Tao Liu [1], Duanzhi Duan [1], Lai-ting Ho [3] and Zhuangde Jiang [1]

1 State Key Laboratory for Manufacturing Systems Engineering, Xi'an Jiaotong University, 28 Xianning West Road, Xi'an 710049, China
2 Research Institute of Xi'an Jiaotong University, Hangzhou 311200, China
3 State Key Laboratory of Ultra-Precision Machining Technology, Department of Industrial and Systems Engineering, The Hong Kong Polytechnic University, Hung Hom, Kowloon, Hong Kong 999077, China
* Correspondence: shuming.yang@mail.xjtu.edu.cn (S.Y.); benny.cheung@polyu.edu.hk (C.F.C.)

Abstract: Silicon carbide (SiC) is a typical, difficult-to-machine material that has been widely used in the fabrication of optical elements and structural and heat-resistant materials. Parallel grinding has been frequently adopted to produce a high-quality surface finish. Surface generation is a vital issue for assessing surface quality, and extensive modeling has been developed. However, most of the models were based on a disc wheel with a cylindrical surface, whereas the surface topography generation based on an arc-shaped tool has been paid relatively little attention. In this study, a new theoretical model for surface generation in ultra-precision parallel grinding has been established by considering the arc-shaped effect, synchronous vibration of the wheel, and cutting profile interference in the tool feed direction. Finally, the ground surface generation mechanism and grinding ductility were analyzed in the grinding of SiC ceramics. The results showed that the spiral and straight-line mode vibration patterns were the main feature of the machined surface, and its continuity was mainly affected by the phase shift. Furthermore, for the in-phase shift condition, the grinding ductility was more significant than for the out-of-phase shift due to the continuously decreasing relative linear speed between the wheel and workpiece.

Keywords: parallel grinding; phase shift; SiC ceramics; surface generation; vibration pattern; ultra-precision machining

Citation: Chen, S.; Yang, S.; Cheung, C.F.; Liu, T.; Duan, D.; Ho, L.-t.; Jiang, Z. Study on the Surface Generation Mechanism during Ultra-Precision Parallel Grinding of SiC Ceramics. *Crystals* **2023**, *13*, 646. https://doi.org/10.3390/cryst13040646

Academic Editors: Francisco M. Morales, Chen Li, Chongjun Wu and Binbin Meng

Received: 19 February 2023
Revised: 24 March 2023
Accepted: 2 April 2023
Published: 9 April 2023

1. Introduction

SiC is an advanced engineered ceramic widely used in optics (hot-press mold and telescope mirror) [1–3], electronics [4], and biomedical fields [5] due to its excellent unique properties, such as extreme hardness, good wear resistance, high intensity, high durability, and excellent thermal stability [6–8]. Despite the salient potential properties of SiC ceramics, it is a typical hard-to-machine material and has poor machinability due to lower fracture toughness and extreme hardness. At present, ultra-precision grinding (a root-mean-square figure accuracy of <0.1 μm and a surface roughness Sq of < 0.01 μm) is the predominant cutting technique to produce a high-quality surface in the machining of hard and brittle materials [9–11]. In ultra-precision grinding, the machined surface quality principally depends on the motion precision and machining method of the CNC grinding machine. A microremoval process of target materials and the fabrication of curved surface shapes can be realized by precisely controlling the movement path of the grinding wheel [12–14]. According to the relative motion relationship between the grinding wheel spindle and the workpiece rotation spindle, the methods can be divided into the parallel grinding method [15], transverse grinding method [16], and oblique grinding method [17]. Oblique axis grinding can be further divided into spherical grinding wheel grinding, point grinding, normal grinding, and generating grinding according to the contact state and motion mode between the grinding wheel and workpiece [18]. Even with the same processing parameters, different grinding methods will produce completely different surface morphologies. This is mainly because grinding follows the "copy" nature, and different grinding surface morphologies are formed due to the different trajectories of abrasive particle movement and the different degrees and modes of mutual interference in different paths. Parallel grinding is the most commonly used complex surface machining method. The workpiece spindle and grinding wheel spindle are controlled by the x–z slides to machine various rotary symmetric surfaces. In this method, the cutting direction of the grinding wheel is consistent with the rotation direction of the workpiece on the cutting zone between the tool and the workpiece, so the grinding surface is formed along the radial direction of the spiral trajectory. In this grinding mode, a diamond arc grinding wheel or spherical grinding wheel is generally adopted.

To achieve a "damage free" surface, extensive research has been carried out on surface generation mechanisms in the grinding of SiC ceramics. There are two types of surface generation mechanisms in grinding: brittle fracture or ductile mode [19]. To obtain a good surface finish, the ductile grinding mode needs to be the dominant mechanism [20,21]. Therefore, much research has been conducted on the influence of physical parameters on the ductile-to-brittle transition. The nanoindentation test [22–24] and single abrasive scratch [25–28] experiment were widely employed in the study of crack damage and the surface formation mechanism in the grinding of SiC ceramics. The primary goal of these studies was to determine the threshold for the ductile-to-brittle transition in order to control the grinding load below the critical value, so as to achieve crack-free machining with a good surface finish. This research provided the quantitative relationship between the load and crack size; however, the surface generation resulted from the combined effect of the wheel geometry and tool dynamic characteristics has received little m. In grinding, surface generation is a function of the wheel shape and trajectory path and hence is affected by the tool motion error and tool geometry. In fact, the wheel spindle vibration oriented primarily perpendicular to the workpiece surface is the sensitive direction of the surface topography generation, which has a great impact on the machining accuracy and surface roughness [32–35] Chen et al. [36] investigated the influence of wheel vibration on surface roughness, and a theoretical kinematic model for surface generation was established by simplifying the wheel into a circular section. Cao et al. [37] developed a topography generation model by considering unbalanced wheel vibration, and the influence of vibration amplitude, grit size, and machining parameters on surface roughness and surface microwaviness were investigated. However, the waviness generation was considered as a motion copy of the tool oscillation locus and ignored the influence of the wheel shape. Since a 2D cross section of the wheel (cylinder grinding wheel) was picked to simulate the surface generation, it cannot be used to evaluate the impact of vibration phase shift on surface topography generation. Chen et al. [38] investigated the influence of grinding parameters on surface generation and found that the phase shift had a dominant impact on both surface roughness and waviness features. Based on the authors' experimental results, the distribution of cutting points in parallel grinding was studied, and a waviness pattern model was developed. The results showed that the phase shift had a remarkable effect on waviness geometry evolution, and a strategy was proposed to suppress the surface waviness [39]. Pan et al. [40] proposed a new method to improve the ground surface quality and uniformity in the grinding of a complex optical surface with a non-integer rotation speed ratio (phase shift effect).

However, most of the previous modeling work for ground surface generation was based on a disc wheel with a cylindrical surface, and the cutting profile interference in the circumferential direction of the workpiece was considered. In fact, the surface generation was more sensitive in the cross-feed direction when a grinding wheel with an arc cutting edge was adopted. The surface generation process is governed by many factors, such as wheel vibration, material properties, wheel geometry, and phase shift, which are essential in achieving surface quality control and grinding process optimization. In order to predict the surface generation more accurately, in this study, a new model for surface generation in parallel grinding was developed by comprehensively taking into account the effect of phase shift, tool vibration, and wheel cutting nose radius. The phase shift effect involved in the surface generation in the grinding of SiC ceramics was analyzed.

2. Modeling of Surface Generation in Ultra-Precision Grinding Considering Tool Vibration

In parallel grinding, a diamond grinding wheel with a circular arc on the cutting edge profile is mostly adopted. The cutting direction of the grinding wheel is tangentially parallel to the rotation direction of the workpiece, but the direction of their linear velocities are opposite to each other (reverse grinding). Figure 1 is a typical schematic diagram of parallel grinding. When the wheel rotates, it moves toward the center of the workpiece at a constant feed rate. Since the rotation speed and feed rate of the workpiece remain constant, the distance the grinding wheel moves toward the center of the workpiece is constant for each cycle the workpiece rotates; thus the tool path of parallel grinding on the workpiece surface is a spiral. In the grinding process, the linear velocity of the workpiece is different at different cutting points of the grinding wheel, and the volume of the material removed when the grinding wheel completes one circle is proportional to the radial position of the grinding point on the workpiece surface. Therefore, the relative motion error of the grinding wheel, geometrical shape of the grinding wheel, and machining parameters will have an important influence on the ground surface topography generation. In ultra-precision parallel grinding, the workpiece spindle is vertical to the grinding wheel spindle, and the wheel performs a linear feed motion from the edge of the workpiece to the rotating center along a spiral path to realize the material removal process, as shown in Figure 1.

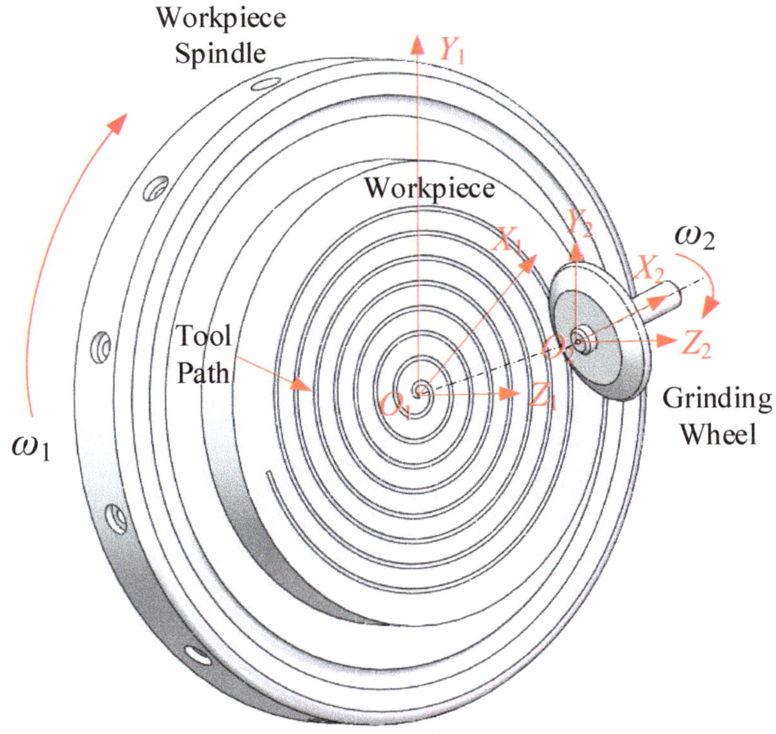

Figure 1. Schematic diagram of ultra-precision parallel grinding.

In parallel grinding, an extremely tiny depth of cut and a low rotational speed of the workpiece spindle combined with a high-speed grinding wheel are generally selected to achieve the microvolume of material removal. In this process, the grinding wheel cutting profile is constantly fed horizontally along the wheel axis direction (ignoring abrasive grains) to generate the micromachined topography. As shown in Figure 2, the spindle speed of the grinding wheel (ω_2) and the tool feed speed (v_f) together generate the tool interference in the adjacent cutting path interval, in which the surface residual profile (radial direction) forms. The angular position of the grinding wheel cutting point (θ_i) on the workpiece is equal to the angle of rotation of the workpiece. The grinding operation parameters can determine the relative cutting position and tool path of the grinding wheel with respect to the workpiece, which can affect the surface topography generation of the workpiece. Therefore, the machining parameters, the geometry of the grinding wheel cutting profile, and the relative position of and error between the grinding wheel and the workpiece have a great influence on the formation of the ground surface.

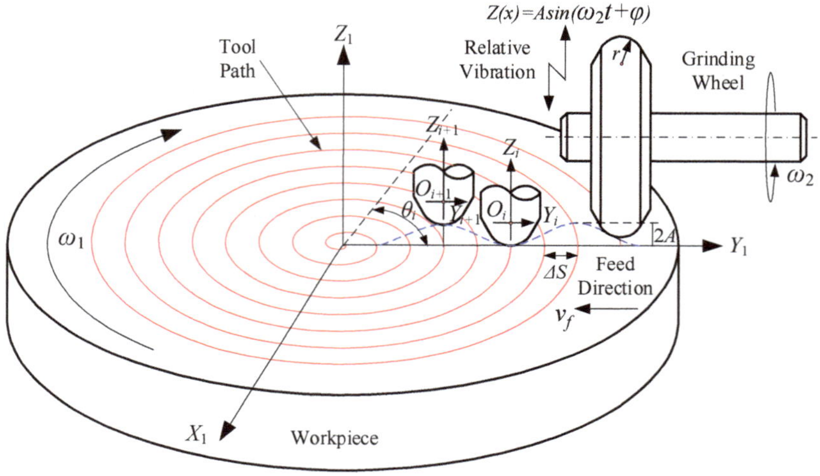

Figure 2. Schematic of locus of the grinding wheel movement in ultra-precision parallel grinding (red solid line indicates tool path and blue dash line represents wheel vibration locus).

The ground surface generation strongly depends on the relative travel trajectory of the cutting tool, which is determined by the operation parameters, tool geometry, and movement errors. The most significant of these is the relative motion error of the grinding wheel in the normal direction, which causes the varying engaged depth of the tool. This leads to the increment in surface residual height and degrades the surface roughness. In the case of non-constant grinding conditions due to the motion errors acting on the grinding wheel, the resultant machined surface is a function of the tool cutting edge geometry and the relative movement of the tool.

In grinding, the relative motion of the wheel is the result of the superposition of the motion of each axis, which transfers characteristics of the wheel cutting profile to the part surface and determines the final machined surface generation. Therefore, the geometric modeling of surface topography generation can be calculated according to the motion process of each axis. Figure 2 schematically shows the motions of the wheel and workpiece in ultra-precision parallel grinding, which include the rotation of the workpiece and grinding wheel as well as the linear movement of the machine table along the feed direction (x axis). However, the vibration with the same frequency as the wheel rotation is the dominant mode of tool vibration, which is responsible for the ground surface profile variation and degrades the surface quality.

According to the geometric relation of the relative motion of the grinding wheel with respect to the workpiece, the coordinates of the circular arc contour of the grinding wheel in each radial position of the workpiece and the intersection point of the adjacent circular arc contour were obtained, with which the residual surface profile can be calculated to reconstruct the finished surface. The coordinate system X_1-Y_1-Z_1 was employed to calculate the workpiece surface profile generation. The discrete workpiece surface can be established in the radical and circumferential directions, in which it is divided into a certain number of segments along the circumferential and radical directions of the workpiece with an equal angle interval ($\Delta\theta$) and tool path interval (S), respectively. The number of cutting positions on the workpiece surface can be expressed as:

$$\begin{cases} N_r = \frac{2\pi}{\Delta\theta} \\ N_c = \frac{R_s}{S},\ S = \frac{v_1}{v_f} \end{cases} \tag{1}$$

where $\Delta\theta$—the discrete angle value (rad);

R_s—the radius of the workpiece (mm);

S—the feed rate (mm/r);

v_1—the rotation speed of the workpiece (r/min);

v_f—the feed speed (mm/min).

In parallel grinding, the workpiece spindle rotates with a slow angular velocity, and the grinding wheel slowly moves towards the center of the workpiece transversely with a constant feed speed, whereby the tool trajectory is a spiral in the XY plane. In order to accurately predict the surface topography of the workpiece under the condition of tool vibration, it is first necessary to evaluate the motion trajectory of the cutting profile in the workpiece coordinate system. In polar coordinates, the movement trajectory of the wheel in discrete form can be calculated as follows:

$$\begin{cases} \theta_w(i) = i \cdot \Delta\theta + j \cdot 2\pi,\ i = 0, 1, 2 \cdots, N_r - 1 \\ R_w(i) = R_s - \left(\frac{\Delta\theta}{2\pi} \cdot i + j\right) \cdot S, j = 0, 1, 2 \cdots, N_c - 1 \\ Z_w(i) = A \sin(\omega_2 \frac{\Delta\theta i + 2\pi j}{\omega_1} + \varphi) \end{cases} \tag{2}$$

where A—the vibration amplitude of the grinding wheel (mm);

ω_1—the angular speed of the workpiece (rad/min);

ω_2—the angular speed of the wheel (rad/min);

φ—phase angle (rad).

In ultra-precision grinding, the residual height generation on the workpiece surface is tracked by the wheel location movement. The discrete tool path model represents a series of wheel loci. For a given pair of tool positions, there is an intersection between the two cutting profiles of the grinding wheel due to the fine feed rate in ultra-precision parallel grinding. In order to calculate the intersected segment, a cutting plane is defined, which includes the lowest engagement point of the wheel and leaves the deepest scratch on the machined surface, as shown in Figure 3. Therefore, the machined surface can be obtained by calculating the intersection of two adjacent cutting profiles.

There is an arc edge on the cutting profile of the grinding wheel, which moves in a single path of approximately a circle, which is expressed as:

$$(y - iS)^2 + [z + Z_w(i)]^2 = r^2 \tag{3}$$

In the process of ground surface generation, the tool cutting profile of the grinding wheel moves along the spiral tool path, and the tool–workpiece interference in the feed direction envelopes the final machined surface. In actual grinding, the wheel axis changes due to the existence of vibration, which in turn results in a variable residual profile on the machined surface. Therefore, the surface topography of a certain section of processing, the influence of the previous feed, and the subsequent feed should be fully considered. The specific consideration method is as follows:

$$\begin{cases} [y-(i-1)S]^2+\left\{z+A\sin[\omega_2\frac{\Delta\theta i+2\pi(j-1)}{\omega_1}+\varphi]\right\}^2=r^2 \\ (y-iS)^2+\left[z+A\sin(\omega_2\frac{\Delta\theta i+2\pi j}{\omega_1}+\varphi)\right]^2=r^2 \\ [y-(i+1)S]^2+\left\{z+A\sin[\omega_2\frac{\Delta\theta i+2\pi(j+1)}{\omega_1}+\varphi]\right\}^2=r^2 \end{cases} \tag{4}$$

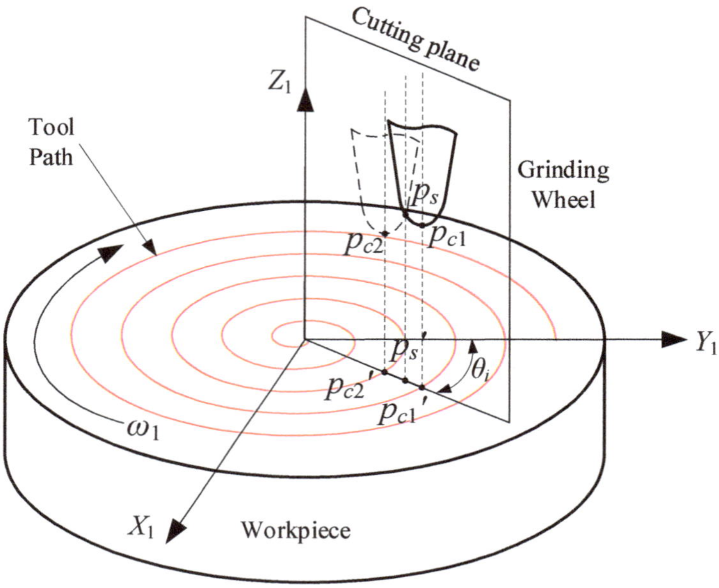

Figure 3. Schematic diagram of surface generation modeling in grinding.

The scallop height is calculated at a cutting point by comparing the height value of neighboring tool cutting profiles and selecting the minimum value, which is responsible for the final ground surface generation. The surface residual height of each cutting point can be expressed as:

$$z_e=\min\{z_{i,j-1}(\rho,\theta),z_{i,j}(\rho,\theta),z_{i,j+1}(\rho,\theta)\} \tag{5}$$

The experiments were performed on a four-axis grinding machine (Moore Nanotech 450UPL). The ultra-precision grinding machine consists of two linear guides (x axis and z axis) and two rotation spindles (workpiece spindle and wheel spindle), as shown in Figure 4. The samples were mounted on an aluminum fixture, which was placed on the vacuum chuck of the workpiece spindle. Before grinding, dynamic balancing of the two spindles was carried out to make the radial runout less than 10 nm. In order to obtain the amplitude of the wheel vibration, a laser displacement sensor with high resolution was employed. A resin-bonded diamond grinding wheel was selected with a 1500# mesh size. The parallel grinding tests were performed on reaction-sintered SiC to investigate surface generation and verify the ground surface model. Before the test, all the samples were ground to obtain a relatively flat surface to reduce the effect of tool wear, and we then preformed the wheel dressing operation before each experiment. A Taylor Hobson roughness profiler (Form TalySurf PGI 1240; measuring error: 0.8 nm vertical resolution) was used to measure the machined surface after the tests.

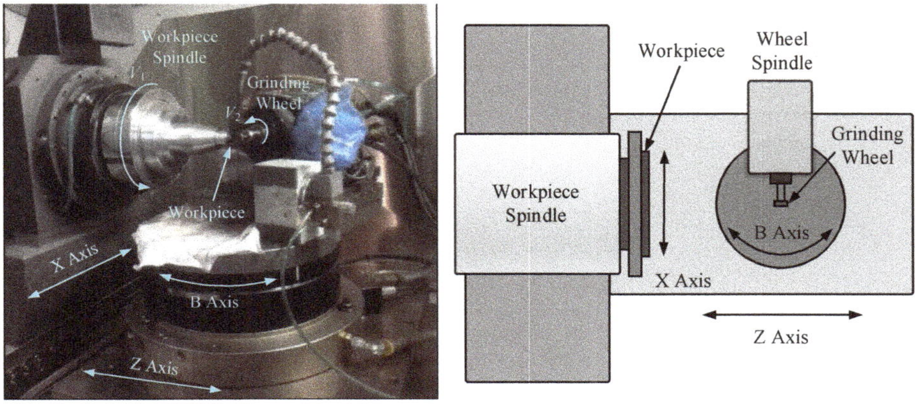

(a) (b)

Figure 4. Configurations of 4-axis ultra-precision grinding machine: (**a**) ultra-precision machine tool for grinding experiments; (**b**) structural layout of the grinding machine.

Table 1 shows the grinding conditions and operation parameters adopted for the machining tests. A diamond arc-shaped wheel was chosen with a 0.5 mm nose radius. The influence of phase shift with respect to wheel synchronous vibration can have a significant effect on surface topography generation in ultra-precision parallel grinding, which can be selected by adjusting the grinding wheel speed.

Table 1. Grinding operating conditions.

Contents	Parameter Setting
Diamond grinding wheel	Resin bond, thickness: 6 mm, diameter: 20 mm, nose radius: 0.5 mm, grit size: 1500#
Workpiece material	RB-SiC, thickness: 5 mm, diameter: 12 mm
Operating parameters	Wheel speed: 39,000 rpm–40,350 rpm, workpiece speed: 1500 frpm, depth of cut: 10 μm, feed speed: 10 mm/min

3. Experiment Results and Discussion

3.1. Experimental Results and Validations

To verify the surface generation model, a series of grinding tests under different phase shifts were conducted. A laser displacement sensor was used to measure the wheel spindle error motion. The amplitude of the tool vibration was obtained by recording 5 values under variable phase shifts, and it was found that the average amplitude value was about 2.5 μm in all ranges of wheel speeds. The cross section profiles of the ground surface were measured under different phase shifts: Figure 5(a1,a2) phase = 0, Figure 6(a1,a2) m, Figure 6(b1,b2) phase = 0.3, Figure 6(c1,c2) phase = 0.5, Figure 6 (d1,d2) phase = 0.7, and Figure 6(e1,e2) phase = 0.9. Figure 5 shows a comparison between the simulated surface profile and the measured results, in which the surface profile height and period are highly consistent. The machined surface had different waviness patterns and evolved continuously with the processing cycles, and it finally developed a series of microwaviness marks around the rotation center of the workpiece. In the process of machining, microwaviness stripes were formed on the workpiece surface by the grinding wheel vibration, which was affected by the grinding wheel speed, feed speed, and workpiece speed, resulting in the formation of surface stripes with different geometric patterns.

The continuous waviness structure formed when the phase shift was 0, 0.1, and 0.9, as shown in Figure 5(a1) and Figure 6(a1,e1); however, a discontinuously structured wave pattern was generated when the phase shift was 0.3, 0.5, and 0.7, as shown in

Figure 6(b1,c1,d1). This indicates that the phase shift has a significant effect on the surface finish, in which the wheel speed only has a minor change (for example, 0.385% for the wheel speed change from 39,000 rpm to 39,150 rpm in Figure 5(a2) and Figure 6(a2), respectively). The theoretical surface profile height in Figure 5(a2) shows a maximum at the phase shift of 0 (out-of-phase shift), in which the wheel speed is equal to an integer multiple of the workpiece speed (V_1/V_2 = 39,000/1500 = 26). With the further increase in phase shift (in-phase shift), however, the machined surface profile height presents a dramatic decrease (height amplitude decreases from 2.5μm to 0.5μm), and the speed ratio is not an integer number (such as V_1/V_2 = 39,150/1500 = 26.1 in Figure 6(a2)). With the development of the phase shift, the discontinuity of the waviness pattern occurs at the phase shift of 0.3, 0.5, and 0.7, as shown in Figure 6(b1,c1,d1). In this case, there is a larger difference in the tool depth of cut at adjacent tool tracks in the presence of grinding wheel vibration. As the phase shift increases from 0 to 0.5, the cross section profile height of the machined surface decreases; however, the profile height increases when the phase shift further increases from 0.7 to 0.9 (the phase shift is close to 1). This shows that the phase shift can contribute to reducing the surface scallop height and improving the machined surface quality. These results showed that the theoretical model could accurately predict the resulting surface generation, including the surface waviness and profile. In comparison to the conventional modeling approaches that take into account the surface generation in the wheel cutting direction, the proposed modeling algorithm, by considering the cutting profile interference in the tool feed direction, is more accurate in predicting the surface waviness profile generation in the parallel grinding operation, in which the effects of the phase shift, wheel vibration, and wheel cutting nose on surface generation are comprehensively calculated.

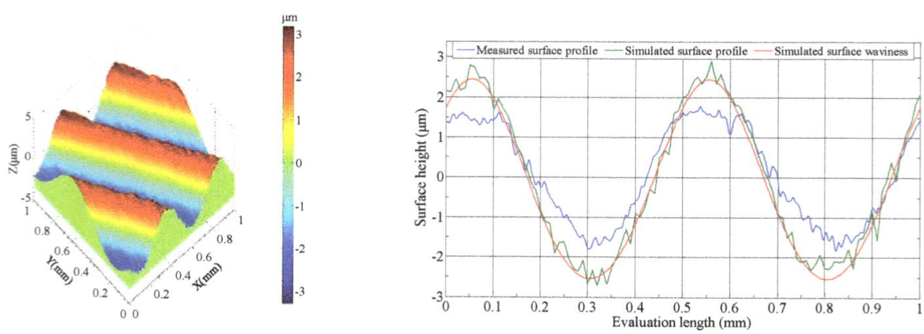

(a1) **(a2)**

Figure 5. Simulated surface and measured surface under out-of-phase shift (phase = 0): (**a1**) simulated surface topography; (**a2**) surface profile height.

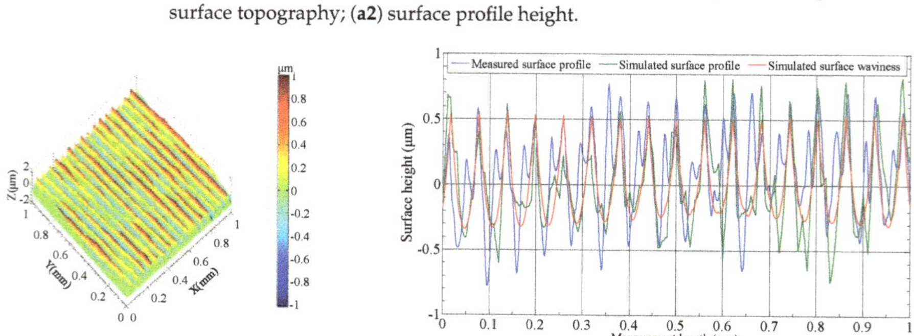

(a1) **(a2)**

Figure 6. *Cont*

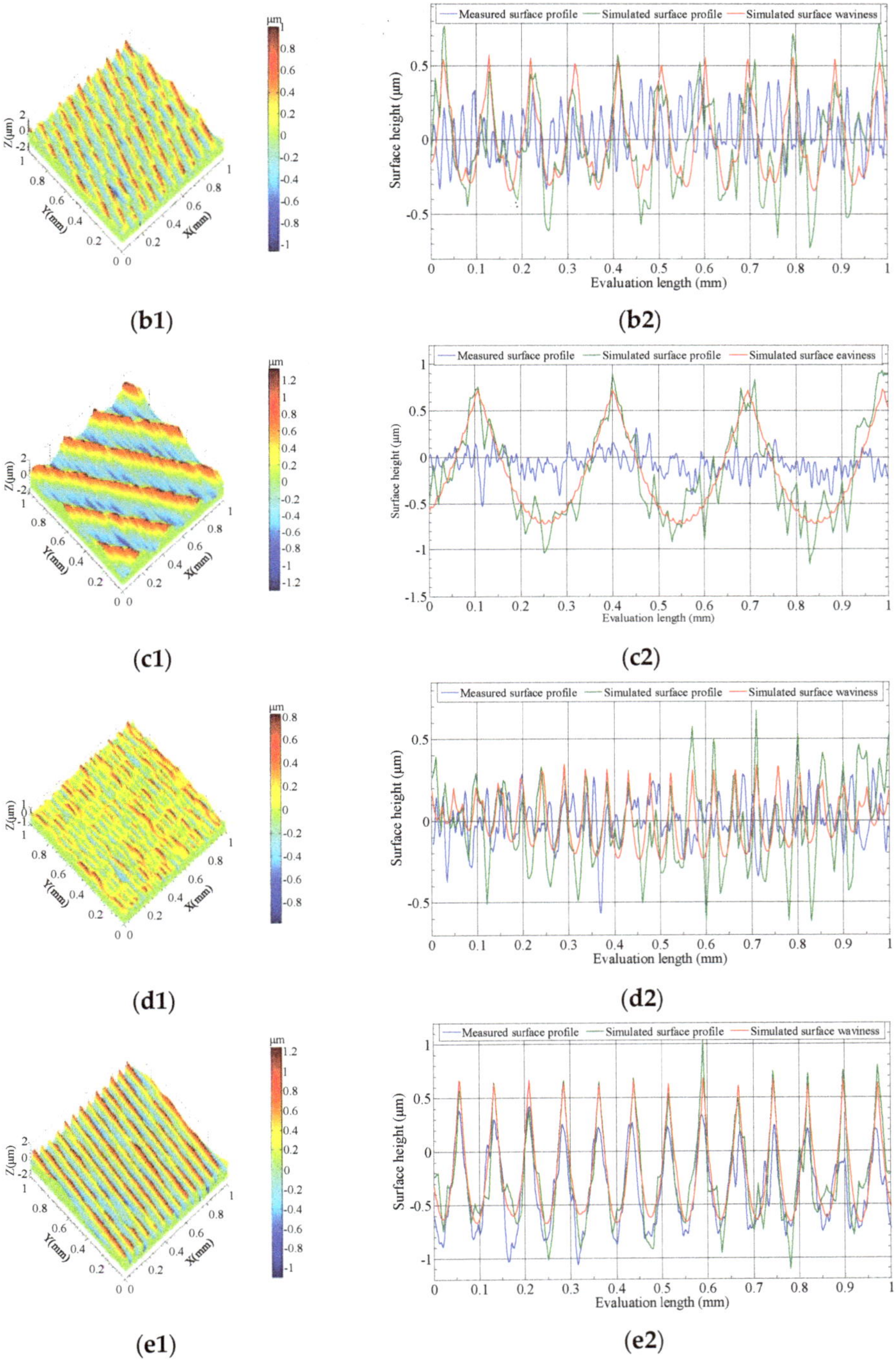

Figure 6. Simulated surface and measured surface under different phase shifts: (**a1**–**e1**) simulated surface topography; (**a2**–**e2**) surface profile height.

The model is capable of estimating the ground surface generation precisely and determining the role of the phase shift in the ultra-precision parallel grinding as well. However, the deviation from the actual experiments was attributed to the wheel wear in the grinding of the SiC ceramics.

From the results above, it is found that the surface waviness is the main characteristic of machined surface generation, which is caused by the synchronous vibration of the grinding wheel. The phase shift plays a key role in governing the surface waviness profile, in which the phase shift farther away from an integer is more conducive to generating denser waviness and a lower surface profile height.

3.2. Experimental Analysis of Surface Generation

The machined surface images of the samples under different phase conditions (out of phase and in phase) are shown in Figure 7. It can be clearly observed that a spiral mark formed on the ground surface, and the mark period changed with the phase shift development which is caused by the vibration of the grinding wheel. In the central area, the vibration mark evolved into a straight-line mode, which acted along the tangent orientation of a circular section (uncut area). The straight-line pattern for the vibration marks were attributed to the gradually decreasing linear velocity when the cutting point between the tool and workpiece moved from the outer regime to the centers of the rotation for the part spindle. We found a more ductile area for the ground surface generation in the center area in comparison with the edge regime for parallel grinding of SiC. This shows a higher grinding ductility in the center area, which is caused by the gradually decreasing chip size under constant decreasing liner cutting speed. Furthermore, the ground surface condition improved with the appearance of the phase shift both for the edge and center regions, in which a less fractured area was formed due to smaller chips generated by a higher degree of wheel cutter trajectory interference, as shown in Figure 7a,b. This shows that the phase shift is an important factor both for the evolution of vibration marks and grinding ductility. The main reason for this is the increase in interference on adjacent cutting paths under in-phase shift conditions, which results in the decrease in surface residual height. This indicates that the radial interference of adjacent grinding paths has a significant effect on the machined surface quality and cannot be ignored. The phase shift alters the relative grinding wheel position between neighboring tool paths with respect to the workpiece, which directly influences the relative depth of cut for abrasive grains distributed on the wheel and in turn effects the ductility for material removal. The phase shift effect can enhance the interference of adjacent abrasive grains' cutting paths and help to reduce the surface residual height, which can significantly improve the surface quality of grinding.

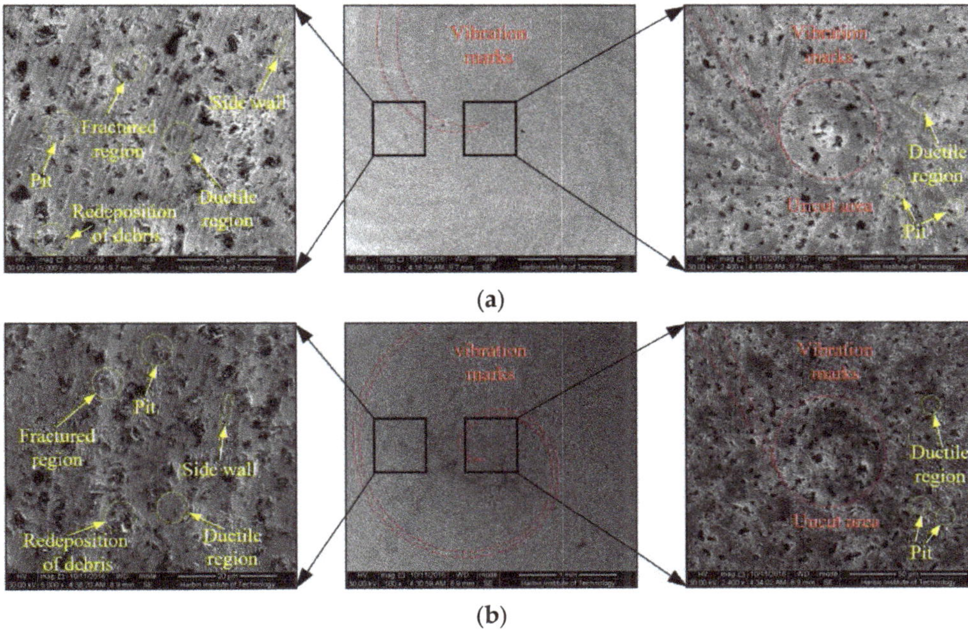

Figure 7. SEM micrographs of RB-SiC ground surface: (**a**) out of phase (phase shift = 0); (**b**) in phase (phase shift = 0.1).

4. Conclusions

In this study, the surface generation mechanism in ultra-precision parallel grinding of RB-SiC ceramics was investigated, and a model for surface generation was established by considering wheel synchronous vibration to calculate the evolution mechanism of vibration marks on the ground surface. The surface generation model was established by considering multiple factors including the arc-shaped effect, tool synchronous vibration, and cutting profile interference in the tool feed direction, in which the effect of the phase shift on the surface topography generation was validated. The main conclusions are as follows:

(1) Wheel synchronous vibration has a dominant impact on surface generation, in which slight changes in the rotational speed of the grinding wheel leads to great waviness in pattern variation. This is due to the huge difference in depth of cut for tool cutting profiles which interferes with neighboring cutting edges in the feed direction.

(2) The phase shift has a great influence on the continuity of waviness generation; the phase (phase = 0, 0.1, and 0.9) closer to an integer tends to generate continuous waviness, whereas the phase (phase = 0.3 and 0.5) farther away from an integer tends to generate discontinuous waviness patterns. This is caused by the relative changes in the depth of cut between adjacent tool paths.

(3) In parallel grinding, the surface generation by way of the ductile mode is more susceptible to appear in the central location of the ground surface due to the decreasing linear velocity. The phase shift has an effect on grinding ductility; for in phase, a ductile removal surface is more likely formed.

There is a good agreement between the novel, developed model and the real experimental results. The theoretical model and experimental results for surface generation provide an effective method to control surface quality and suppress the surface microwaviness in the parallel grinding of ceramic materials.

Author Contributions: Conceptualization, S.C. and C.F.C.; methodology, S.C.; software, T.L.; validation, S.C., S.Y. and D.D.; formal analysis, Z.J.; investigation, S.C.; resources, L.-t.H. and D.D.; data curation, S.C.; writing—original draft preparation, S.C.; writing—review and editing, S.C.; visualization, S.C.; supervision, S.Y. and C.F.C.; project administration, Z.J.; funding acquisition, S.C. All authors have read and agreed to the published version of the manuscript.

Funding: This work was supported by the Major Program of the National Natural Science Foundation of China (52293404), Major Program of the National Natural Science Foundation of China (52293405), National Natural Science Foundation of China (52105481), National Natural Science Foundation of China (52005397), Natural Science Foundation of Zhejiang Province (LQ21E050010), National Key R&D Program of China "strategic science and technology innovation cooperation" project (SQ2022YFE020500), and China Postdoctoral Science Foundation (2019M663681). The authors would like to thank the Key Research and Development Program of Shaanxi Province (2021ZDLGY12-06) and the Fundamental Research Funds for the Central Universities (XZD012022068) of China for their financial support.

Data Availability Statement: Not applicable.

Conflicts of Interest: The authors declare no conflict of interest.

References

1. Beaucamp, A.; Simon, P.; Charlton, P.; King, C.; Matsubara, A.; Wegener, K. Brittle-ductile transition in shape adaptive grinding (SAG) of SiC aspheric optics. *Int. J. Mach. Tool Manufact.* **2017**, *115*, 29–37. [CrossRef]
2. Jalluri, T.D.P.V.; Gouda, G.M.; Dey, A.; Rudraswamy, B.; Sriram, K.V. Development and characterization of silicon dioxide clad silicon carbide optics for terrestrial and space applications. *Ceram. Int.* **2022**, *48*, 96–110. [CrossRef]
3. Huang, P.; Zhang, J.Q. Strain rate effect on the ductile brittle transition in grinding hot pressed SiC ceramics. *Micromachines* **2020**, *11*, 545. [CrossRef]
4. Tang, Z.R.; Gu, L.; Ma, H.P.; Dai, K.F.; Luo, Q.; Zhang, N.; Huang, J.Y.; Fan, J.J. Study on the surface structure of N-Doped 4H-SiC homoepitaxial layer dependence on the growth temperature and C/Si ratio deposited by CVD. *Crystals* **2023**, *13*, 193. [CrossRef]
5. Saddow, S.E. Silicon carbide technology for advanced human healthcare applications. *Micromachines* **2022**, *13*, 346. [CrossRef]
6. Xie, Y.L.; Deng, D.X.; Guan, P.; Huang, X.; Zhao, C.Y. Fabrication of silicon carbide microchannels by thin diamond wheel grinding. *Int. J. Adv. Manuf. Technol.* **2020**, *111*, 309–323. [CrossRef]
7. Zhang, T.Y.; Liu, F.; Liu, Y.; Wu, C.G.; Liang, S.Y. Ultraviolet nanosecond laser-ablated groove analysis of 2.5D Cf/SiC composites. *Crystals* **2023**, *13*, 223. [CrossRef]
8. Huang, S.Q.; Wei, Z.Z.; Rao, X.S.; Li, C.; Zhang, F.H. Optimization of processing parameters during electrical discharge diamond grinding of RB- SiC ceramics based on grey relational theory. *Diam. Abras. Eng.* **2021**, *41*, 56–62.
9. Li, C.; Hu, Y.X.; Zhang, F.H.; Geng, Y.Q.; Meng, B.B. Molecular dynamics simulation of laser assisted grinding of GaN crystals. *Int. J. Mech. Sci.* **2023**, *239*, 107856. [CrossRef]
10. Li, C.; Piao, Y.C.; Zhang, F.H.; Zhang, Y.; Hu, Y.X.; Wang, Y.F. Understand anisotropy dependence of damage evolution and material removal during nanoscratch of MgF_2 single crystals. *Int. J. Extreme Manuf.* **2023**, *5*, 015101. [CrossRef]
11. Li, C.; Piao, Y.C.; Meng, B.B.; Hu, Y.X.; Li, L.Q.; Zhang, F.H. Phase transition and plastic deformation mechanisms induced by self-rotating grinding of GaN single crystals. *Int. J. Mach. Tool Manufact.* **2022**, *172*, 103827. [CrossRef]
12. Yu, S.M.; Yao, P.; Huang, C.Z.; Chu, D.K.; Zhu, H.T.; Zou, B.; Liu, H.L. On-machine precision truing of ultrathin arc-shaped diamond wheels for grinding aspherical microstructure arrays. *Precis. Eng.* **2022**, *73*, 40–50. [CrossRef]
13. Kuriyagawa, T.; Zahmaty, M.S.S.; Syoji, K. A new grinding method for aspheric ceramic mirrors. *J. Mater. Process. Technol.* **1996**, *62*, 387–392. [CrossRef]
14. Heike, K.F.; Taghi, T.; Bahman, A. Material removal mechanism in ultrasonic-assisted grinding of Al_2O_3 by single-grain scratch test. *Int. J. Adv. Manuf. Technol.* **2017**, *91*, 2949–2962.
15. Zhong, Z.W.; Venkatesh, V.C. Recent developments in grinding of advanced materials. *Int. J. Adv. Manuf. Technol.* **2009**, *41*, 468–480. [CrossRef]
16. Weck, M.; Hennes, N.; Schulz, A. Dynamic behaviour of cylindrical traverse grinding processes. *CIRP Ann.* **2001**, *50*, 213–216. [CrossRef]
17. Yang, Y.C.; Wu, Y.-R.; Tsai, T.-M. An analytical method to control and predict grinding textures on modified gear tooth flanks in CNC generating gear grinding. *Mech. Mach. Theory* **2022**, *177*, 105023. [CrossRef]
18. Chen, B.; Li, S.C.; Deng, Z.H.; Guo, B.; Zhao, Q.L. Grinding marks on ultra-precision grinding spherical and aspheric surfaces. *Int. J. Precis. Eng. Manuf. Green Tech.* **2017**, *4*, 419–429. [CrossRef]
19. Chen, S.S.; Cheung, C.F.; Zhang, F.H. An experimental and theoretical analysis of surface generation in the ultra-precision grinding of hard and brittle materials. *Int. J. Adv. Manuf. Technol.* **2018**, *97*, 2715–2729. [CrossRef]
20. Dong, Z.C.; Cheng, H.B. Ductile mode grinding of reaction-bonded silicon carbide mirrors. *Appl. Opt.* **2017**, *56*, 7404–7412. [CrossRef] [PubMed]

21. Antwi, E.K.; Liu, K.; Wang, H. A review on ductile mode cutting of brittle materials. *Front. Mech. Eng.* **2018**, *13*, 251–263. [CrossRef]
22. Datye, A.; Schwarz, U.D.; Lin, H.T. Fracture toughness evaluation and plastic behavior law of a single crystal silicon carbide by nanoindentation. *Ceramics* **2018**, *1*, 198–210. [CrossRef]
23. Page, T.F.; Oliver, W.C.; McHargue, C.J. The deformation behavior of ceramic crystals subjected to very low load nanoindentations. *J. Mater. Res.* **1992**, *7*, 450–473. [CrossRef]
24. Rao, X.S.; Zhang, F.H.; Luo, X.C.; Ding, F.; Cai, Y.K.; Sun, J.N.; Liu, H.T. Material removal mode and friction behaviour of RB-SiC ceramics during scratching at elevated temperatures. *J. Eur. Ceram. Soc.* **2019**, *39*, 3534–3545. [CrossRef]
25. Shen, X.T.; Song, X.; Wang, X.C.; Sun, F.H. Grinding characteristics of CVD diamond grits in single grit grinding of SiC ceramics. *Int. J. Adv. Manuf. Technol.* **2021**, *114*, 2783–2797. [CrossRef]
26. Zheng, Z.D.; Huang, K.; Lin, C.T.; Zhang, J.G.; Wang, K.; Sun, P.; Xu, J.F. An analytical force and energy model for ductile-brittle transition in ultra-precision grinding of brittle materials. *Int. J. Mech. Sci.* **2022**, *220*, 107107. [CrossRef]
27. Giridhar, D.; Sakthivel, G.; Vijayaraghavan, L.; Krishnamurthy, R.; Kumar, M.S.; Gangadhar, K.; Kannan, T. Characterization of single-grit grooving process of silicon carbide ceramic using multisensory approach. *Silicon* **2022**, *14*, 5563–5575. [CrossRef]
28. Gao, S.; Wang, H.X.; Huang, H.; Kang, R.K. Molecular simulation of the plastic deformation and crack formation in single grit grinding of 4H-SiC single crystal. *Int. J. Mech. Sci.* **2023**, *247*, 108147. [CrossRef]
29. Tao, H.F.; Liu, Y.H.; Zhao, D.W.; Lu, X.C. Effects of wheel spindle vibration on surface formation in wafer self-rotational grinding process. *Int. J. Mech. Sci.* **2022**, *232*, 107620. [CrossRef]
30. Czapla, T.; Fice, M.; Niestrój, R. Experimental identification of wheel-surface model parameters: Various terrain conditions. *Sci. Rep.* **2022**, *12*, 16015. [CrossRef] [PubMed]
31. Yin, T.F.; To, S.; Du, H.H.; Zhang, G.Q. Effects of wheel spindle error motion on surface generation in grinding. *Int. J. Mech. Sci.* **2022**, *218*, 107046. [CrossRef]
32. Chen, S.S.; Cheung, C.F.; Zhang, F.; Zhao, C.Y. Three-dimensional modelling and simulation of vibration marks on surface generation in ultra-precision grinding. *Precis. Eng.* **2018**, *53*, 221–235. [CrossRef]
33. Huo, F.W.; Kang, R.K.; Li, Z.; Guo, D.M. Origin, modeling and suppression of grinding marks in ultra-precision grinding of silicon wafers. *Int. J. Mach. Tool Manufact.* **2013**, *66*, 54–65. [CrossRef]
34. Inasaki, I.; Karpuschewski, B.; Lee, H.S. Grinding chatter–origin and suppression. *CIRP Ann. Manuf. Technol.* **2001**, *50*, 515–534. [CrossRef]
35. XU, L.X.; Li, H.; Cai, X.T.; Zhou, P.X.; Chen, Y.W.; Wu, J.F. Study on surface quality in ultrasonic vibration grinding of SiC ceramics with small diameter grinding wheel. *Diam. Abras. Eng.* **2020**, *40*, 67–77.
36. Chen, J.B.; Fang, Q.H.; Li, P. Effect of grinding wheel spindle vibration on surface roughness and subsurface damage in brittle material grinding. *Int. J. Mach. Tool Manufact.* **2015**, *91*, 12–23. [CrossRef]
37. Cao, Y.L.; Guan, J.Y.; Li, B.; Chen, X.L.; Yang, J.X.; Gan, C.B. Modeling and simulation of grinding surface topography considering wheel vibration. *Int. J. Adv. Manuf. Technol.* **2013**, *66*, 937–945. [CrossRef]
38. Chen, S.S.; Yang, S.M.; Cheung, C.F.; Ho, L.T.; Zhang, F.H. Suppression strategy of micro-waviness error in ultra-precision parallel grinding. *Nanomanuf. Metrol.* **2022**, *5*, 423–429. [CrossRef]
39. Chen, S.S.; Cheung, C.F.; Zhao, C.Y.; Zhang, F.H. Simulated and measured surface roughness in high-speed grinding of silicon carbide wafers. *Int. J. Adv. Manuf. Technol.* **2017**, *91*, 719–730. [CrossRef]
40. Pan, Y.C.; Zhao, Q.L.; Guo, B.; Chen, B.; Wang, J.H. Suppression of surface waviness error of fresnel micro-structured mold by using non-integer rotation speed ratio in parallel grinding process. *Micromachines* **2020**, *11*, 652. [CrossRef]

crystals

MDPI

Article

Mechanical Characteristics Generation in the Workpiece Subsurface Layers through Cutting

Michael Storchak

Institute for Machine Tools, University of Stuttgart, Holzgartenstraße 17, 70174 Stuttgart, Germany; michael.storchak@ifw.uni-stuttgart.de

Abstract: The cutting process generates specific mechanical characteristics in the subsurface layers of the shaped parts. These characteristics have a decisive influence on the working properties and product durability of these parts. The orthogonal cutting process of structural heat-treated steel's effect on the mechanical properties of the machined subsurface layers was evaluated by instrumented the nanoindentation method and sclerometry (scratch) method. As a result of this study, the relationship between the specific work in the tertiary cutting zone and the total deformation work during indenter penetration during the instrumented nanoindentation was established. The dependence of the indenter penetration depth during sclerometry of the machined subsurface layers of the workpiece was also studied. The orthogonal cutting process was carried out at different cutting speeds and tool rake angles. The cutting speed increase and the increase in the tool rake angle cause an increase in the indenter penetration work during the instrumented nanoindentation and an increase in the maximum indenter penetration depth during sclerometry. Simultaneously, the measured microhardness of the machined surfaces decreases with both an increase in cutting speed and an increase in the tool rake angle.

Keywords: cutting; subsurface layers; cutting specific work; nanoindentation; sclerometry; indenter penetration depth; total deformation work

Citation: Storchak, M. Mechanical Characteristics Generation in the Workpiece Subsurface Layers through Cutting. *Crystals* **2023**, *13*, 761. https://doi.org/10.3390/cryst13050761

Academic Editors: Chen Li, Chongjun Wu, Binbin Meng and Shanshan Chen

Received: 20 March 2023
Revised: 21 April 2023
Accepted: 24 April 2023
Published: 3 May 2023

1. Introduction

The main objectives of the cutting process used in shaping and finishing operations are to ensure the required geometric shape and dimensional accuracy of the part, as well as the specified quality of the machined surfaces. By performing this objective, the cutting process realizes the tool penetration into the machined material. In this case, the machined subsurface layers of the workpiece are affected by thermomechanical loads as a result of this penetration. The consequences of these loads are plastic strain and subsequent damage of the machined layers with the formation of chips. These processes contribute significant changes in the machined subsurface layers of the workpiece. As a result, specific mechanical characteristics are generated in them. These characteristics have a fundamental effect on the working properties and longevity of the part being shaped in the cutting process. To evaluate such mechanical characteristics, the hardness characteristics of the machined subsurface layers [1] and the total residual stresses formed in them [2] are usually used. The methods used to measure these characteristics have been well established for quite a long time, providing an evaluation of the subsurface state of the parts. However, these methods ensure just a macro estimate of the mechanical characteristics of the machined surfaces. They are insufficiently sensitive to the conditions of thermomechanical effects from the cutting process on the machined surfaces of the workpiece. High sensitivity in evaluating the mechanical characteristics of the subsurface layers of shaped parts using different machining processes is ensured by micro- and nanometric control methods: instrumented nanoindentation and sclerometry (scratch test).

The presented paper is devoted to the evaluation of integral mechanical characteristics in the workpiece subsurface layers generated as a result of the cutting process, and the study of the relationship between these characteristics and the factors influencing them from the orthogonal cutting process.

2. Methods for the Determination of Machined Layers Mechanical Properties

In the past few decades, micro- and nanometric methods have been used to evaluate the mechanical characteristics of materials and parts. These methods include instrumented nanoindentation [3,4] and sclerometry (scratching or scratch test) [5]. The techniques [6] used in the application of these evaluation methods provide extended possibilities for determining the mechanical characteristics of the material's subsurface layers both in instrumented nanoindentation [7] and in sclerometry [8]. The prototype of instrumental nanoindentation as a process of continuous penetration of an indenter into a test material, which undergoes elastic loading, plastic strain, and subsequent damage, may be the study of Atkins and Tabor [9]. The method of material testing by continuously indenter loading and then unloading was further refined by developing methods and devices to measure the indenter load while simultaneously measuring the depth of its penetration into the tested material [10]. At the same time, the indenter penetration depth was narrowed to the nano-scale range. The result of this research was the creation of a widely used instrumented nanoindentation method. Nanoindentation is a well-established method for studying the mechanical properties of various materials. This method is widely represented in well-known publications: see, for example, [11].

Methods for analyzing the indenter penetration diagram were originally presented in the studies of Doerner and Nix [12] as well as Oliver and Pharr [13]. In particular, they presented methods for estimating hardness and Young's modulus based on the initial slope of the unloading curve. The different results of determining the hardness value obtained using the nanoindentation method were analyzed in a study by Vargas and colleagues [14]. Kang et al. studied various optimization methods to determine the material mechanical properties through instrumented nanoindentation [15]. Guillonneau and colleagues proposed a method for determining the mechanical properties based on the analysis of contact stiffness fluctuations in the tested materials [16]. The friction effect in the contact between the indenter and the studied material was investigated by Harsono and colleagues [17]. The relationship between indentation modes and mechanical properties of materials, taking into account friction and the inclination angle value of the indenter's working part is presented in Wang's study [18]. Sivaram and colleagues studied the effects of the pile-up and sink-in effect on the material's mechanical properties during instrumental nanoindentation of thin coatings and integrated circuit materials [19].

When evaluating the mechanical properties of the studied materials and coatings, the researchers mostly concentrate on determining the microhardness of the subsurface layers (see, e.g., [20]) and the residual stresses formed in them (see, e.g., [21]). This is especially the case when evaluating the mechanical properties of various steels and alloys. Furthermore, plasma hardened wheel steels [22], additively manufactured stainless steels [23], as well as steels used to manufacture gear wheels [24] and many others have been investigated. Moon and colleagues studied the contribution of the steel microstructure components to its hardening [25]. The estimation of dislocation density by measuring the microhardness of heterogeneous materials was studied in the work of Mendas et al. [26] and Ameri and colleagues [27]. The application of the instrumented nanoindentation method to evaluate the residual stresses (see, e.g., [21,28,29]) provided the possibility to obtain important information about the mechanical properties of the subsurface layers of the tested materials and coatings. This is particularly important for the characterization of surfaces processed by various mechanical processes, in particular by cutting, plastic deformation, heat treatment, etc., and the tool working surfaces used to realize such processes. Moreover, the method of instrumented nanoindentation is also used to study and evaluate the stress–strain state of the workpiece's subsurface layers. Using nanoindentation, the strain degree of additively manufactured stainless steels [30], Fe-ion-irradiated steels [31], structural steels under tensile loads [32], the strain degree, fracture process, and microstructure changes of austenitic steels [33,34] and others have been evaluated. Wagner et al. studied the behavior of non-metallic particles in 42CrMo4 alloyed steel [35]. Zhou and colleagues studied the effect of indenter loading modes and conditions during nanoindentation of hardened alloy steel [36].

Instrumented nanoindentation has been further advanced due to the recent rapid development of numerical modeling for this method. A nanoindentation three-dimensional model of tool steel coatings is presented in a study by Bobzin et al. [37]. Khan and his colleagues developed a finite-element model of the nanoindentation process of corrosion-resistant aluminum coating [38]. The material model parameters were also determined using inverse fitting via experimental macromechanical testing of the coating material. The same inverse method of determining the parameters of the thin-film metallic glass material model was applied by Han et al. [39]. In this case, the determination of material model parameters is additionally supported using an artificial neural network. The evaluation of the mechanical properties of thin metal coatings using indentation simulation is devoted to the work of Wittler et al. [40]. The Pöhl study presents a method for determining the basic mechanical properties of the test material [41]. Recommendations are provided on the possibilities of determining these properties from the simulated process of indenter penetration into the studied material.

An important part of the instrumental nanoindentation method studies is the investigations devoted to the improvement of instruments and devices for this method implementation. The accuracy and repeatability of results obtained with nanoindentation are ensured by various methods of these devices' calibration: see, e.g., [30]. Peng and colleagues developed a handheld device to implement the nanoindentation method [42]. The indentation force applied to the indenter is detected by a voice coil motor, and the indenter movement is detected using an eddy current transducer. Li and colleagues developed a method for instrumental nanoindentation device calibration [43] based on the use of an optical sonde interferometer and a step height reference. The effect of environmental temperature on the results of nanoindentation tests directly during these tests was studied by Fritz and Kiener [44].

The scratch test (sclerometry) was developed almost in parallel with the instrumented nanoindentation method [45]. The scratch test is well suited for the qualitative characterization of thin, hard coatings applied by different methods [46,47]. This test has been widely used for many years in studies of coating–substrate contact adhesion [48,49]. The scratch test is characterized by its relative simplicity and cost-effectiveness, as well as by the fact that no thorough preparation of the test specimen surface is necessary [46]. At the same time, the scratch test is subject to some uncertainty, especially when determining the fracture load of the examined surface [50]. To reduce the uncertainty in determining the mechanical characteristics of the tested subsurface layers of materials using the scratch test, the multi-pass test has been used in recent decades [51,52]. The multi-pass scratch test is characterized by multiple scratches on the same trace with the progressive or constant load on the indenter. The resulting subsurface fractures are determined either by evaluating the scratch images [51] or by analyzing the characteristics of the scratch test: the tangential load on the indenter or the indenter penetration depth [52]. In the last few decades, the scratch test has also been used to study the mechanical properties of subsurface layers from materials, mainly metals processed using cutting or plastic deformation, and to study the tribomechanical characteristics of contacting surfaces during application [53]. Aurich and Steffes studied the mechanical properties of the subsurface layers during grinding using single-grain scratching [54]. The effect of cutting speed on the plastic strain degree of AISI 4140H steel and the formation of pile-up was established. Fan and colleagues studied the plastic mechanisms of dislocations during the scratch test [55]. The effect of scratching modes with a single grain of titanium alloy Ti-6Al-4V on the indenter penetration depth and pile-up height was studied by Pratap et al. [56]. Thus, the scratch test can be successfully applied to evaluate the effect of material machining processes on the mechanical characteristics of the subsurface layers subjected to a thermo-mechanical impact during processing.

The instrumented nanoindentation method and the scratch test (sclerometry) ensure the measurement not only of individual mechanical characteristics of the tested specimens, such as microhardness, indenter penetration depth, residual stress values, etc., but also of the integral mechanical characteristics of the subsurface layers, in particular the indenter penetration work (energy), its elastic and plastic parts. In this regard, the mechanical state of the subsurface layers subjected to thermomechanical impact through various machining processes is of special interest [57]. The effect of machining conditions on the strain energy of the subsurface layers is presented in a study by Bezyazychny and colleagues [58]. Yamamoto et al. studied the relationship between the elastic and plastic components, as well as the total energy of the indentation, and the microhardness of the tested metals [59]. Ren and Liu [60] and Wang with colleagues [61] devoted their research to studying the relationship between the mechanical properties of hard-working metals and alloys and the cutting process conditions.

The analysis of instrumental nanoindentation and sclerometry methods has shown the essential possibilities of these techniques for micro- and nanometric analysis in determining the mechanical characteristics generated in the shaping process of the subsurface layers of various parts. Thus, these methods can be successfully used to evaluate the influence of generalized cutting process parameters on the formed integral mechanical characteristics of the subsurface layers.

3. Materials and Methods

The cutting process has a significant impact on the machined surface of the shaped part. In addition to ensuring a predetermined geometric shape of the part and microgeometry of the machined surface, the cutting process also generates the physical and mechanical characteristics of the workpiece's subsurface layers. The cutting process's effect on these subsurface layer properties of the machined workpiece is multifaceted and often confrontational. A significant role in this influence is played by the multiplicity and complexity of physical and mechanical processes occurring in the cutting zones of real spatial machining processes such as drilling, milling, threading, etc. To neutralize the secondary effects influence of deformation and subsequent damage of the machined material that takes place in real three-dimensional machining processes, the influence of the cutting process on the formation of physical and mechanical characteristics of the machined surfaces was studied in the conditions of a two-dimensional (free orthogonal) cutting process.

3.1. Materials

Experimental studies of the formation of the physical and mechanical characteristics of the subsurface layers through the machining process were performed on a special stand, reproducing the orthogonal cutting process [62,63]. The experimental stand ensures rectilinear movement of the table with the workpiece relative to the fixed tool with a stepless adjustable cutting speed from 0 to 200 m/min [64]. A well-known representative of heat-treated structural steel AISI 1045 was selected as the machined material. The chemical composition of AISI 1045 steel is specified in Table 1. The mechanical and thermal properties of the steel AISI 1045 are listed in Table 2.

The workpieces of the machined material were pre-grinded to ensure the necessary accuracy, in particular the side parallelism. The final workpiece size was 170 mm × 60 mm × 3 mm. The workpieces were then subjected to annealing to ensure a uniform material structure and the absence of residual stresses. The workpiece hardness after annealing was 180 HB. SNMG-SM-1105 (ISO SNMG 15 06 12-SM 1105) replacement carbide inserts, Sandvik Coromant, Sandviken, Sweden, were used as cutting elements. Table 2 shows the mechanical and thermal properties of this carbide cutting insert. The cutting wedge geometry required for a stable cutting process was ensured by installing a replaceable carbide plate in the tool body at a pre-defined inclination angle and grinding the tool clearance face [64]. The tool clearance angle was α = 8° and the curvature radius of the cutting edge was ρ = 20 µm. Grinding wheels and grinding pastes [65,66] made of synthetic superhard materials were used for grinding and subsequent polishing of carbide inserts. The cutting forces for subsequent determination of the cutting power were measured on the workpiece [62,64]. For this, the workpiece was clamped in a three-component Kistler type 9121 dynamometer using a special clamping device—Figure 1a. Experimental studies were performed at different cutting speeds and different tool rake angles.

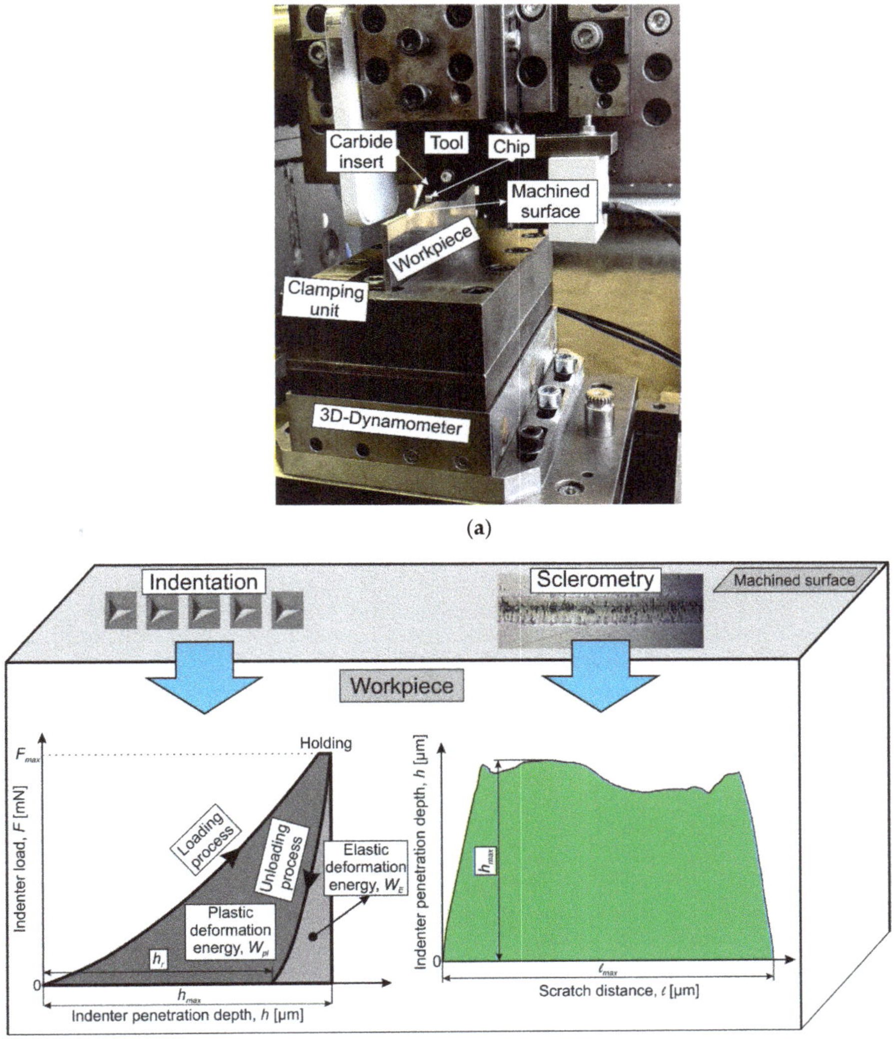

(a)

(b)

Figure 1. Experimental setup for the realization of the orthogonal cutting process and scheme for determination of an instrumented nanoindentation and sclerometry of machined surfaces: (**a**) experimental setup for cutting forces measurement; (**b**) instrumented indentation and sclerometry diagrams.

Table 1. Chemical composition of the steel AISI 1045 [67].

Material	Fe	C	Si	Mn	P	S	Cr
AISI 1045	98.51–98.98%	0.420–0.50%	≤0.04%	0.50–0.90%	≤0.03%	≤0.035%	≤0.4%

Table 2. Mechanical and thermal properties of the steel AISI 1045 steel and carbide insert [67,68].

Material	Strength (MPa)		Elastic Modulus (GPa)	Elongation (%)	Hard-ness	Poisson's Ratio	Specific Heat (J/kg·K)	Thermal Expansion (μm/m·°C)	Thermal Conductivity (W/m·K)
	Tensile	Yield							
AISI 1045	690	620	206	12	HB 180	0.29	486	14	49.8
SNMG-SM-1105	-	-	650	-	HRC 76	0.25	251	-	59

The cutting speed V_C was set at four levels: 48 m/min (0.8 m/s), 96 m/min (1.6 m/s), 144 m/min (2.4 m/s), and 192 m/min (3.2 m/s). The tool rake angle had the following values: $-10°$, $0°$, and $10°$. To determine the cutting forces on the tool clearance face, the depth of cut (undeformed chip thickness) was varied in three levels and was 0.1 mm, 0.2 mm, and 0.3 mm. The cutting process was repeated for each value of cutting speed and tool rake angle at least 5 times. The measuring results of cutting forces were averaged over these repetitions. The maximum uncertainty in measuring the cutting forces was not more than 10%.

The influence of the cutting process on the mechanical characteristics of the machined workpieces' subsurface layers was evaluated by instrumented nanoindentation and sclerometry of the machined surfaces—Figure 1b. To perform the above analysis, thin sections of the workpiece part with a machined surface were prepared [69]. Micromechanical studies of the mechanical characteristics of the machined subsurface layers were carried out on the "Micron-gamma" device [52,70]. The indenter is loaded via an electromagnetic loader. The indenter penetration depth is determined by a differentially switched inductive sensor [70]. This ensures minimal temperature drift. The accuracy of the indenter depth measurement is 5 nm. Instrumented nanoindentation of the machined surfaces was performed at a maximum indenter load of 350 mN. The accuracy of the indenter load setting is 0.02 mN. The change rate of the indenter load was 30 mN/s. After reaching the maximum load on the indenter, the nanoindentation process was delayed at this load and then unloading was performed. The delay time was 5 s.

To perform instrumental nanoindentation and sclerometry, a Berkovich indenter was used. The total penetration energy W as well as its plastic W_{pl} and elastic W_e parts were determined from the instrumented nanoindentation diagram—Figure 1b. The instrumented nanoindentation process was repeated at least 12 times for each set value of cutting speed and tool rake angle. The result of determining the indenter penetrating energy was averaged over the tests performed. The measurement uncertainty was no more than 9%. Sclerometry of machined workpiece surfaces was performed at an indenter load of 100 mN. The indenter movement speed was 20 μm/s. The scratch length was about 630 μm. The maximum indentation depth along the scratch length was determined from the sclerometry diagram—Figure 1b.

The sclerometry process of machined workpiece surfaces was repeated at least 10 times for each value of cutting speed and tool rake angle. The measurement results were then averaged. The uncertainty in determining the maximum indenter penetration depth along the length of the scratch was no more than 12%. An optical profilometer "Micron-alpha" was used to analyze the scratches applied on the "Micron-gamma" device [52]. The scanning principle of the profilometer is based on white light interferometry. This provides a non-contact study of machined surface microtopography with a vertical resolution of 2 nm.

3.2. Methods

As a result of the elastic–plastic interaction of the tool with the machined material and subsequent damage of this material through chip formation, along with the geometric characteristics of the machined subsurface layers, certain physical and mechanical characteristics are also formed in the cutting process. The value and parameters of these characteristics' distribution with constant materials of the contact pair (the pairing tool—machined material) are substantially determined by the conditions of the cutting process: cutting conditions, tool geometry, the presence or absence of lubrication and cooling, etc. Depending on these conditions, different mechanical properties of the subsurface layers are generated, generally characterized by their hardening or softening [62,71,72]. Whether the conditions of adiabatic hardening or isothermal softening prevail in the cutting zones depends on whether the hardened or softened layer of the machined material is formed [73,74]. This process is determined by the contact conditions between the tool and the machined material in the primary and tertiary cutting zones [74], wherein the contact conditions in the tertiary cutting zone have an overwhelming influence [62]. Therefore, the influence of the contact conditions of the tool with the workpiece in this cutting zone is considered further. This influence is evaluated by the energy impact of the tool on the machined material, forming a certain stress–strain state in this material. The formed stress–strain state generates, in turn, certain physical and mechanical characteristics of the subsurface layers. These characteristics remain after the cutting process is finished and the thermomechanical loads are removed. In the cutting process, the contact of the tool clearance face with the workpiece surface is characterized by two stresses: q_{sz}—stress in the contact of the stagnation zone with the workpiece at length h_{sz} and q_{fw}—stress in the contact of the tool clearance face chamfer with the workpiece at length h_{fw}—Figure 2.

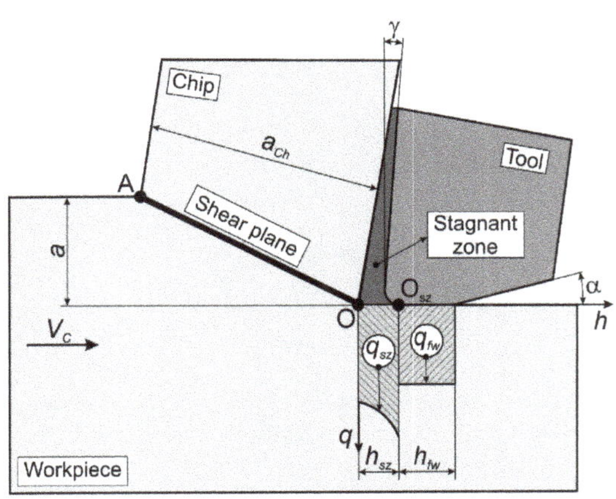

Figure 2. Scheme for the interaction of the tool with the workpiece in the tool clearance face.

The energy interaction of the tool clearance face with the workpiece surface that results in the stresses q_{sz} and q_{fw} can be determined by three different methods:

- according to the cutting work in the tertiary zone, similar to the determination of the specific deformation work in the primary cutting zone [71–74];
- according to the stresses acting on the tool clearance face during the cutting process [71,73,75];
- according to the cutting power on the tool clearance face [71,75].

In this study, the energy interaction was estimated using the third method, as the product of the cutting forces acting on the tool clearance face and the relative speed between the tool and the workpiece:

$$P_{ccf} = F_{ccf} \cdot V_C, \quad (1)$$

$$P_{tcf} = F_{tcf} \cdot V_C, \tag{2}$$

where P_{ccf} is the cutting power on the tool clearance face, P_{tcf} is the thrust power on the tool clearance face, F_{ccf} is the cutting force on the tool clearance face, F_{tcf} is the thrust force on the tool clearance face, and V_C is the cutting speed.

The cutting forces acting during machining on the tool clearance face (in the tertiary cutting zone) were determined using the value of total cutting forces at the undeformed chip thickness (depth of cut for orthogonal cutting), equal to zero [71,75]. To determine these forces, the dependences of the total cutting forces on the depth of cut were extrapolated to a cutting depth equal to zero [62,76]:

$$\underset{\mathbb{R}^k}{\forall} F \underset{\mathbb{R}^k}{\exists} \left(F_{ccf} \cup F_{tcf} \right) : F_{a=0} = extrap(F_C \cup F_T), \tag{3}$$

where $\mathbb{R}$ is the existence space of kinetic characteristics in the cutting process, k is the dimensionality of the existence space, a is the depth of cut (unformed chip thickness), F_C is the total cutting force, and F_T is the total thrust force.

The mechanical characteristics of the workpiece subsurface layers generated as a result of the cutting process are evaluated in this study based on the indenter penetration work as a result of the instrumented nanoindentation and the maximum indenter penetration depth during sclerometry (see Section 3.1). Since these characteristics are generated as a result of the interaction of the tool clearance face with the workpiece surface (see above), the correspondence of the indenter penetration work and its maximum penetration depth with the cutting power at the tool clearance face is considered. This is postulated by the following statement:

➢ *Thermomechanical interaction of the tool with the workpiece (mainly in the tertiary cutting zone) is evaluated based on the cutting power at the tool clearance face in proportion to the indenter penetration work in the machined surface of the workpiece, estimated using instrumented nanoindentation and in proportion to the maximum indenter penetration depth in the subsurface layers, and estimated using sclerometry of the machined subsurface layers:*

$$\underset{\Re}{\forall} S_C \in \Re \overset{S_{Cn}}{\underset{S_{C1}}{\exists}} P_{ccf} \vee P_{tcf} : P_{cf} \propto W_{IN}, \tag{4}$$

$$\underset{\Re}{\forall} S_C \in \Re \overset{S_{Cn}}{\underset{S_{C1}}{\exists}} P_{ccf} \vee P_{tcf} : P_{cf} \propto h_{S\,\max}, \tag{5}$$

where $\Re$ is the existence space of cutting process states (conditions), S_C is the cutting process state, W_{IN} is the total indenter penetration work by instrumented nanoindentation of the machined surface, and $h_{S\,max}$ is the maximum indenter penetration depth during sclerometry of the machined surface.

4. Results

The thermomechanical impact of the tool on the machined workpiece during cutting is accompanied by significant plastic deformation of the machined material and its subsequent failure with chip formation. This impact generates in the machined subsurface layers of the workpiece's specific mechanical characteristics and the corresponding material microstructure. In this case, in different cutting zones, different mechanical characteristics and different microstructures of the machined material corresponding to its stress–strain state in each cutting zone are observed. Based on the generated mechanical characteristics of the subsurface layers and the formed microstructure of the machined material, the degree of the material thermomechanical loading in each cutting zone can be estimated [71–73]. This loading degree is ensured through specific conditions (characteristics) of the cutting process, acting mainly in the tertiary cutting zone (see Section 3.2).

4.1. Cutting Process Characteristic

Figure 3 shows a scheme of the free orthogonal cutting process and the microstructure of the machined material in different cutting zones.

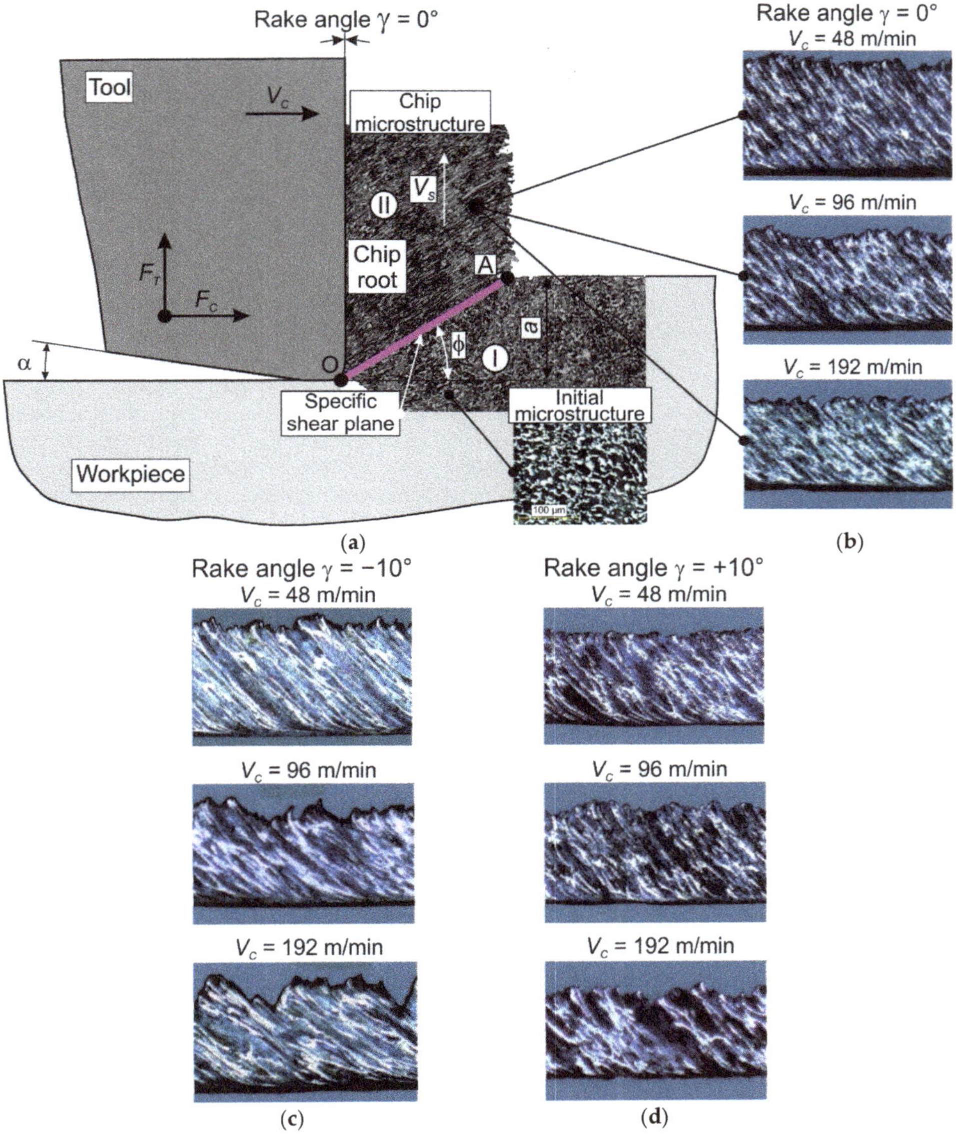

Figure 3. Orthogonal cutting scheme with chip microstructure images at different cutting speeds and tool rake angles: (**a**) orthogonal cutting scheme, combined with the chip root and an image of the initial microstructure; (**b**) chip microstructure images at a tool rake angle $\gamma = 0°$ and different cutting speeds; (**c**) chip microstructure images at a tool rake angle $\gamma = -10°$ and different cutting speeds; (**d**) chip microstructure images at a tool rake angle $\gamma = +10°$ and different cutting speeds. (The following symbols are used in the figure: V_S—chip speed; φ—shear angle; F_C and F_T—cutting and thrust force, respectively).

Cutting process scheme aligned with the chip root—Figure 3a. In the primary cutting zone I, characterized by insignificant plastic strain and low cutting temperatures, the initial microstructure of the machined material is preserved almost up to the specific shear plane (see Figure 3a). The initial microstructure of the machined material (AISI 1045 steel) corresponds to the microstructure of structural heat-treated steels after annealing, which has been widely reported in publications [67,71,72]. The microstructure of workpieces subjected to the subsequent cutting process had a ferrite–perlite structure. The image of the initial microstructure shows light ferrite grains and dark pearlite grains (see Figure 3a). In the secondary cutting zone II, after passing the specific shear plane, the machined material is subjected to significant plastic deformation and significant cutting temperature. This cutting zone is characterized by a significant gradient in the strain degree and cutting temperature [62,71,73]. As a result, the material microstructure (formed chips) in the secondary cutting zone is textured [77]. Images of the chip microstructure at different cutting speeds for the tool rake angle $\gamma = 0^\circ$, shown in the cutting process scheme (see Figure 3a), are presented in Figure 3b. The chip microstructure images for the tool rake angles $\gamma = -10^\circ$ and $+10^\circ$ at the different studied cutting speeds are shown in Figure 3c,d, respectively.

With increasing the cutting speed V_C, the inclination of the chip deformation texture increases for all studied tool rake angles (see Figure 3b–d). This is explained by the fact that with increasing the cutting speed V_C, the movement speed of the chip material layers V_S increases proportionally. At the same time, increasing the tool rake angle causes a decrease in the inclination angle of the chip deformation texture. In all likelihood, this is due to the significantly greater strain degree of the machined material at negative tool rake angles (see Figure 3c). Decreasing the tool rake angle also increases shear strain in the chips of the machined material. With a significant value of cutting speed V_C and negative tool rake angles, the shear strain of the chip material increases significantly. This results in the formation of a shear band (see Figure 3c, V_C = 192 m/min). As a result of this stress–strain state of the machined material in the secondary cutting zone, there is a transformation from the formation of flow chips to the formation of segmented chips [77].

To determine the cutting power during the interaction of the tool clearance face with the workpiece machined surface (tertiary cutting zone), the cutting forces on the tool clearance face are identified when the cutting speed and the tool rake angle are changed. These forces are determined by extrapolating the dependence of cutting forces on the depth of cut (undeformed chip thickness) to zero depth of cut (see Section 3.2). Figure 4 shows the dependencies of cutting force and thrust force, as well as the cutting power of these forces, on the cutting speed and the tool rake angle. With increasing the cutting speed, the forces decrease slightly (see Figure 4a). The reason for the decrease in cutting forces on the clearance face is the increase in cutting temperature with an increase in the cutting speed, which in turn can lead to softening of the machined material. This is realized if the process of machined material softening as a result of increased cutting speed prevails over the process of machined material hardening as a result of the plastic strain of this material and the hardening effect of the strain rate on the machined material. In contrast to the changing in the character of cutting forces, the cutting power increases significantly with increasing cutting speed—Figure 4b. The significant range of cutting speed variation has a predominant influence here. The influence of the tool rake angle on the cutting forces and cutting power (see Figure 4c,d) is expected and coincides with similar dependencies known from the cutting theory [71,72]. With a decrease in the tool rake angle, the cutting forces and cutting powers increase significantly. The same change in the character of cutting forces and cutting powers also remains for other cutting speeds. The reason for this is that the machined material deformation increases considerably with decreasing tool rake angles, especially with negative tool rake angles.

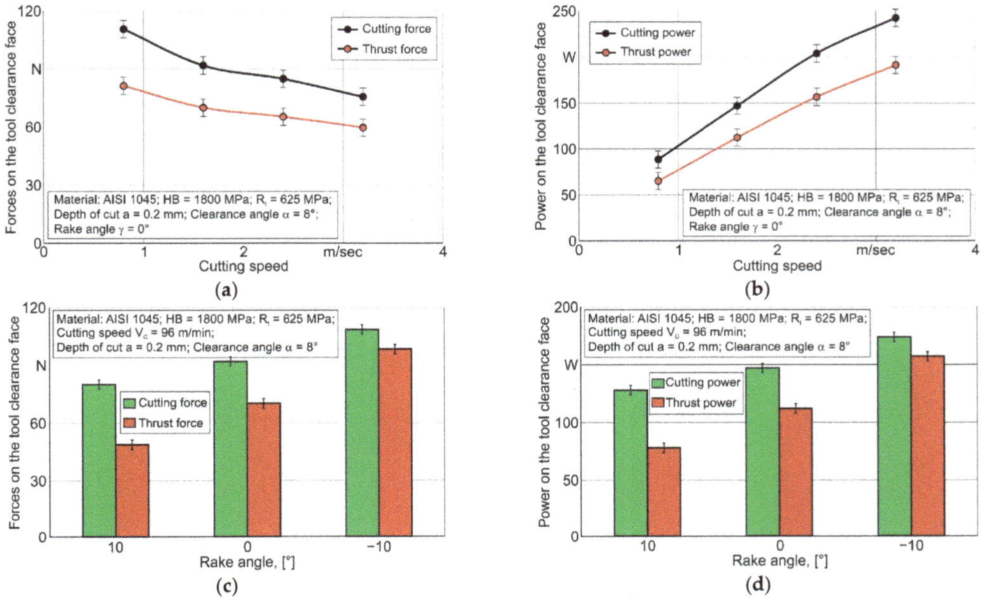

Figure 4. Cutting forces and cutting power on the tool clearance face depending on the cutting speed and the tool rake angle: (**a**) cutting forces depending on the cutting speed; (**b**) cutting power depending on the cutting speed; (**c**) cutting forces depending on the rake angle; (**d**) cutting power depending on the rake angle.

4.2. Mechanical Characteristics of Machined Subsurface Layers

As the mechanical characteristics of the workpiece subsurface layers are generated through the cutting process, the total indentation work W and the maximum indentation depth h_{max} are considered (see Section 3). The diagrams of multiple indenter penetration during instrumented nanoindentation of the workpiece machined surface at cutting speed $V_C = 96$ m/min (1.6 m/s) and the tool rake angle $\gamma = 0°$ exemplarily are shown in Figure 5. Similar indentation diagrams were obtained by testing the workpiece surfaces machined with all the studied values of cutting speeds and tool angles. Such diagrams are used to determine the total indentation work (see Section 3.1), which is then used to match the cutting power.

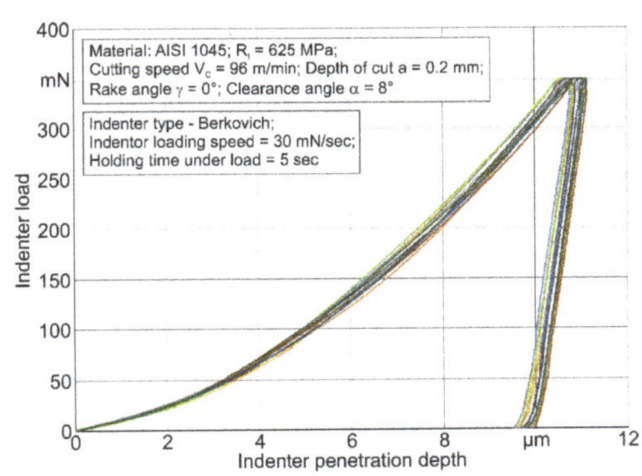

Figure 5. Instrumented nanoindentation diagram.

The maximum indenter penetration depth is determined by a sclerometry test on the surface of the machined material. Figure 6 shows exemplarily the results of this test for a machined surface with a cutting speed of $V_C = 96$ m/min (1.6 m/s) and a tool rake angle of $\gamma = 0°$. When the sclerometry test is carried out, the indenter is oriented with the tip forward along the sclerometry trace. The sclerometry diagram family at the specified parameters that is shown as an example in Figure 6a,b shows a part of the space image from one of the scratch families and the profile of this image. According to the sclerometry results, the greatest averaged depth of indentation over the family of scratches is determined (see also Section 3.1). This value is used later to match the cutting power at the tool clearance face.

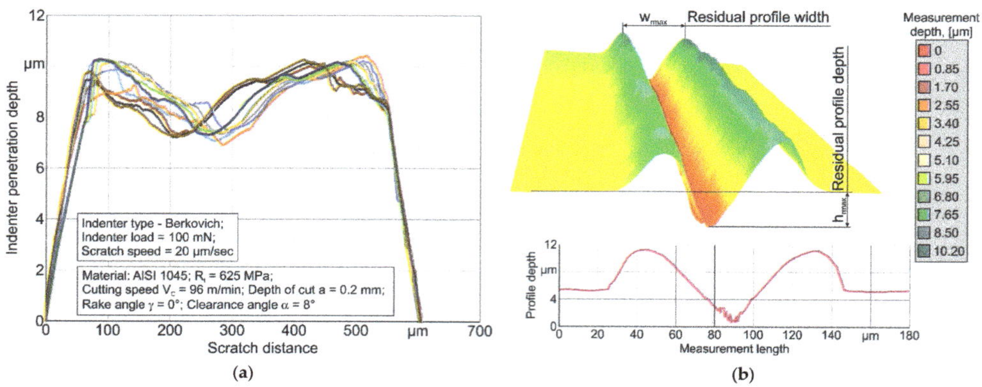

Figure 6. Sclerometry results of the machined surface: (**a**) indenter penetration depth along the sclerometry distance; (**b**) image of the scratch profile (w_{max}—maximum scratch width).

The instrumented nanoindentation test described above (see Figure 5) was used to determine the total indenter work W for all studied values of cutting speeds and tool rake angles. In parallel, the microhardness of the workpiece surfaces processed using the orthogonal cutting process was determined. The dependences of the indentation work value and microhardness of the machined surface on the cutting speed and the tool rake angle are shown in Figure 7.

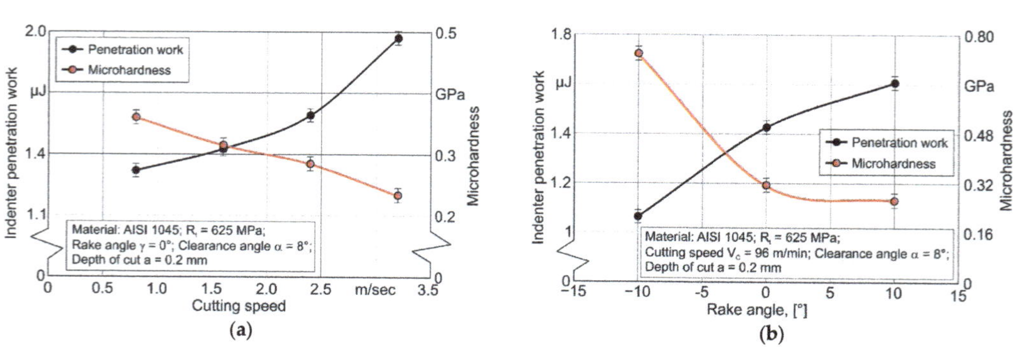

Figure 7. Total indentation work and microhardness of machined surfaces: (**a**) depending on the cutting speed; (**b**) depending on the tool rake angle.

The indentation work increases with increasing cutting speed from 0.8 m/s to 2.4 m/s almost linearly—Figure 7a. A significantly greater increase in work W is observed at a cutting speed of 3.2 m/s. In all probability, this increase is explained by a significant increase in the cutting temperature at cutting speed $V_C = 3.2$ m/s compared to the cutting temperature at lower cutting speeds. This leads to some softening in the subsurface layers of the machined workpiece. This is also evidenced by a decrease in the microhardness of the workpiece's machined surface, the changed behavior of which, with increasing cutting speed, is similar to the changed behavior of the indenter introduction work. With the softening of the machined surface, the indenter penetration depth increases during instrumental nanoindentation of the studied surfaces. In turn, this leads to an increase in the penetration work *W*. Increasing the tool rake angle γ from -10° to 10° leads to an increase in the indentation work—Figure 7b. This changing character of the indentation work is explained by a decrease in the strain degree of the machined material with an increase in the tool rake angle. Reducing the strain degree of the machined material contributes to an increase in the indentation depth during instrumented indentation of the machined workpiece surface and, accordingly, increases the indentation work. With increasing tool rake angle, the microhardness of the machined workpiece surface decreases significantly (see Figure 7b). This also corresponds to the changed behavior of the strain degree in the subsurface layers of the machined material.

The sclerometry test of machined workpiece surfaces with the determination of the maximum indentation depth h_{max} (see Figure 6) was performed by testing the workpiece surfaces machined with all the studied values of cutting speeds and tool rake angles. The dependence of the maximum indentation depth on the cutting speed and the tool rake angle is shown in Figure 8. The maximum indenter penetration depth during sclerometry significantly increases with increasing cutting speed—Figure 8a. This change in the character of the indenter penetration depth corresponds to a certain softening of the machined material due to an increase in cutting temperature with an increase in the cutting speed. The maximum indenter penetration depth increases with the increasing tool rake angle—Figure 8b. This is quite logical since an increase in the tool rake angle causes a decrease in the strain degree of the machined subsurface layers. A smaller strain degree in the subsurface layers causes an increase in the indenter penetration depth when the scratch is applied to the workpiece's machined surface.

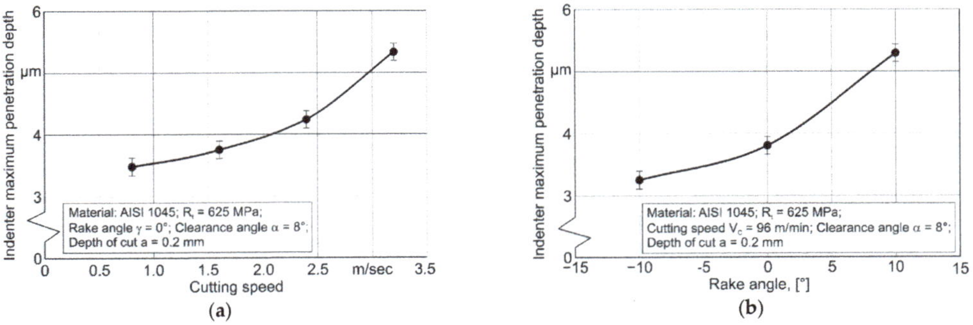

Figure 8. Indenter penetration depth during sclerometry: (**a**) depending on cutting speed; (**b**) depending on tool rake angle.

The dependence shape of the total indentation work in instrument nanoindentation on the cutting speed (see Figure 7a) is similar to the dependence shape of the maximum indentation depth in sclerometry on the cutting speed (see Figure 8a). In all likelihood, this can be explained by the fact that the basis of these two dependencies is the same physical process of machined material softening with increasing the cutting speed. The change range of cutting speed is maintained for both of the studied mechanical characteristics: the total indentation work and its maximum penetration depth. To the same degree, the dependence shape of these two characteristics on the cutting speed is saved. The numerical dependence of these characteristics on the cutting speed is, of course, different. This difference is caused naturally by the different physical nature of the studied characteristics.

5. Discussion

Analyzing the results of instrumental nanoindentation and sclerometry of workpiece surfaces machined at different cutting speeds and using different tool rake angles indicates a significant influence of cutting conditions on the mechanical characteristics of the machined material's subsurface layers. This gives the basis for further use of the postulate formulated above (see Section 3.2) in evaluating the effect of the cutting process on the mechanical characteristics of the machined material's subsurface layers. For this purpose, a comparison of the cutting power in the tertiary cutting zone with the indenter penetration work and its penetration depth was performed.

Figure 9 shows the dependencies of indentation work as a result of the instrumented nanoindentation of machined workpiece surfaces at different cutting speeds and different tool rake angles. The increase in cutting power causes a directly proportional increase in the indenter penetration work in the region of increasing cutting speed from 0.8 m/s to 2.4 m/s. When the cutting power is further increased, in the region of cutting speed equal to 3.2 m/s, the indenter penetration work increases to a significantly greater extent—Figure 9a. The microhardness of the workpiece's machined surfaces is related to the increase in cutting power in the tertiary cutting zone by an inversely proportional relationship (see Figure 9a). When the cutting power increases due to the growth of cutting forces on the tool clearance face caused by a decrease in the tool rake angle, the indenter penetration work decreases accordingly. In parallel, the microhardness of machined workpiece surfaces increases with increasing cutting power—Figure 9b.

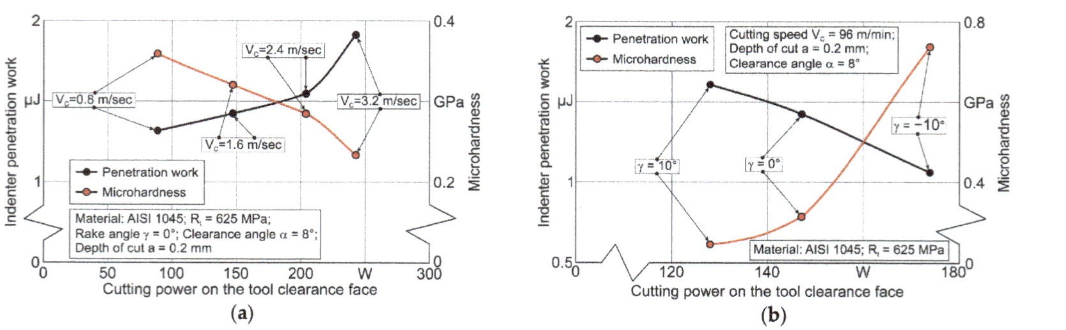

Figure 9. The coincidence of the total indenter penetration work with the cutting energy on the tool clearance face: (**a**) depending on the cutting speed; (**b**) depending on the tool rake angle.

Dependences of the maximum indenter penetration depth during scratching (sclerometry) on the cutting power on the tool clearance face are shown in Figure 10. With increasing the cutting power due to increasing cutting speed, the maximum depth of indenter penetration increases—Figure 10a. This increase is almost the same over the entire range of cutting speeds. An increase in cutting power due to an increase in cutting forces, which in turn is caused by a decrease in the tool rake angle, leads to a decrease in the maximum penetration depth of the indenter—Figure 10b.

When analyzing the influence of cutting power in the tertiary zone on the indentation work as a result of instrumental nanoindentation, as well as when analyzing the influence of cutting power on the maximum indentation depth as a result of scratching the machined surface, it is necessary to take into account the cutting conditions that cause cutting power changes in each specific case. For example, it is necessary to consider whether the increase in cutting power is caused by an increase in cutting speed or by an increase in cutting forces.

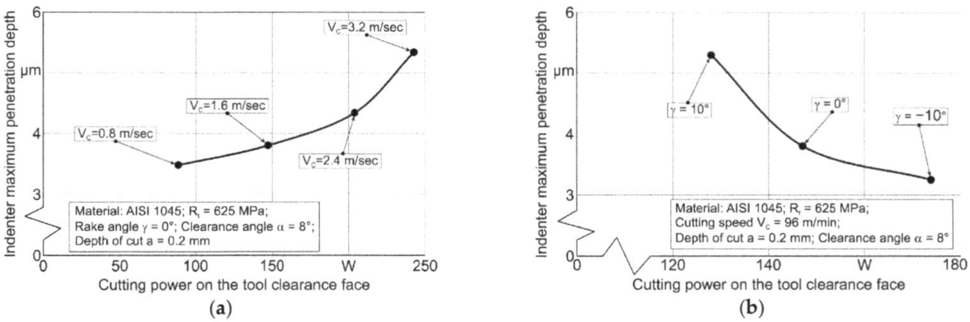

Figure 10. The coincidence of indenter penetration depth during sclerometry with the cutting energy on the tool clearance face: (**a**) depending on the cutting speed; (**b**) depending on the tool rake angle.

Thus, if it is necessary to determine the cause of changes in the integral mechanical characteristics of the workpiece's subsurface layers processed through the cutting process, thermomechanical analysis of the interaction between the tool and the workpiece should be performed in parallel with micro- and nanomechanical tests of the machined subsurface layers.

6. Conclusions

The indenter penetration work during instrumented nanoindentation and the maximum indenter penetration depth during sclerometry are proposed to be applied as mechanical characteristics of the machined subsurface layers.

Increasing the cutting speed and increasing the tool rake angle causes an increase in the indenter penetration work during instrumented nanoindentation and an increase in the maximum indenter depth during sclerometry. Increasing the cutting power in the tertiary cutting zone (on the tool clearance face) also causes an increase in the indenter penetration work during instrumented nanoindentation and its maximum penetration depth during sclerometry. In parallel, the measured microhardness of the machined surfaces decreases with both an increase in cutting speed and an increase in the tool rake angle.

The reasons for cutting power changes may be the basis for interpreting the mechanical properties' behavior of the machined workpiece subsurface layers with changes in the cutting process conditions and evaluating this effect.

The considered integral characteristics of the machined subsurface layers can be used to evaluate the influence of machining conditions on the physical–mechanical state of these layers. This possibility is ensured due to the close connection between the thermomechanical interaction of the tool and the machined material with the physical–mechanical state of the material generated as a result of the cutting process.

Funding: This study was funded by the German Research Foundation (DFG) in the project HE-1656/153-1 "Development of a Concept for Determining the Mechanical Properties of the Cutting Material in Machining".

Data Availability Statement: Not applicable.

Acknowledgments: The author would like to thank the German Research Foundation (DFG) for their support, which is highly appreciated.

Conflicts of Interest: The author declares no conflict of interest.

References

1. Guo, Y.; Saldana, C.; Compton, W.D.; Chandrasekar, S. Controlling deformation and microstructure on machined surfaces. *Acta Mater.* **2011**, *59*, 4538–4547. [CrossRef]
2. Saptaji, K.; Afiqah, S.N.; Ramdan, R.D. A Review on Measurement Methods for Machining Induced Residual Stress. *Indones. J. Comput. Eng. Des.* **2019**, *1*, 106–120. [CrossRef]
3. Pharr, G.M. Recent advances in small-scale mechanical property measurement by nanoindentation. *Curr. Opin. Solid State Mater. Sci.* **2015**, *19*, 315–316. [CrossRef]
4. Voyiadjis, G.Z.; Yaghoobi, M. Review of Nanoindentation Size Effect: Experiments and Atomistic Simulation. *Crystals* **2017**, *7*, 321. [CrossRef]
5. Wredenberg, F.; Larsson, P.-L. Scratch testing of metals and polymers: Experiments and numerics. *Wear* **2009**, *266*, 76–83. [CrossRef]
6. Fischer-Cripps, A.C. Critical review of analysis and interpretation of nanoindentation test data. *Surf. Coat. Technol.* **2006**, *200*, 4153–4165. [CrossRef]
7. Li, X.; Bhushan, B. A review of nanoindentation continuous stiffness measurement technique and its applications. *Mater. Charact.* **2002**, *48*, 11–36. [CrossRef]
8. Kolawole, O.; Ispas, I. Evaluation of geomechanical properties via scratch tests: Where are we and where do we go from here? *SN Appl. Sci.* **2020**, *2*, 1633. [CrossRef]
9. Atkins, A.; Tabor, D. Plastic indentation in metals with cones. *J. Mech. Phys. Solids* **1965**, *13*, 149–164. [CrossRef]
10. Pethicai, J.B.; Hutchings, R.; Oliver, W.C. Hardness measurement at penetration depths as small as 20 nm. *Philos. Mag. A* **1983**, *48*, 593–606. [CrossRef]
11. Fischer-Cripps, A.C. Nanoindentation of thin films and small volumes of materials. In *Nanoindentation*, 2nd ed.; Springer: Berlin/Heidelberg, Germany, 2011; 276p. [CrossRef]
12. Doerner, M.F.; Nix, W.D. A method for interpreting the data from depth-sensing indentation instruments. *J. Mater. Res.* **1986**, *1*, 601–609. [CrossRef]
13. Oliver, W.C.; Pharr, G.M. An improved technique for determining hardness and elastic modulus using load and displacement sensing indentation experiments. *J. Mater. Res.* **1992**, *7*, 1564–1583. [CrossRef]
14. Vargas, A.L.M.; Blando, E.; Hübler, R. Elasto–Plastic materials behavior evaluation according to different models applied in indentation hardness tests. *Measurement* **2019**, *139*, 134–139. [CrossRef]
15. Kang, J.J.; Becker, A.A.; Wen, W.; Sun, W. Extracting elastic-plastic properties from experimental loading-unloading indentation curves using different optimization techniques. *Int. J. Mech. Sci.* **2018**, *144*, 102–109. [CrossRef]
16. Guillonneau, G.; Kermouche, G.; Bec, S.; Loubet, J.-L. Determination of mechanical properties by nanoindentation independently of indentation depth measurement. *J. Mater. Res.* **2012**, *27*, 2551–2560. [CrossRef]
17. Harsono, E.; Swaddiwudhipong, S.; Liu, Z.S. The effect of friction on indentation test results. *Model. Simul. Mater. Sci. Eng.* **2008**, *16*, 065001. [CrossRef]
18. Wang, Y. Effects of indenter angle and friction on the mechanical properties of film materials. *Results Phys.* **2016**, *6*, 509–514. [CrossRef]
19. Sivaram, S.; Jayasinghe, J.A.S.C.; Bandara, C.S. Qualitative Study on Pile-up Effect on Hardness Test by Nano-Indentation. *Eng. J. Inst. Eng. Sri Lanka* **2021**, *54*, 47. [CrossRef]
20. Tsybenko, H.; Farzam, F.; Dehm, G.; Brinckmann, S. Scratch hardness at a small scale: Experimental methods and correlation to nanoindentation hardness. *Tribol. Int.* **2021**, *163*, 107168. [CrossRef]
21. Dean, J.; Aldrich-Smith, G.; Clyne, T.W. Use of nanoindentation to measure residual stresses in surface layers. *Acta Mater.* **2011**, *59*, 2749–2761. [CrossRef]
22. Kanaev, A.T.; Ramazanova, Z.M.; Biizhanov, S.K. Study of plasma-hardened wheel steel using nanoindentation. *Ind. Lab. Diagn. Mater.* **2020**, *86*, 56–60. (In Russian) [CrossRef]
23. England, J.; Uddin, M.J.; Ramirez-Cedillo, E.; Karunarathne, D.; Nasrazadani, S.; Golden, T.D.; Siller, H.R. Nanoindentation Hardness and Corrosion Studies of Additively Manufactured 316L Stainless Steel. *J. Mater. Eng. Perform.* **2022**, *31*, 6795–6805. [CrossRef]
24. Oila, A.; Bull, S.J. Nanoindentation testing of gear steels. *Int. J. Mater. Res.* **2003**, *94*, 793–797. [CrossRef]
25. Moon, J.; Kim, S.; Jang, J.-I.; Lee, J.; Lee, C. Orowan strengthening effect on the nanoindentation hardness of the ferrite matrix in microalloyed steels. *Mater. Sci. Eng. A* **2008**, *487*, 552–557. [CrossRef]
26. Mendas, M.; Benayoun, S.; Miloud, M.H.; Zidane, I. Microhardness model based on geometrically necessary dislocations for heterogeneous material. *J. Mater. Res. Technol.* **2021**, *15*, 2792–2801. [CrossRef]
27. Ameri, A.A.H.; Elewa, N.N.; Ashraf, M.; Escobedo-Diaz, J.P. General methodology to estimate the dislocation density from microhardness measurements. *Mater. Charact.* **2017**, *131*, 324–330. [CrossRef]
28. Wang, H.; Zhu, L.; Xu, B. *Residual Stresses and Nanoindentation Testing of Films and Coatings;* Springer Nature Singapore Pte Ltd. and Science Press: Beijing, China, 2018; 207p, ISBN 978-981-10-7840-8. [CrossRef]
29. Zhang, W.; Wang, X.; Hu, Y.; Wang, S. Predictive modelling of microstructure changes, micro-hardness and residual stress in machining of 304 austenitic stainless steel. *Int. J. Mach. Tools Manuf.* **2018**, *130–131*, 36–48. [CrossRef]

30. Ding, K.; Zhang, Y.; Birnbaum, A.J.; Michopoulos, J.G.; McDowell, D.L.; Zhu, T. Strain gradient plasticity modeling of nanoindentation of additively manufactured stainless steel. *Extrem. Mech. Lett.* **2021**, *49*, 101503. [CrossRef]
31. Lin, P.; Nie, J.; Liu, M. Nanoindentation experiment and crystal plasticity study on the mechanical behavior of Fe-ion-irradiated A508-3 steel. *J. Nucl. Mater.* **2022**, *571*, 154002. [CrossRef]
32. Li, C.; Zhao, H.; Sun, L.; Yu, X. In situ nanoindentation method for characterizing tensile properties of AISI 1045 steel based on mesomechanical analysis. *Adv. Mech. Eng.* **2019**, *11*, 1–11. [CrossRef]
33. Paul, V.; Ameyama, K.; Ota-Kawabata, M.; Ohmura, T. Evaluation of Deformation and Fracture Behavior in 304L Austenitic Steel Harmonic Structures through Nanoindentation. *Steel Res. Int.* **2023**, *94*, 2200354. [CrossRef]
34. Yang, L.; Sun, K.; Peng, W.; Li, X.; Zhang, L. Effects of Grain Boundary Angles on Initial Deformation of 304 Austenitic Stainless Steel under Nanoindentation: A Molecular Dynamics Simulation. *Crystals* **2022**, *12*, 58. [CrossRef]
35. Zhou, G.; Guo, J.; Zhao, J.; Tang, Q.; Hu, Z. Nanoindentation Properties of 18CrNiMo7-6 Steel after Carburizing and Quenching Determined by Continuous Stiffness Measurement Method. *Metals* **2020**, *10*, 125. [CrossRef]
36. Wagner, R.; Lehnert, R.; Storti, E.; Ditscherlein, L.; Schröder, C.; Dudczig, S.; Peuker, U.A.; Volkova, O.; Aneziris, C.G.; Biermann, H.; et al. Nanoindentation of alumina and multiphase inclusions in 42CrMo4 steel. *Mater. Charact.* **2022**, *193*, 112257. [CrossRef]
37. Bobzin, K.; Kalscheuer, C.; Carlet, M.; Schmauder, S.; Guski, V.; Verestek, W.; Tayyab, M. 3D deformation modeling of CrAlN coated tool steel compound during nanoindentation. *Surf. Coat. Technol.* **2023**, *453*, 129148. [CrossRef]
38. Khan, M.K.; Hainsworth, S.V.; Fitzpatrick, M.E.; Edwards, L. A combined experimental and finite element approach for determining mechanical properties of aluminium alloys by nanoindentation. *Comput. Mater. Sci.* **2010**, *49*, 751–760. [CrossRef]
39. Han, G.; Marimuthu, K.P.; Lee, H. Evaluation of thin film material properties using a deep nanoindentation and ANN. *Mater. Des.* **2022**, *221*, 111000. [CrossRef]
40. Wittler, O.; Mrossko, R.; Kaulfersch, E.; Wunderle, B.; Michel, B. Mechanical characterisation of thin metal layers by modelling of the nanoindentation experiment. In Proceedings of the 2nd Electronics System-Integration Technology Conference, Greenwich, UK, 1–4 September 2008; pp. 995–998. [CrossRef]
41. Pöhl, F. Determination of unique plastic properties from sharp indentation. *Int. J. Solids Struct.* **2019**, *171*, 174–180. [CrossRef]
42. Peng, G.; Xu, F.; Chen, J.; Hu, Y.; Wang, H.; Zhang, T. A cost-effective voice coil motor-based portable micro-indentation device for in situ testing. *Measurement* **2020**, *165*, 108105. [CrossRef]
43. Li, Z.; Herrmann, K.; Pohlenz, F. A comparative approach for calibration of the depth measuring system in a nanoindentation instrument. *Measurement* **2006**, *39*, 547–552. [CrossRef]
44. Fritz, R.; Kiener, D. Development and application of a heated in-situ SEM micro-testing device. *Measurement* **2017**, *110*, 356–366. [CrossRef]
45. Randall, N.X. The current state-of-the-art in scratch testing of coated systems. *Surf. Coat. Technol.* **2019**, *380*, 125092. [CrossRef]
46. Li, J.; Beres, W. Scratch Test for Coating/Substrate Systems—A Literature Review. *Can. Met. Q.* **2007**, *46*, 155–173. [CrossRef]
47. Lin, C.K.; Berndt, C.C. Measurement and analysis of adhesion strength for thermally sprayed coatings. *J. Therm. Spray Technol.* **1994**, *3*, 75–104. [CrossRef]
48. Bull, S.J. Failure modes in scratch adhesion testing. *Surf. Coat. Technol.* **1991**, *50*, 25–32. [CrossRef]
49. Zivica, F.; Babic, M.; Adamovic, D.; Mitrovic, S.; Todorovic, P.; Favarob, G.; Pantic, M. Influence of the surface roughness on adhesion of chrome coatings on alloy tool steel x165crmov12. *J. Balk. Tribol. Assoc.* **2012**, *18*, 228–237.
50. Sousa, F.J.P.; Tridapalli, D.; Pereira, M.; Flesch, C.A.; Alarcon, O.E. Evaluation of measurement uncertainties for a scratching tester. *Measurement* **2006**, *39*, 594–604. [CrossRef]
51. Yildiz, F.; Alsaran, A. Multi-pass scratch test behavior of modified layer formed during plasma nitriding. *Tribol. Int.* **2010**, *43*, 1472–1478. [CrossRef]
52. Storchak, M.; Zakiev, I.; Zakiev, V.; Manokhin, A. Coatings strength evaluation of cutting inserts using advanced multi-pass scratch method. *Measurement* **2022**, *191*, 110745. [CrossRef]
53. Farayibi, P.; Hankel, J.; Hassend, F.v.G.; Blüm, M.; Weber, S.; Röttger, A. Tribological characteristics of sintered martensitic stainless steels by nano-scratch and nanoindentation tests. *Wear* **2023**, *512*, 204547. [CrossRef]
54. Aurich, J.C.; Steffes, M. Single Grain Scratch Tests to Determine Elastic and Plastic Material Behavior in Grinding. *Adv. Mater. Res.* **2011**, *325*, 48–53. [CrossRef]
55. Fan, P.; Katiyar, N.K.; Zhou, X.; Goel, S. Uniaxial pulling and nano-scratching of a newly synthesized high entropy alloy. *APL Mater.* **2022**, *10*, 111118. [CrossRef]
56. Pratap, A.; Divse, V.; Goel, S.; Joshi, S.S. Understanding the surface generation mechanism during micro-scratching of Ti-6Al-4V. *J. Manuf. Process.* **2022**, *82*, 543–558. [CrossRef]
57. Liu, H.; Xu, X.; Zhang, J.; Liu, Z.; He, Y.; Zhao, W.; Liu, Z. The state of the art for numerical simulations of the effect of the microstructure and its evolution in the metal-cutting processes. *Int. J. Mach. Tools Manuf.* **2022**, *177*, 103890. [CrossRef]
58. Bezyazichnyy, V.; Rybinsk State Aviation Technical University; Prokofiev, M.; Vinogradova, N. Research of the Influence of Technological Machining Conditions on the Accumulation of Latent Energy Deformation in the surface Parts. *Bull. PNIPU Aerosp. Eng.* **2015**, *43*, 131–144. [CrossRef]
59. Yamamoto, M.; Tanaka, M.; Furukimi, O. Hardness–Deformation Energy Relationship in Metals and Alloys: A Comparative Evaluation Based on Nanoindentation Testing and Thermodynamic Consideration. *Materials* **2021**, *14*, 7217. [CrossRef]

60. Ren, X.; Liu, Z. Influence of cutting parameters on work hardening behavior of surface layer during turning superalloy Inconel 718. *Int. J. Adv. Manuf. Technol.* **2016**, *86*, 2319–2327. [CrossRef]
61. Wang, Q.; Liu, Z.; Wang, B.; Song, Q.; Wan, Y. Evolutions of grain size and micro-hardness during chip formation and machined surface generation for Ti-6Al-4V in high-speed machining. *Int. J. Adv. Manuf. Technol.* **2016**, *82*, 1725–1736. [CrossRef]
62. Kushner, V.; Storchak, M. Determining mechanical characteristics of material resistance to deformation in machining. *Prod. Eng.* **2014**, *8*, 679–688. [CrossRef]
63. Heisel, U.; Kushner, V.; Storchak, M. Effect of machining conditions on specific tangential forces. *Prod. Eng.* **2012**, *6*, 621–629. [CrossRef]
64. Heisel, U.; Storchak, M.; Krivoruchko, D. Thermal effects in orthogonal cutting. *Prod. Eng.* **2013**, *7*, 203–211. [CrossRef]
65. Filatov, Y.D.; Filatov, A.Y.; Syrota, O.O.; Yashchuk, V.P.; Monteil, G.; Heisel, U.; Storchak, M. The influence of tool wear particles scattering in the contact zone on the workpiece surface microprofile formation in polishing quartz. *J. Superhard Mater.* **2010**, *32*, 415–422. [CrossRef]
66. Filatov, Y.D.; Sidorko, V.I.; Filatov, O.Y.; Yaschuk, V.P.; Heisel, U.; Storchak, M. Surface quality control in diamond abrasive finishing. *Proc. SPIE* **2009**, *7389*, 73892O. [CrossRef]
67. Villarrazo, N.; Caneda, S.; Pereira, O.; Rodríguez, A.; López de Lacalle, L.N. The Effects of Lubricooling Ecosustainable Techniques on ToolWear in Carbon Steel Milling. *Materials* **2023**, *16*, 2936. [CrossRef]
68. Storchak, M.; Stehle, T.; Möhring, H.-C. Determination of thermal material properties for the numerical simulation of cutting processes. *Int. J. Adv. Manuf. Technol.* **2021**, *118*, 1941–1956. [CrossRef]
69. Storchak, M.; Jiang, L.; Xu, Y.; Li, X. Finite element modeling for the cutting process of the titanium alloy Ti10V2Fe3Al. *Prod. Eng.* **2016**, *10*, 509–517. [CrossRef]
70. Storchak, M.; Zakiev, I.; Träris, L. Mechanical properties of subsurface layers in the machining of the titanium alloy Ti10V2Fe3Al. *J. Mech. Sci. Technol.* **2018**, *32*, 315–322. [CrossRef]
71. Zorev, N.N. *Metal Cutting Mechanics*; Pergamon Press GmbH: Frankfurt am Main, Germany, 1966; 526p.
72. Oxley, P.L.B. Development and Application of a Predictive Machining Theory. *Mach. Sci. Technol.* **1998**, *2*, 165–189. [CrossRef]
73. Kushner, V.; Storchak, M. Modelling the material resistance to cutting. *Int. J. Mech. Sci.* **2017**, *126*, 44–54. [CrossRef]
74. Kushner, V.; Storchak, M. Determination of Material Resistance Characteristics in Cutting. *Procedia CIRP* **2017**, *58*, 293–298. [CrossRef]
75. Tsekhanov, J.; Storchak, M. Development of analytical model for orthogonal cutting. *Prod. Eng.* **2015**, *9*, 247–255. [CrossRef]
76. Storchak, M.; Möhring, H.-C.; Stehle, T. Improving the friction model for the simulation of cutting processes. *Tribol. Int.* **2022**, *167*, 107376. [CrossRef]
77. Storchak, M.; Mohring, H.-C. Numerical and Experimental Analysis of Chip Formation at Ultrahigh Cutting Speed. *MM Sci. J.* **2019**, *2019*, 3243–3249. [CrossRef]

MDPI
St. Alban-Anlage 66
4052 Basel
Switzerland
www.mdpi.com

Crystals Editorial Office
E-mail: crystals@mdpi.com
www.mdpi.com/journal/crystals

www.ingramcontent.com/pod-product-compliance
Lightning Source LLC
Chambersburg PA
CBHW061504210926
51850CB00008B/1819
* 9 7 8 3 0 3 6 5 8 8 6 6 7 *